Fundamentals of Enhanced Oil Recovery

Fundamentals of Enhanced Oil Recovery

Editor

Marcin Kremieniewski

MDPI • Basel • Beijing • Wuhan • Barcelona • Belgrade • Manchester • Tokyo • Cluj • Tianjin

Editor
Marcin Kremieniewski
Oil and Gas Institute—National Research Institute
Poland

Editorial Office
MDPI
St. Alban-Anlage 66
4052 Basel, Switzerland

This is a reprint of articles from the Special Issue published online in the open access journal *Energies* (ISSN 1996-1073) (available at: https://www.mdpi.com/journal/energies/special_issues/Fundamentals_of_Enhanced_Oil_Recovery).

For citation purposes, cite each article independently as indicated on the article page online and as indicated below:

LastName, A.A.; LastName, B.B.; LastName, C.C. Article Title. *Journal Name* **Year**, *Volume Number*, Page Range.

ISBN 978-3-0365-4375-8 (Hbk)
ISBN 978-3-0365-4376-5 (PDF)

Contents

About the Editor

Marcin Kremieniewski

Dr. Marcin Kremieniewski is an expert in the field of technology, selection and processing of cement slurries and drilling fluid compositions. He carries out research and design work for many entities related to the energy and oil industries. He has extensive knowledge in the field of drilling technology, hydraulic binders, drilling fluids, environmental engineering as well as energy and geoengineering. He deals with issues related to reducing energy consumption, increasing the efficiency of extraction, rational energy management in the oil industry, proper well sealing, rational selection of drilling fluids, increasing the efficiency of oil extraction, reducing investment costs in the oil sector and environmental aspects in the oil sector. Author or co-author of over 100 scientific publications, 8 scientific monographs, as well as many national and international conference presentations. He is also the creator of numerous inventions that have patents at international inventions exhibitions. He is supervisor to doctoral dissertations and a reviewer of many scientific magazines. He is in charge of research teams implementing various projects in the field of drilling technology. He holds the Mining Director, 3rd degree and Meritorious for the Oil Mining and Gas Industry distinction.

energies

MDPI

Editorial

Improving the Efficiency of Oil Recovery in Research and Development

Marcin Kremieniewski

Oil and Gas Institute-National Research Institute, 25A Lubicz Str., 31-503 Krakow, Poland; kremieniewski@inig.pl

Citation: Kremieniewski, M. Improving the Efficiency of Oil Recovery in Research and Development. *Energies* **2022**, *15*, 4488. https://doi.org/10.3390/en15124488

Received: 9 June 2022
Accepted: 16 June 2022
Published: 20 June 2022

1. Introduction

By creating a special edition entitled Fundamentals of Enhanced Oil Recovery, the editors focus on the problem of the global increase in energy demand. In recent years, there has been a trend towards switching to alternative energy sources, but despite this, oil and natural gas will continue to be the main source of energy for the next several decades. However, it should be borne in mind that renewable energy resources have some limitations in terms of quantity. The progressive exploitation of the deposits contributes to the increasing degree of oil depletion, and this makes extraction more and more difficult, even though the deposits are not yet completely empty. Such a situation results in the necessity to search for new methods of increasing oil and gas production. Intensifying oil production is a very rational use of energy that has not yet been fully used. Production intensification methods are used to utilize these energy reserves. Such activities can be implemented and carried out at any stage of the well implementation, from the well design stage, through drilling, to oil and gas production. For this purpose, it is extremely important to conduct research and design works that include issues related to increasing the efficiency of oil extraction. It is also very important to improve the effectiveness of sealing the borehole and the correct selection of cement slurries, drilling fluids and borehole washing liquids. In addition, the issues related to the reduction in energy consumption and energy management in the oil industry are very important at the present time, both from the economic point of view and in terms of the rational use of energy. The focus should also be on innovations that include modern technologies supporting the efficiency of extraction and the latest technical, technological and operational challenges in the oil sector. All these factors are discussed in the international industrial arena, and the most important topics, after a rigorous assessment procedure by eminent specialists from around the world, find their way to prestigious magazines, such as MDPI *Energies* Special Issue.

2. A Short Review of the Contributions in This Issue

Many articles can be grouped under the title Fundamentals of Enhanced Oil Recovery. As for the improvement of oil recovery, in the first article [1], the authors discuss the problem of gas migration through fresh cement slurry and hardened cement slurry [2]. Gas transfer is an unfavorable phenomenon that can be minimized and sometimes even eliminated by using an appropriate cement slurry recipe [3]. According to the authors, the appropriate selection of the quantity and quality of components enables the design of slurries with the required static strength values [4]. Additionally, the cementitious sheath of such anti-migration slurry has a low porosity and a very low proportion of large pore spaces. Furthermore, the mechanical parameters do not deteriorate during long-term deposition in borehole conditions [5]. The authors designed a solution in the form of a cement slurry, the cement sheath of which has high corrosion resistance. The presence of appropriate additives and admixtures favors the design of the slurry, which contributes to the improvement of gas recovery by eliminating the negative phenomenon of gas migration.

The next article presents the technology of preparing drilling fluids with a high content of internal phase [6]. The correct selection of drilling fluid is very important from the point of view of improving oil recovery [7]. Various types of drilling fluids are used for drilling boreholes, but most often, due to the economic aspect, a water-based mud is used [8]. However, for drilling in difficult geological conditions, an inversion mud is used, in which the oil phase to the water phase most often occurs in the range from 70/30 to 90/10 [9]. The authors in publication [6] present a solution where the oil to water ratio is 50/50 to 20/80, and such a mud can be used to drill a borehole in HPHT conditions. According to the authors, the new drilling mud solution contributes to the improvement of the improved oil recovery. Additionally, inversion drilling muds are characterized by high electrical parameters; the ES stability is above 300 V [10]. They also have stable rheological parameters and low filtration. Due to the reduced oil content, the developed drilling fluid system is more economical and has limited toxicity.

Issues related to the geological sequestration of acid gases, including CO_2, are now an increasingly common solution to prevent the progressive changes in the Earth's climate. In the publication [11], the authors focus on the analysis of the research carried out in the Borzęcin sequestration area [12]. The area is located in western Poland. The tests are aimed at examining the migration paths of the injected acid gases (mainly mixtures of CO_2 and H_2S) to the aquifers under the natural gas deposit [13]. As part of their research, the authors conduct two well sampling actions, during which they take samples of the waters lying under the Borzęcin reservoir and then examine their physicochemical parameters. Such tests have not been conducted so far. Tests of reservoir waters from selected wells were also carried out, including isotope analyzes [14]. The work carried out was aimed at broadening the state of knowledge, which is valuable for the risk assessment of the acid gas sequestration process taking place on a specific example, and to improve the efficiency of gas injection.

During the improvement of oil recovery, one of the basic problems accompanying the use of water in EOR processes is the precipitation of inorganic sediments [15]. As the authors write in the publication [16], these are the most common deposits of calcium carbonate and deposits of calcium, barium and strontium sulphates, with calcium carbonate being the most common one [17,18]. In order to eliminate this unfavorable phenomenon, appropriate types of inhibitors are used, which was presented in the publication [16]. However, in order to carry out such tests, it is necessary to develop the research technology and build a test stand. Such a solution was presented by the authors of the publication [16] and they discussed the most important research results.

Fundamentals of Enhanced Oil Recovery is strongly associated with the improved adhesion of cement to pipes and rock formation [19]. On the other hand, the adhesion is strongly influenced by proper preparation of the borehole and cleaning of the annular space after drilling and before cementing [1,20]. It is connected with the fact that low values of adhesion of the cement sheath to the rock formation and to the surface of the casing cause the formation of uncontrolled gas flows [4]. Additionally, the lack of adhesion also reduces the stabilization of the pipe in the borehole [21]. In [22], the authors focus on determining the impact of cleaning the annular space on the adhesion of the cement sheath to the rock formation. The issues related to the preparation of the borehole for cementation by appropriate cleaning of the rock formation from mud cake residues contribute to the improvement of oil recovery, which contributes to the improvement of drilling works [23].

The increase in oil production translates into shell gas exploration [24,25]. Therefore, in the publication [26], the authors discuss the developed method of finding similarities between particular geological structures in terms of their hydrocarbon-generating properties and hydrocarbon resources. The measurements and geochemical studies of holes located in the Ordovician, Silurian and Cambrian formations of the Polish part of the East European Platform are used here [27,28]. The classification of objects is based on a cluster analysis, where the focus is on the issue of generating clusters that are grouped into samples in gas, condensate and oil windows [29]. The characteristic geochemical properties of the samples classified into selected clusters are also determined [30]. Researchers successfully classified the samples into individual windows and determined their percentages in Silurian, Ordovician and Cambrian units. Doing so is some sort of empirical challenge towards improving oil and gas recovery.

In work [31], the authors rightly state that flooding technology is an important measure for increasing the recovery of crude oil in oil fields. In this paper, the authors used the direct numerical simulation (DNS) method, which is based on the Navier–Stokes equation and the fluid volume (VOF) method, to investigate the dynamic behavior of oil-water flow in a low-permeability pore structure [32,33]. On the basis of the results obtained from the research, the authors concluded that the variability in the non-uniformity of the viscosity action results from the difference in the viscosity of oil and water. On the other hand, the complex dynamic behavior of the two-phase oil-water flow on the pore scale, demonstrated by capillarity, play a decisive role in determining the area of spatial sweep and the final index of oil recovery [34,35]. From the work, they conclude that the absolute viscosity of oil and water has a significant influence on the degree of oil recovery by adjusting the relative importance between the action of viscosity and capillarity.

Enhanced Oil Recovery is also about the correct selection of drilling fluids. Therefore, an appropriate design of the cement slurry to seal the borehole is important [36]. For this purpose, the additives and admixtures for the cement slurry should be properly selected in order to obtain the required parameters of the hardened cement slurry [37]. Such requirements depend on the geological conditions of the drill hole. There are very high regimes in deep boreholes, which must be met by cement slurries; therefore, the necessity to use advanced, innovative measures is implied [38]. An example of the above is the possibility of using the addition of nanosilica, as discussed in [39], in order to improve the technological parameters of both the liquid cement slurry and the resulting cement sheath. The authors of the work presented the results of research on the mechanical parameters for cement stone with the addition of silica nanoparticles. The samples are deposited in an environment of elevated temperature of 90 °C. The cement sheath made of cement slurry, which contains an admixture of nanosilica, shows an improvement in mechanical properties. This is manifested by an increase in compressive strength. It is very important from the point of view of undertaking further works in the borehole and improving the efficiency of production.

The issue of the tightness of the borehole is of key importance for its long-term durability; thus, it is part of the broad thematic spectrum of Enhanced Oil Recovery. To ensure durability for many years, the column of pipes is sealed with cement slurry [40]. However, slurry that is pumped down the mud, if it comes into contact with the mud cake, may not seal well the annular space of the borehole [41]. Therefore, it is important to properly clean the hole, which is a big problem, because there are many variables affecting its stability [42]. The contact time of the well with the scrubber is important. On the one hand, insufficient contact time does not guarantee the correct removal of the mud cake. On the other hand, long contact times can damage the borehole wall. Therefore, in [43], the researchers conducted a study to evaluate the impact of the contact time of the washer on cleaning the annular space. The study of the degree of mud cake removal depending on the contact time is based on the determination of the adhesion of the cement sheath to the rock formation [44]. By comparing the obtained adhesion with a reference sample, the researchers determine the effectiveness of the deposit removal. On the basis of these studies, they determined the optimal contact time with the cleaner.

In [45], the author deals with the issue of the influence of temperature on the gradation of oil pipes. Corrosion in the oil industry is an important aspect when extracting natural gas from a deposit with high temperatures [46]. According to the author [47], the water in the tank is often in the form of steam with a pressure of up to several dozen MPa. As a result of its extraction, it cools, which causes [48] condensation. Condensed water in contact with the acidic components of the gas causes corrosion, especially in the presence of aggressive gases. Therefore, a very important issue is to determine the effect of water condensation as a result of the changes in the temperature of gases containing CO_2 and H_2S [49,50] on the corrosion of steel, which is in contact with extraction pipes and casing pipes. On the basis of the obtained results, the author answers the question of what effect temperature, gas components and pressure have on the corrosion of the borehole construction material, and indicates the selection of the borehole material to prevent corrosion in aggressive environments.

Fundamentals of Enhanced Oil Recovery is also an operational challenge. These include the new analytical procedure for the preselection of gas wells for water shut-off procedures, as discussed in the next paper [51], based on the available results of the integrated geological and deposit data analysis. The basis for assessing the possibility of cross-flow formation is the assessment of the presence or absence of impermeable barriers at intervals, supplemented by perforation [52,53]. The authors used data from wells obtained in different years from measurements with different types of probes. Based on the modified quantitative and qualitative interpretation techniques, permeable and impermeable layers are distinguished in the analyzed drilling sections. The process of verifications carried out for eight boreholes initially selected by the operator located in the Carpathian Foredeep in Poland is discussed [54,55].

A properly designed cement slurry is one of the most important elements in the Fundamentals of Enhanced Oil Recovery through improved borehole sealing [56]. Therefore, the publication [57] discusses the influence of the Hblock fine-grained material on selected parameters of the cement slurry. The fine-grained additive used shortens the setting time and the transition time from the value of the initial setting time to the final setting time [58]. It shortens the time needed to bind the cement slurry and proceed to further work [59]. Moreover, such action helps to eliminate the possibility of the gaseous medium penetrating into the structure of the liquid cement slurry [60]. The publication discusses the effect of the additive on the technological parameters of the cement slurry and the possibility of using fine-grained material as an innovative technology in fluids for oil wells.

Innovative technologies in oil wells are part of the Special Issue Fundamentals of Enhanced Oil Recovery. One of these technologies is the use of jet pumps for the utilization of the associated petroleum gas, as discussed in [61]. This is a very important issue because the combustion of this gas causes environmental degradation and poses a potential threat to the human body [62]. The possibility of simultaneous use of a suction-piston pump driven by a shuttle and jet machine in the oil well was proposed after prior estimation of the pressure distribution along the borehole. The authors used the well-known methods of Poettman-Carpenter and Baksendel. In order to carry out the work, the researchers developed a methodology for the practical use of these equations in order to calculate the parameters of the jet pump based on the independent parameters of the oil well [63,64]. The final result of the work [61] is a list of recommendations for the selection of the location of an oil-gas ejector inside a selected oil well and a generalization of the principles of selecting the ideal location of such ejectors for other shafts. The second result is a reasonable method to rationally determine the location of the ejector in the oil well and calculate its geometry, which ensures complete selection of the petroleum gas released into the oil well ring. Such activities will contribute to the intensification of oil extraction from boreholes and the improvement of the environment in the oil field.

The correct selection of drilling fluids is also an issue that contributes to Enhanced Oil Recovery. Therefore, in the article [65], the authors presented the influence of enzymatic and oxidizing factors on polymers used in drilling mud technology. The concentration of calcium hypochlorite, urea peroxide, sodium persulfate, amylase and cellulase was determined to reduce the rheological parameters of drilling fluids. Additionally, researchers developed a method of treating drilling fluid prior to cementation, and developed a drilling fluid containing enzymatic or oxidizing agents to prepare the borehole for cementation [66,67]. On the basis of the work carried out, the authors obtained positive results regarding the possibility of diluting the drilling fluid immediately before cementing, and it is possible to use oxidants and enzymes in the composition of the drilling fluids [68,69].

3. Conclusions

The present publication focuses on issues related to improved oil recovery technology. The articles presented in this Special Issue show various related challenges, including the following: increasing the efficiency of oil recovery, improved borehole sealing, correct selection of drilling fluids, reducing energy consumption, appropriate energy management in the oil industry, new technologies supporting the efficiency of recovery, technical and technological challenges, operational challenges and innovative technologies in oil drilling. The works included in this Special Issue cover important elements not only of the exploitation of oil and gas fields, but most of all the most important aspects, which include the elements of borehole implementation. The published submissions cover a broad spectrum of technologies that address the fundamentals of improved oil recovery. The task of editing and selecting articles for this collection was both stimulating and rewarding. We would like to thank the staff and reviewers very much for their efforts and contributions. Thanks to their hard work, a very interesting Special Issue has been created, which will contribute to increasing knowledge and to the further development of technology to improve oil recovery.

Conflicts of Interest: The author declares no conflict of interest.

References

1. Kremieniewski, M.; Wiśniowski, R.; Stryczek, S.; Orłowicz, G. Possibilities of Limiting Migration of Natural Gas in Boreholes in the Context of Laboratory Studies. *Energies* **2021**, *14*, 4251. [CrossRef]
2. Kremieniewski, M. O konieczności prowadzenia serwisowych badań parametrów technologicznych zaczynów uszczelniających. *Nafta-Gaz* **2019**, *1*, 48–55. [CrossRef]
3. Rogers, M.J.; Dillenbeck, R.L.; Eid, R.N. Transition time of cement slurries, definitions and misconceptions, related to annular fluid migration. In Proceedings of the SPE Annual Technical Conference and Exhibition, Houston, TX, USA, 26–29 September 2004; Society of Petroleum Engineers: Richardson, TX, USA, 2004.
4. Kremieniewski, M. Recipe of Lightweight Slurry with High Early Strength of the Resultant Cement Sheath. *Energies* **2020**, *13*, 1583. [CrossRef]
5. Kremieniewski, M.; Stryczek, S.; Wiśniowski, R.; Rzepka, M.; Gonet, A. Influence of bentonite addition on parameters of fresh and hardened cement slurry. *AGH Drill. Oil Gas* **2017**, *34*, 335–348. [CrossRef]
6. Błaż, S.; Zima, G.; Jasiński, B.; Kremieniewski, M. Invert Drilling Fluids with High Internal Phase Content. *Energies* **2021**, *14*, 4532. [CrossRef]
7. Błaż, S. Badania laboratoryjne nad opracowaniem składu płuczki inwersyjnej. *Nafta-Gaz* **2015**, *3*, 54–63.
8. Błaż, S. Analiza właściwości technologicznych płuczki inwersyjnej w warunkach HPHT. *Nafta-Gaz* **2016**, *6*, 403–412. [CrossRef]
9. Elkatatny, S. Mitigation of barite sagging during the drilling of high-pressure high-temperature wells using an invert emulsion drilling fluid. *Powder Technol.* **2019**, *352*, 325–330. [CrossRef]
10. Askø, A.; Alsvik, E.T.; Danielsen, T.H.; Haga, M.A. Low-Density Invert Emulsion Drilling Fluid Enables Recovery of Oil Reserves in Extremely Depleted Reservoirs: A Case History from Valhall, Norway. In Proceedings of the IADC/SPE International Drilling Conference and Exhibition, Galveston, TX, USA, 3–5 March 2020.
11. Warnecki, M.; Wojnicki, M.; Kuśnierczyk, J.; Szuflita, S. Study of the Long Term Acid Gas Sequestration Process in the Borzęcin Structure: Measurements Insight. *Energies* **2021**, *14*, 5301. [CrossRef]
12. Lubaś, J.; Szott, W. 15-year experience of acid gas storage in the natural gas structure of Borzęcin—Poland. *Nafta-Gaz* **2010**, *66*, 333–338.
13. Lubaś, J.; Szott, W.; Jakubowicz, P. Effects of Acid Gas Reinjection on CO2 Concentration in Natural Gas Produced from Borzęcin Reservoir. *Nafta-Gaz* **2012**, *68*, 405–410.
14. Chadwick, R.A.; Marchant, B.; Williams, G.A. CO_2 storage monitoring: Leakage detection and measurement in subsurface volumes from 3D seismic data at Sleipner. *Energy Procedia* **2014**, *63*, 4224–4239. [CrossRef]
15. McGlade, C.; Sondak, G.; Han, M. Whatever Happened to Enhanced Oil Recovery? International Energy Agency. 2018. Available online: https://www.iea.org/commentaries/whatever-happened-to-enhanced-oil-recovery (accessed on 10 January 2022).
16. Wojnicki, M.; Lubaś, J.; Gawroński, M.; Szuflita, S.; Kuśnierczyk, J.; Warnecki, M. An Experimental Investigation of WAG Injection in a Carbonate Reservoir and Prediction of the Recovery Factor Using Genetic Programming. *Energies* **2022**, *15*, 2127. [CrossRef]
17. Mogensen, K.; Masalmeh, S. A review of EOR techniques for carbonate reservoirs in challenging geological settings. *J. Pet. Sci. Eng.* **2020**, *195*, 107889. [CrossRef]

18. Afzali, S.; Rezaei, N.; Zendehboudi, S. A comprehensive review on Enhanced Oil Recovery by Water Alternating Gas (WAG) injection. *Fuel* **2018**, *227*, 218–246. [CrossRef]
19. Kremieniewski, M. Cleaning of the casing string before cementation, based on research using a rotational viscometer. *Nafta-Gaz* **2018**, *74*, 676–683. [CrossRef]
20. Lavrov, A.; Torsæter, M. *Physics and Mechanics of Primary Well Cementing*; Springer International Publishing: Houston, TX, USA, 2016. Available online: https://www.springer.com/gp/book/9783319431642 (accessed on 9 August 2021); ISBN 978-3-319-43165-9.
21. Stryczek, S.; Wiśniowski, R.; Gonet, A.; Rzyczniak, M.; Sapińska-Śliwa, A. Wpływ wybranych superplastyfikatorów na właściwości reologiczne zaczynów cementowych stosowanych podczas cementowania kolumn rur okładzinowych w otworach wiertniczych. *Przemysł Chem.* **2018**, *97*, 903–905. [CrossRef]
22. Kremieniewski, M.; Błaż, S.; Stryczek, S.; Wiśniowski, R.; Gonet, A. Effect of Cleaning the Annular Space on the Adhesion of the Cement Sheath to the Rock. *Energies* **2021**, *14*, 5187. [CrossRef]
23. Jasiński, B. Ocena wpływu cieczy przemywającej na jakość zacementowania rur w otworze wiertniczym po użyciu płuczki glikolowo-potasowej. *Nafta-Gaz* **2016**, *6*, 413–421. [CrossRef]
24. Mandal, P.P.; Rezaee, R.; Emelyanova, I. Ensemble Learning for Predicting TOC from Well-Logs of the Unconventional Goldwyer Shale. *Energies* **2021**, *15*, 216. [CrossRef]
25. Ma, L.; Slater, T.; Dowey, P.J.; Yue, S.; Rutter, E.; Taylor, K.G.; Lee, P.D. Hierarchical integration of porosity in shales. *Sci. Rep.* **2018**, *8*, 11683. [CrossRef] [PubMed]
26. Kwilosz, T.; Filar, B.; Miziołek, M. Use of Cluster Analysis to Group Organic Shale Gas Rocks by Hydrocarbon Generation Zones. *Energies* **2022**, *15*, 1464. [CrossRef]
27. Piesik-Buś, B.; Filar, B. Analysis of the current state of natural gas resources in domestic deposits and a forecast of domestic gas production until 2030. *Nafta-Gaz* **2016**, *6*, 376–382.
28. Puskarczyk, E. Application of Multivariate Statistical Methods and Artificial Neural Network for Facies Analysis from Well Logs Data: An Example of Miocene Deposits. *Energies* **2020**, *13*, 1548. [CrossRef]
29. Mroczkowska-Szerszeń, M.; Ziemianin, K.; Brzuszek, P.; Matyasik, I.; Jankowski, L. The organic matter type in the shale rock samples assessed by FTIR-ART analyses. *Nafta-Gaz* **2015**, *6*, 361–369.
30. Łętkowski, P.; Gołąbek, A.; Budak, P.; Szpunar, T.; Nowak, R.; Arabas, J. Determination of the statistical similarity of the physicochemical measurement data of shale formations based on the methods of cluster analysis. *Nafta-Gaz* **2016**, *72*, 910–918. [CrossRef]
31. Ning, T.; Xi, M.; Hu, B.; Wang, L.; Huang, C.; Su, J. Effect of Viscosity Action and Capillarity on Pore-Scale Oil–Water Flowing Behaviors in a Low-Permeability Sandstone Waterflood. *Energies* **2021**, *14*, 8200. [CrossRef]
32. Su, J.; Wang, L.; Gu, Z.; Zhang, Y.; Chen, C. Advances in Pore-Scale Simulation of Oil Reservoirs. *Energies* **2018**, *11*, 1132. [CrossRef]
33. Kamal, M.S.; Hussein, I.A.; Sultan, A.S. Review on Surfactant Flooding: Phase Behavior, Retention, IFT, and Field Applications. *Energy Fuels* **2017**, *31*, 7701–7720. [CrossRef]
34. Tsuji, T.; Jiang, F.; Christensen, K. Characterization of immiscible fluid displacement processes with various capillary numbers and viscosity ratios in 3D natural sandstone. *Adv. Water Resour.* **2016**, *95*, 3–15. [CrossRef]
35. Guo, Y.; Zhang, L.; Zhu, G.; Yao, J.; Sun, H.; Song, W.; Yang, Y.; Zhao, J. A Pore-Scale Investigation of Residual Oil Distributions and Enhanced Oil Recovery Methods. *Energies* **2019**, *12*, 3732. [CrossRef]
36. El-Gamal, S.M.; Hashem, F.S.; Amin, M.S. Influence of carbon nanotubes, nanosilica and nanometakaolin on some morphologicalmechanical properties of oil well cement pastes subjected to elevated water curing temperature and regular room air curing temperature. *Constr. Build. Mater.* **2017**, *146*, 531–546. [CrossRef]
37. Khalil, M.; Jan, B.M.; Tong, C.W.; Berawi, M.A. Advanced nanomaterials in oil and gas industry: Design, application and challenges. *Appl. Energy* **2017**, *191*, 287–310. [CrossRef]
38. Lau, H.C.; Yu, M.; Nguyen, Q.P. Nanotechnology for oilfield applications: Challenges and impact. *J. Pet. Sci. Eng.* **2017**, *157*, 1160–1169. [CrossRef]
39. Pikłowska, A.; Ziaja, J.; Kremieniewski, M. Influence of the Addition of Silica Nanoparticles on the Compressive Strength of Cement Slurries under Elevated Temperature Condition. *Energies* **2021**, *14*, 5493. [CrossRef]
40. Kremieniewski, M. Influence of Graphene Oxide on Rheological Parameters of Cement Slurries. *Energies* **2020**, *13*, 5441. [CrossRef]
41. Błaż, S. Nowe rodzaje cieczy przemywających osady z płuczki inwersyjnej przed zabiegiem cementowania otworów wiertniczych. *Nafta-Gaz* **2017**, *5*, 302–311. [CrossRef]
42. Wiśniowski, R.; Skrzypaszek, K.; Małachowski, T. Selection of a suitable rheological model for drilling fluid using applied numerical methods. *Energies* **2020**, *13*, 3192. [CrossRef]
43. Kremieniewski, M.; Kędzierski, M.; Błaż, S. Increasing the Efficiency of Sealing the Borehole in Terms of Spacer Pumping Time. *Energies* **2021**, *14*, 6702. [CrossRef]
44. Kremieniewski, M. Hybrid Washer Fluid for Primary Cementing. *Energies* **2021**, *14*, 1295. [CrossRef]
45. Bęben, D. The Influence of Temperature on Degradation of Oil and Gas Tubing Made of L80-1 Steel. *Energies* **2021**, *14*, 6855. [CrossRef]

46. Zhang, Y.; Pang, X.; Qu, S.; Gao, X.; Li, K. The relationship between fracture toughness of CO_2 corrosion scale and corrosion rate of X65 pipeline steel under supercritical CO2 condition. *Int. J. Greenh. Gas Control* **2011**, *5*, 1643–1650. [CrossRef]
47. Bęben, D. Badania skuteczności działania wybranych inhibitorów korozji stosowanych okresowo w przemyśle wydobywczym. *Ochr. Przed Korozją* **2019**, *62*, 376–381.
48. Mahmoodian, M.; Qingi, C. Failure assessment and safe life prediction of corroded oil and gas pipelines. *J. Pet. Sci. Eng.* **2017**, *151*, 434–438. [CrossRef]
49. Yougui, Z. Electrochemical Mechanism and Model of H_2S Corrosion of Carbon Steel. Ph.D. Thesis, Ohio University, Athens, GA, USA, 2015.
50. Stachowicz, A. Korozja rur wydobywczych odwiertów gazowych zawierających CO_2. *Nafta-Gaz* **2011**, *11*, 395–400.
51. Falkowicz, S.; Urbaniec, A.; Stadtműller, M.; Majkrzak, M. A New Strategy for Pre-Selecting Gas Wells for the Water Shut-Off Treatment Based on Geological Integrated Data. *Energies* **2021**, *14*, 7316. [CrossRef]
52. Lakatos, I.J.; Lakatos-Szabo, J.; Szentes, G.; Jobbik, A.; Vago, A. Application of Self-Conforming well Stimulation technology in Oil and Gas Fields—Fundamentals and Case Histories. In Proceedings of the IOR 2017—19th European Symposium on Improved Oil Recovery, Stavanger, Norway, 24–27 April 2017.
53. Alfarge, D.K.; Wei, M.; Bai, B. Numerical simulation study of factors affecting relative permeability modification for water-shutoff treatments. *Fuel* **2017**, *207*, 226–239. [CrossRef]
54. Myśliwiec, M. Poszukiwania złóż gazu ziemnego w osadach miocenu zapadliska przedkarpackiego na podstawie interpretacji anomalii sejsmicznych—podstawy fizyczne i dotychczasowe wyniki. *Prz. Geol.* **2004**, *52*, 299–306.
55. Dziadzio, P. Sekwencje depozycyjne w utworach badenu i sarmatu w SE części zapadliska przedkarpackiego. *Prz. Geol.* **2000**, *48*, 1124–1138.
56. Stryczek, S.; Wiśniowski, R.; Gonet, A.; Złotkowski, A. Wpływ rodzaju cementu na właściwości reologiczne zaczynów uszczelniających stosowanych w technologiach wiertniczych. *Wiert. Naft. Gaz* **2010**, *27*, 721–739.
57. Kremieniewski, M. Influence of Hblock Fine-Grained Material on Selected Parameters of Cement Slurry. *Energies* **2022**, *15*, 2768. [CrossRef]
58. Stryczek, S.; Małolepszy, J.; Gonet, A.; Wiśniowski, R.; Kotwica, Ł. *Wpływ Dodatków Mineralnych na Kształtowanie Się Właściwości Technologicznych Zaczynów Uszczelniających Stosowanych w Wiertnictwie i Geoinżynierii Wydawnictwo*; S.C.M.R: Kraków, Poland, 2011; pp. 1–164. Available online: http://katalog.nukat.edu.pl/lib/item?id=chamo:4133282&fromLocationLink=false&theme=nukat (accessed on 21 May 2015).
59. Kremieniewski, M.; Stryczek, S. Zastosowanie cementu wysokoglinowego do sporządzania zaczynów uszczelniających w technologiach wiertniczych. *Cem. Wapno Beton* **2019**, *22*, 215–226.
60. Stryczek, S.; Brylicki, W.; Małolepszy, J.; Gonet, A.; Wi´sniowski, R.; Kotwica, Ł. Potential use of fly ash from fluidal combustion of brown coal in cementing slurries for drilling and geotechnical works. *Arch. Min. Sci.* **2009**, *54*, 775–786.
61. Bazaluk, O.; Dubei, O.; Ropyak, L.; Shovkoplias, M.; Pryhorovska, T.; Lozynskyi, V. Strategy of Compatible Use of Jet and Plunger Pump with Chrome Parts in Oil Well. *Energies* **2022**, *15*, 83. [CrossRef]
62. Le Billon, P.; Kristoffersen, B. Just cuts for fossil fuels? Supplyside carbon constraints and energy transition. *Environ. Plan. A Econ. Space* **2020**, *52*, 1072–1092. [CrossRef]
63. Pavlychenko, A.; Kovalenko, A. The investigation of rock dumps influence to the levels of heavy metals contamination of soil. In *Mining of Mineral Deposits*; CRC Press: Boca Raton, FL, USA, 2013; pp. 237–238.
64. Skitsa, L.; Yatsyshyn, T.; Liakh, M.; Sydorenko, O. Ways to improve safety of a pumping-circulatory system of a drilling rig. *Min. Miner. Depos.* **2018**, *12*, 71–79. [CrossRef]
65. Uliasz, M.; Zima, G.; Błaż, S.; Jasiński, B. Enzymatic and Oxidizing Agents for the Decomposition of Polymers Used in the Composition of Drilling Muds. *Energies* **2021**, *14*, 5032. [CrossRef]
66. Błaż, S. Dobór środków chemicznych do degradacji polimerów i koloidów ochronnych w płuczkach wiertniczych. *Nafta-Gaz* **2009**, *5*, 371–383.
67. Jasiński, B. Wpływ oksydantów na wielkość sedymentacji fazy stałej w zasolonych płuczkach wiertniczych. *Nafta-Gaz* **2012**, *9*, 602–610.
68. Uliasz, M.; Zima, G.; Błaż, S.; Jasiński, B. Ocena właściwości cieczy wiertniczych w aspekcie zapobiegania migracji gazu w otworach na przedgórzu Karpat. *Nafta-Gaz* **2015**, *1*, 11–17.
69. Nasr-El-Din, H.A.; Al-Otaibi, M.B.; Al-Qahtani, A.A.; Samuel, M.M. An Effective Fluid Formulation to Remove Drilling Fluid Mud Cake in Horizontal and Multi-LateralWells. *SPE Drill. Complet.* **2007**, *22*, 26–32. [CrossRef]

energies

Article

Possibilities of Limiting Migration of Natural Gas in Boreholes in the Context of Laboratory Studies

Marcin Kremieniewski [1], Rafał Wiśniowski [2], Stanisław Stryczek [2] and Grzegorz Orłowicz [2,*]

1 Oil and Gas Institute, National Research Institute, 25A Lubicz Street, 31-503 Kraków, Poland; kremieniewski@inig.pl
2 Department of Drilling and Geoengineering, Faculty of Drilling, Oil and Gas, AGH University of Science and Technology, 30 Mickiewicza Ave., 30-059 Kraków, Poland; wisniows@agh.edu.pl (R.W.); stryczek@agh.edu.pl (S.S.)
* Correspondence: orlowiczg@agh.edu.pl; Tel.: +48-517-693-870

Abstract: Gas migration through fresh and hardened cement slurry is an ongoing problem in the oil industry. In order to eliminate this unfavourable phenomenon, research is being conducted on new compositions of slurries for gas wells. The article presents the results of research for slurries with low and high resistance to gas migration. The proper selection of the quantity and quality of components makes it possible to design slurry with the required static structural strength values. In addition, the cement sheath of such anti-migration slurry has low porosity and a very low proportion of large pore spaces. Additionally, the mechanical parameters do not decrease during long-term deposition in borehole-like conditions. By obtaining these results, it was possible to design slurry whose cement sheath has high corrosion resistance. The new slurry has a lower water-cement ratio. Additionally, GS anti-migration copolymer, anti-filter additive and latex are used. The presence of n-SiO2 aqueous solution and microcement allows for sealing the microstructure of the hardened cement slurry. Such modifications significantly improve the technological parameters of the cement slurry and the cement coat formed from it.

Keywords: gas migration; well cementing; cement slurry; cement sheath; corrosion resistance; gas outflows

Citation: Kremieniewski, M.; Wiśniowski, R.; Stryczek, S.; Orłowicz, G. Possibilities of Limiting Migration of Natural Gas in Boreholes in the Context of Laboratory Studies. *Energies* **2021**, *14*, 4251. https://doi.org/10.3390/en14144251

Academic Editor: Rouhi Farajzadeh

Received: 7 June 2021
Accepted: 12 July 2021
Published: 14 July 2021

1. Introduction

Limitation of uncontrolled natural gas flows from the ring and/or annular space of a borehole is one of the crucial aspects during sealing of casing in boreholes drilled in formation with a higher risk of gas migration (the so-called exhalation). It relates to the main objective of casing sealing, i.e., ensuring durability and tightness between the lowered casing, on the one hand, and the drillhole wall and the previously cemented casing string, on the other [1–5]. As there are cases of gas invasion into the structure of the bonding slurry, resulting in leaks in the structure of the cement coating, it is necessary to carry out research and development work aimed at continually improving the quality of cement slurries formulations and the slurry injection technology [2,5–7]. During cement slurry bonding, in the process of its hydration, gel structure is created and at that time a clear decline of hydrostatic pressure in the ring space of a drillhole is observed. When slurry changes its structure, it is very likely that gas moves in the gelled slurry structure, which may consequently result in gas migration both through fresh and hardened cement slurry [2,3,5,8–13].

2. Uncontrolled Flows of Fluids in a Borehole

Interhorizontal flows of drilling fluids (natural gas, crude oil, brine, drilling fluid) are caused by the lack of balance between formation and hydrostatic pressures in the ring space of a wellbore.

Uncontrolled natural gas flows within particular horizons being drilled are called migrations (drillhole no. 1, Figure 1), whereas gas flow from rock formation on the ground surface is called exhalation (drillhole no. 2, Figure 1).

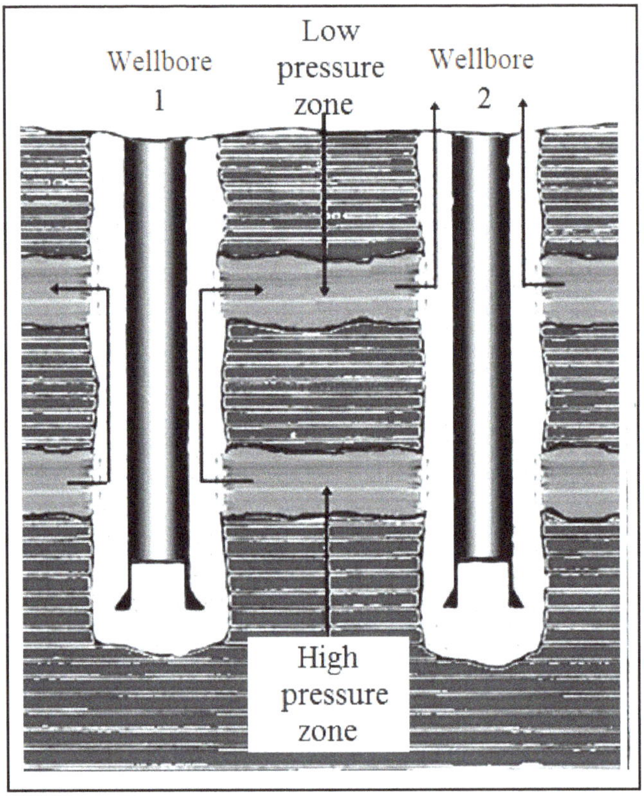

Figure 1. Example of gas migration and exhalation.

The first issues of this kind were observed in the 1960s when making a borehole for gas storage. The topic of uncontrolled natural gas outflows from ring and/or annular spaces of a wellbore is very important in the context of the influence on safety, environmental protection, in terms of legal regulations, with regard to suitable management and exploitation of the borehole area. The results of studies conducted by oil companies (Schlumberger, Halliburton) prove that ca. 15% of failed cementing of a casing string in wellbores are caused by gas migration [4,14].

Presently, this problem occurs to a much higher extent, mainly on some gas formations and in boreholes prepared for gas storage. The above is confirmed by the analysis of the state of cementation of boreholes carried out (in the years 2008 to 2011) in the area of the Carpathian Foredeep and in the Carpathians, which is presented in Figure 2 [15].

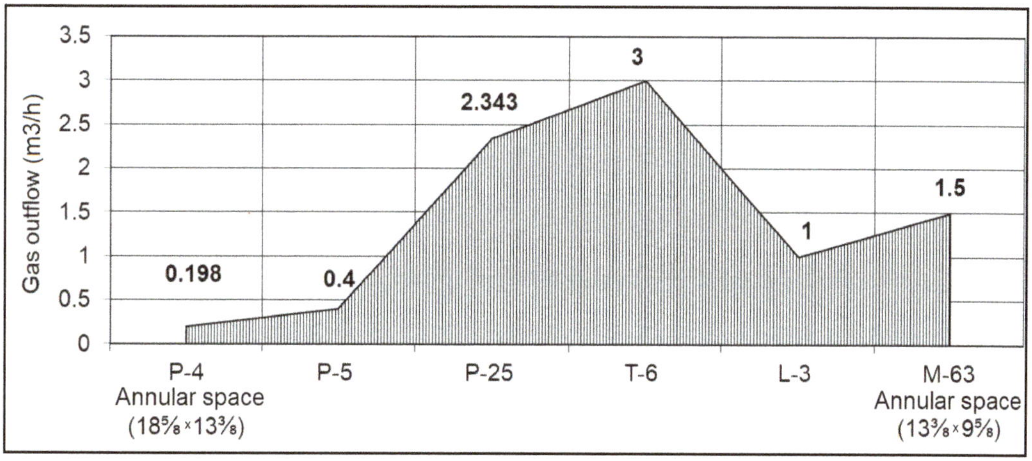

Figure 2. Gas outflows from the ring space in boreholes drilled in the Carpathian Foredeep and Carpathians in the years 2008—2011. P-4—Pogwizdów 4, P-5—Pogwizdów 5, P-25—Pruchnik 25, T-6—Terliczka 6K, L-3—Łapanów 3, M-63—Mirocin 63.

In extreme situations, where an abnormally high pressure occurs after cementing of the casing on a wellhead, it may be necessary to decommission the borehole. In most cases, however, this problem is solved by using different techniques to cement the leaking zone in order to eliminate the gas flow or to reduce the pressure to a value that allows further work to be carried out in compliance with the applicable standards. Such procedures, however, are very costly, so the optimum solution is to minimise uncontrolled gas flows from the design stage of the formulation and the technological parameters of the slurry, while conducting laboratory tests [3,16–20].

3. Minimizing of Uncontrolled Gas Flows

The sealing slurry changes its physical state during hydration from the liquid phase during injection, through the transition phase—gel (slurry bonding), to the solid phase (bonded) slurry. Taking the technological properties of the sealing slurry as a criterion, it can be noted that the potential factors minimising the exhalation of natural gas after cementing include:

- Achieving the end of the slurry bonding before the hydrostatic pressure drops below the formation pressure; reducing the possibility of gas intrusion into the structure of the bonding slurry (Figure 3). Apparatus from Figure 4 was used for measurements.
- Ensuring the shortest possible transition time from 50 Pa to 250 Pa during the build-up of the static structural strength of the slurry (Figure 5). This results in the tightening of possible gas channels in the structure of the bonding slurry [3,9,12,14,18]. The apparatus from Figure 6 was used for measurements.

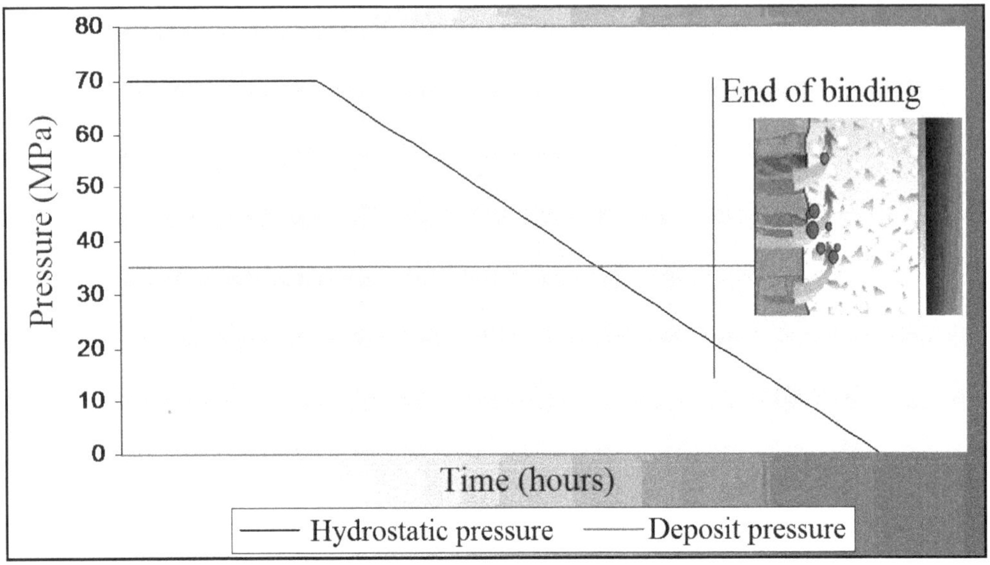

Figure 3. Diagram of hydrostatic pressure reduction below formation pressure.

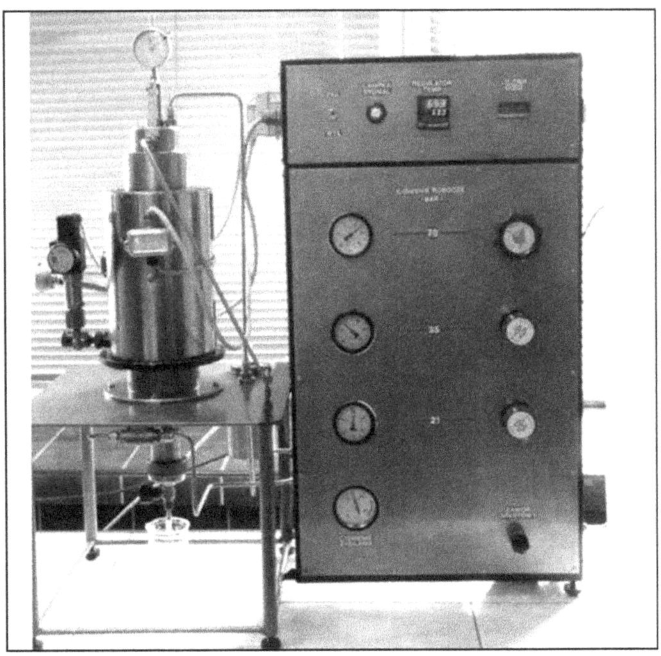

Figure 4. Apparatus for testing gas migration during slurry bonding—hydraulic press (OGI-NRI design).

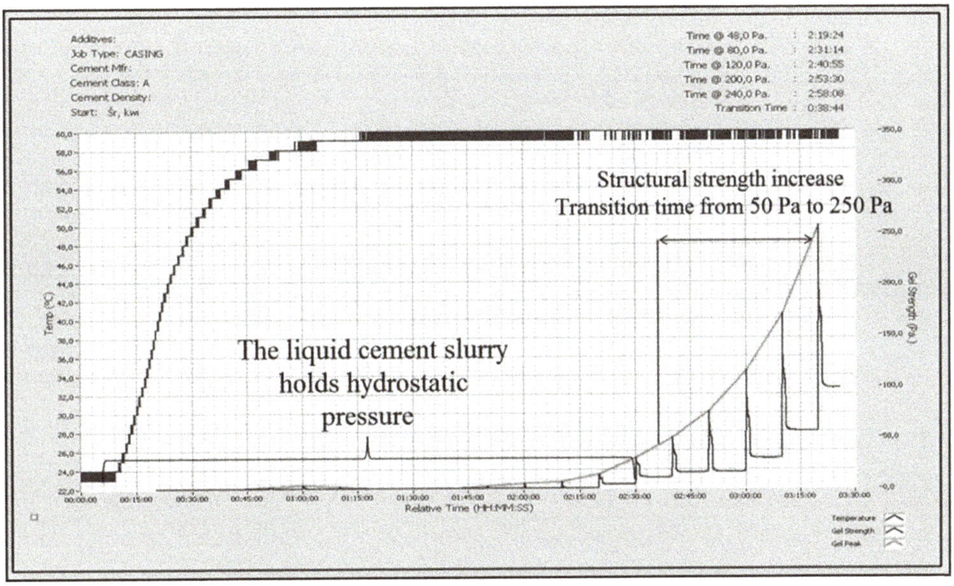

Figure 5. Diagram for the build-up of static structural strength and the formation of a transition period.

Figure 6. Apparatus for testing the build-up of static structural strength in slurry—two-chamber UCA SGSM Ofite model 120-53.

When developing sealing slurry formulations, various additives and admixtures are used in order to reduce the possibility of uncontrolled gas flows in boreholes through the setting slurry and after it has set, i.e., microcements, microsilicas, nanosilicas, which tighten the microstructure, as well as latex and cationic polymers, which shorten the gelling time of the slurry and thus reduce the possibility of natural gas flowing through the hardened cement slurry [2,13,21–24].

4. Research Findings

Laboratory tests conducted by the Oil and Gas Institute, the National Research Institute and AGH University of Science and Technology show significant differences in effectiveness of removing uncontrolled gas outflows as a result of comparing technological parameters of the base slurry, on the one hand, and the slurry after modification with selected additives and admixtures, on the other. Table 1 shows the control slurry formulation (Contr slurry) and the modified formulation (Mdf slurry). The technological parameters of the fresh slurries are summarised in Table 2. There was a significant reduction in the filtration of the slurry from a value of 150 $cm^3/30$ min to just over 2 $cm^3/30$ min. The values of rheological parameters have also decreased. In order to minimise the possibility of uncontrolled gas outflows at the contact between the cement coating—rock formation and the cement coating—casing pipes, the presence of gravitational water was eliminated. The changes made to the slurry formulations result in short thickening times from 30 Bearden Consistency units (Bc) to 100 Bc, thus minimizing the occurrence of gas migration during slurry bonding.

Table 1. Base slurry and slurry with selected additives and admixtures.

Parameter	Kind of Slurry	
	Contr	**Mdf**
w/c water	0.54	0.45
Defoaming additive	0.30	0.50
Liquefying additive	0.20	0.20
Anti-filtration additive	0.20	0.20
GS additive	—	4.0
C_2 anti-filtration additive	—	2.0
Latex	—	6.0
Latex stabilizer	—	2.0
n-SiO_2 aqueous solution	—	6.0
Microcement	—	10.0
G High Sulphate Resistant (G-HSR) oil well cement	100.0	100.0
Bulking additive	0.30	0.30

Table 2. Technological parameters of the fresh slurries.

Parameter		Kind of Slurry	
		Contr	**Mdf**
Density (g/cm^3)		1.78	1.81
Flow properties (mm)		250	300
Filtration ($cm^3/30$ min)		150	2.2
Plastic viscosity (mPa·s)		220	61.5
Flow limit (Pa)		23.5	2.1
Structural strength (Pa)		5.8	3.6
Water separation (%)		3.0	0.0
	Bc value	Time: min.	
Thickening time	30 Bc	2–48	2–19
	100 Bc	4–43	2–29
From 30 to 100 Bc (min)		115	10

Figures 7 and 8 show the gas migration test and static structural strength build-up tests for the control slurry. This slurry obtained the end of bonding time after lowering the hydrostatic pressure value below the formation pressure. The consequence of this can be gas intrusion into the structure of the cement coating and the subsequent creation of gas migration paths. Additionally, a large amount of filtrate in the upper part of the measuring cylinder indicates a low anti-migration efficiency of the slurry in question. During the static structural strength test (Figure 8), no transition period was obtained from a value of 50 Pa to 250 Pa.

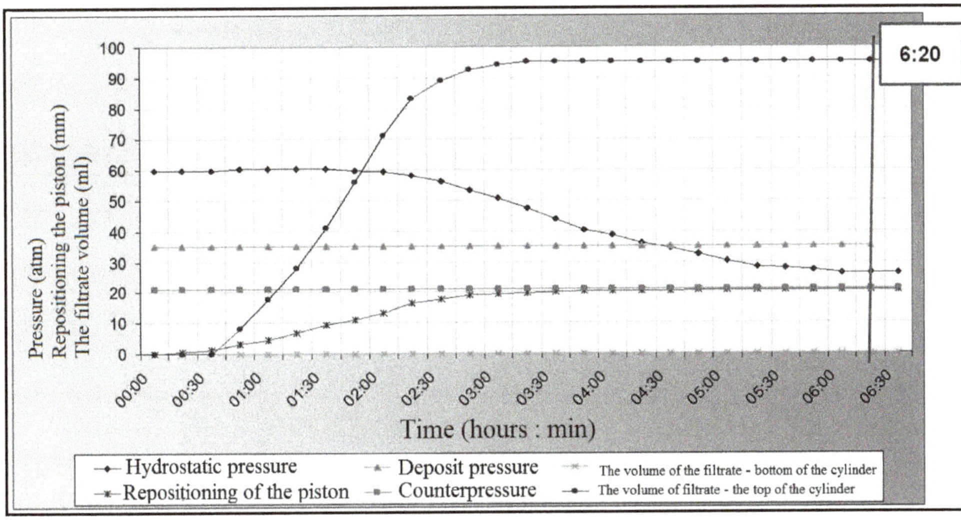

Figure 7. Course of the gas migration test during bonding for slurry with low anti-migration resistance.

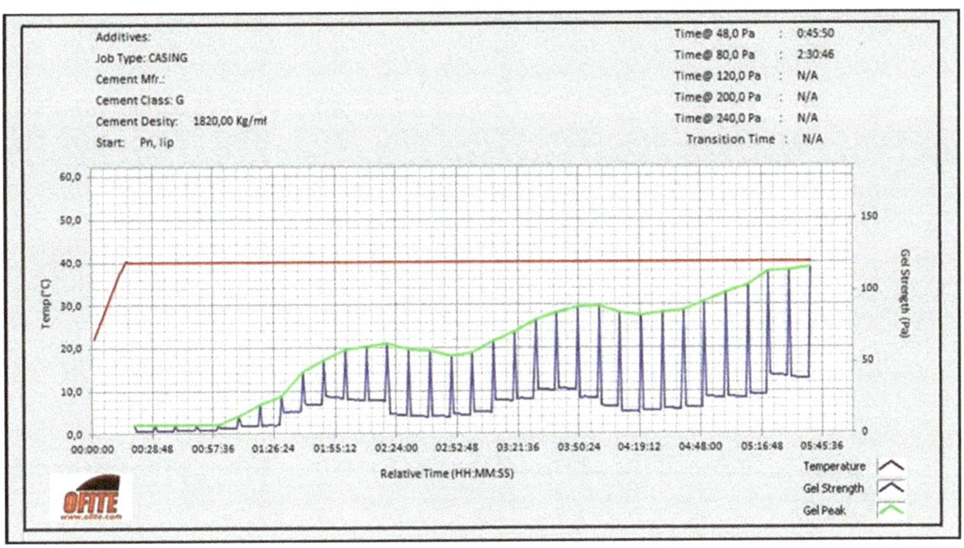

Figure 8. Static structural strength build-up for slurry with low anti-migration resistance.

Figures 9 and 10 show the results of tests on modified slurry which is guaranteed to reduce the possibility of uncontrolled gas flow. This slurry reaches the end of bonding (EB) before the hydrostatic pressure drops below the formation pressure. Thus, it shows a very good ability to maintain the hydrostatic pressure at a high level (Figure 9). Slurry filtration at the top of the cylinder was low, confirming the effectiveness of this slurry in counteracting gas migration. Furthermore, a short transition time of 52 min from 50 Pa to 250 Pa and a steady increase in structural strength were obtained (Figure 10).

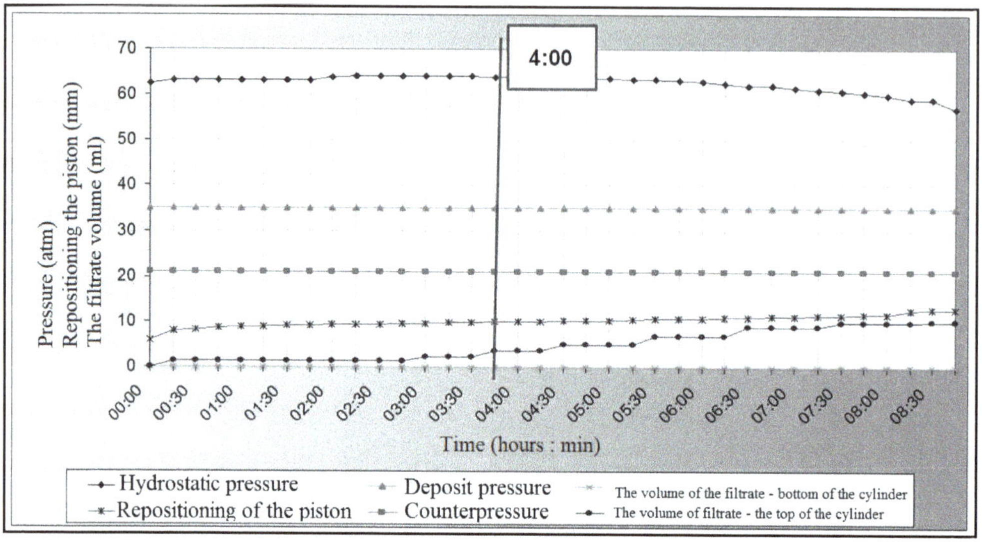

Figure 9. Course of the gas migration test during setting for slurry with high anti-migration resistance.

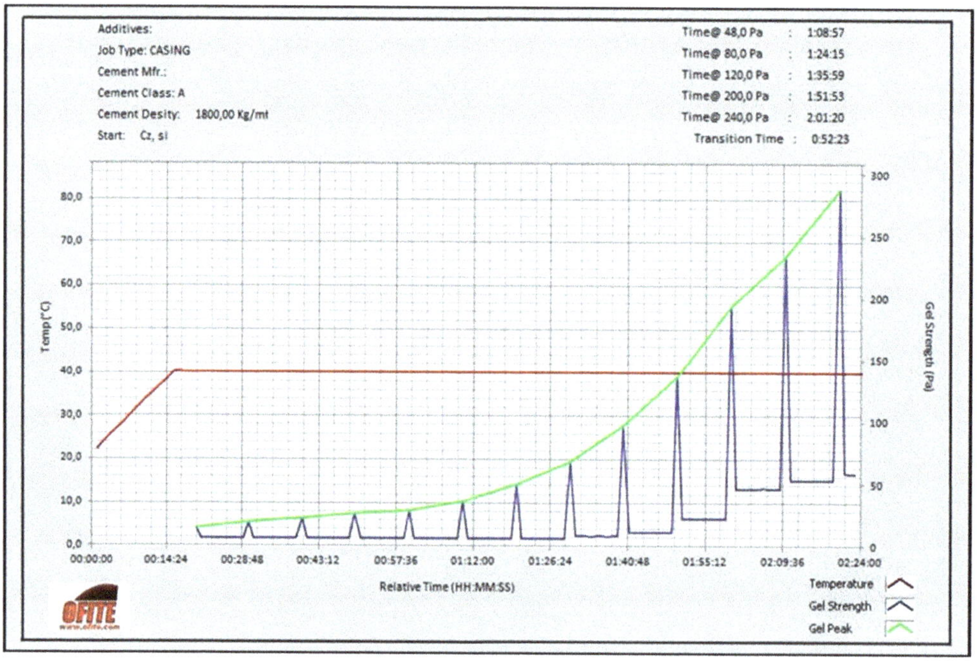

Figure 10. Static structural strength build-up for slurry with high anti-migration resistance.

5. Analysis of the Quality and Durability of the Cement Coating with Regard to the Possibility of Uncontrolled Gas Flows

When analysing the problem of gas outflow through the cement coating, corrosion of the hardened slurry in the annular space must also be taken into account. Test results available in the literature indicate that after 30 years, the cement coating has corroded in approx. 60% of the boreholes and deteriorated its tightness, as shown graphically in Figure 11. The main factors causing corrosion of cement coating in a borehole are formation brine, high temperature and pressure at a given depth [1,11]. The combination of this groups of factors results in chemical compounds with reduced mechanical strength and sometimes no bonding ability [4]. Therefore, in order to design the sealing slurry formulation with improved corrosion resistance, it is necessary to:

- Ensure the lowest possible content of $Ca(OH)_2$ (portlandite) and C_3A phase in the slurry.
- Provide the highest possible amount of calcium silicate hydrate (CSH) phase building up the cement matrix.
- Ensure that the permeability and porosity of the slurry is as low as possible (taking into account that the proportion of gel pores is greater than the proportion of capillary pores) [7,16,17,19,25].

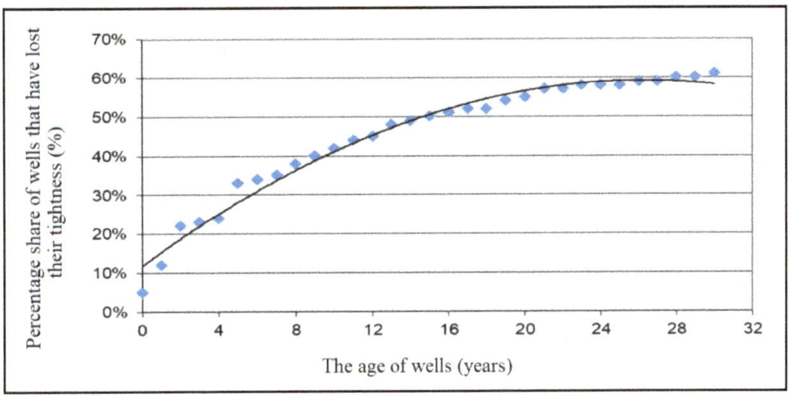

Figure 11. Influence of well ageing on the emergence of pressure in the annular space [26].

The destructive effect of ions present in reservoir brines on hardened cement slurries can be represented by the following chemical reactions [11,19,25]:

- Double exchange between magnesium and calcium cations, e.g.,

Magnesium ions (Mg^{2+}).

- Transformation of portlandite into brucite

$$Ca(OH)_2 + MgCl_2 \rightarrow Mg(OH)_2 + CaCl_2$$

- Extraction of calcium ions Ca^{2+} from the CSH phase and formation of brucite

$$C_{1,75}SH_n + MgCl_2 \rightarrow C_{1,75\text{-}x}SH_n + x\,CaCl_2 + x\,Mg(OH)_2$$

- Formation of expansive hydrated basic magnesium chloride

$$2Ca(OH)_2 + 3MgCl_2 + 2H_2O \rightarrow Mg_3Cl_2\,(OH)_4 \cdot 2H_2O + 2CaCl_2$$

$$C_{1,75}SHn + 3MgCl_2 \rightarrow Mg_3Cl_2\,(OH)_4 \cdot 2H_2O + C_{1,75\text{-}x}SH_n$$

Sulphate ions (SO_4^{2-}) present in the pores of hardened cement slurry cause the formation of salts that are difficult to dissolve. This process is accompanied by a significant increase in the volume of the slurry, and consequently the formation of high tensile stresses in the cement matrix. The swelling compounds formed are gypsum dihydrate and tricalcium aluminate trisulphate hydrate called ettringite.

The hydration products most frequently attacked by sulphate ions are portlandite and C_3A and hydrated calcium aluminates:

$$Ca(OH)_2 + SO_4{}^{2-} \rightarrow CaSO_4 \cdot 2H_2O + 2OH^-$$
Gypsum dihydrate

$$3\,CaO \cdot Al_2O_3 + SO_4{}^{2-} + n\,H_2O + Ca(OH)_2 \rightarrow 3\,CaO \cdot Al_2O_3 \cdot 3CaSO_4 \cdot 32H_2O + m\,H_2O$$
Ettringite

On the basis of an analysis of long-term laboratory test results on corrosion of hardened cement slurries, the following conclusions can be drawn [3,9,16,17,25]:

- In order to limit magnesium corrosion, measures are taken to achieve the lowest possible permeability and porosity of the hardened cement slurry, which reduces the penetration of aggressive Mg^{2+} ions into the cement matrix. Good results are obtained by using mineral additives (primarily granulated blast furnace slag containing small amounts of calcium compounds or fly ash). Therefore, increased resistance to magnesium ion attack is achieved by using metallurgical or ash cements.
- Prevention of sulphate corrosion is based on the use of cements with increased resistance to sulphate aggression (mainly High Sulphate Resistant (HSR) type oil well cements). Polymeric additives are also used to create a micro-membrane between the cement grains, as well as fine mineral fillers.

Tests to determine the durability of hardened sealing slurries are based on the following procedures [9]:

1. Selection of suitable cement slurry formulations for testing, taking into account borehole conditions.
2. Preparation of corrosive environments depending on the drilling area.
3. Preparation of sealing slurry and making of samples of hardened slurries (hardening of samples in autoclaves, Figure 12).
4. Exposing the samples to the selected corrosive environments for a defined exposure time.
5. Periodic visual assessment of samples, photographic documentation and testing of strength, porosity and gas permeability, etc.
6. Periodic replacement of reservoir brines (every six months).

Samples of hardened slurries are deposited under corrosive conditions for a defined period of time. In autoclaves the temperature reaches 120 °C and the maximum pressure can reach 20 MPa. A reference sample is stored in one chamber of the autoclave in tap water, which represents the reference environment.

Figure 13 shows the values of parameters describing the pore microstructure, while Figure 14 compares the mechanical parameters of samples 1 and 2. Sample 1 is a hardened cement slurry with very low corrosion resistance. This sample has a porosity factor (ratio of the volume of space between the grains to the volume of the entire sample) of 40.8%, a large proportion of the largest pore spaces, and its mechanical parameters decrease significantly during deposition. This sample deteriorated almost completely after 18 months of deposition under conditions of increased corrosivity. Sample no. 2, which was modified to achieve improved corrosion resistance, has a low share of macropores (Figure 13), a low porosity index of 28.7% and a high compressive strength, which is almost constant over time.

Figure 12. Autoclaves for storing samples of cement slurries.

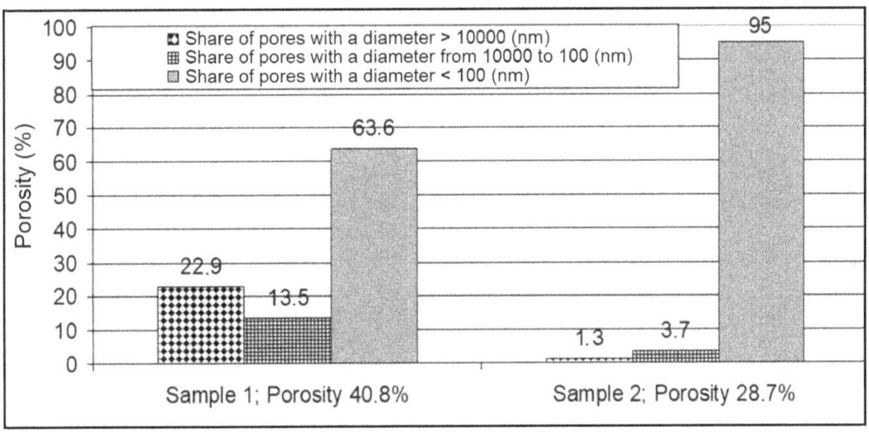

Figure 13. Porosity coefficient values and shares of individual pores for samples 1 and 2.

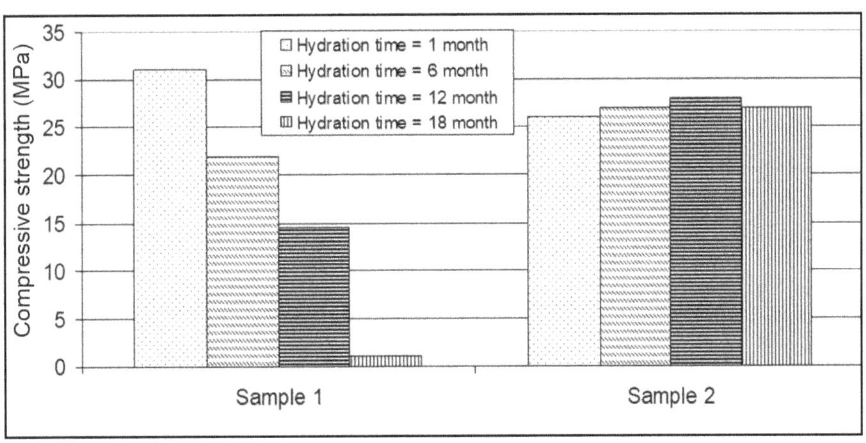

Figure 14. Compressive strength values for samples 1 and 2.

6. Conclusions

When developing sealing slurries to minimise uncontrolled gas leaks, it is important to modify them at the design stage with additives such as microcement, microsilica, latex, high molecular weight cationic polymers, nanosilica or expansive additives.

The modified formulations discussed in this publication are characterised by appropriate densification times for the given conditions. Such values for the time of thickening allow the slurry to be pumped efficiently into the annular space to be sealed, within the time necessary to carry out the procedure. This time is increased by a safety buffer in the event of complications arising in the preparation of the slurry in the field. The designed cement slurry is characterised by a rapid transition from a consistency value of 30 Bc to a value of 100 Bc during the thickening time test. This short transition time of 10 min goes a long way towards minimising the possibility of gas migration through the binding cement slurry. After injection, the slurry begins to gel and a gas-impermeable cement sheath is formed. The designed slurry with anti-migration properties also has the required rheological parameters, which allow it to be pumped efficiently into the cemented annular space, and to efficiently displace the drilling fluid previously present in the borehole. The newly developed slurry for cementing boreholes with an increased risk of gas migration has a low filtration value and does not show any water retention. As a result, the setting and thickening time is maintained at the required level, because the slurry does not "give off" water in the absorption zones that are often found in the area where gas migration occurs. In addition, a slurry with low filtration and no water retention has a homogeneous structure in the liquid state and does not fractionate. The slurry designed for sealing gas wells is resistant to gas migration during setting, as confirmed by the Static Gel Strength Measurement (SGSM) test. With a transition time of 52 min, it can be concluded that the gas will not be able to penetrate the structure of the cement slurry once it is injected.

When analysing the results of tests on the structure of the cement sheath made of slurry with increased resistance to gas migration, it is found that such a formulation has a significantly lower porosity than standard slurries, and a smaller proportion of pore spaces with the largest diameters. Such conditions confirm the required sealing of the borehole and the elimination of microflows through the cement sheath structure. The mechanical strength values obtained for 18 months of deposition under borehole conditions confirm the erosion resistance of the designed slurry. The samples have comparable strength throughout the storage period, with strengths of over 25 MPa. In contrast, the strength of the control sample decreased due to erosion. The strength of this control sample decreases until, after a final period of exposure to erosive conditions, the sample is destroyed.

Long-term tests of changes in mechanical strength, gas permeability, porosity and microstructure are very important in designing an anti-migration slurry. Such tests allow an objective assessment of the course of destructive processes occurring in the cement sheath during long-term exposure in borehole conditions. Research into the durability of hardened cement slurries is required to improve formulas and develop new slurries for sealing boreholes in various geological and technical conditions.

Author Contributions: Conceptualization, M.K.; methodology, M.K.; validation, R.W. and S.S.; formal analysis, R.W. and S.S.; investigation, M.K., R.W. and S.S.; resources, M.K., R.W., S.S. and G.O.; data curation, R.W.; writing—original draft preparation, R.W.; writing—review and editing, G.O.; visualization, S.S.; supervision, S.S.; project administration, M.K., R.W. and S.S.; funding acquisition, M.K. All authors have read and agreed to the published version of the manuscript.

Funding: The paper was financially supported by Ministry of Science and Higher Education Warsaw (Internal order Oil and Gas Institute—National Research Institute Project No. 0015/KW/21).

Data Availability Statement: Not applicable.

Conflicts of Interest: The authors declare no conflict of interest.

References

1. Aadnoy, B.S.; Cooper, I.; Miska, S.Z.; Mitchell, R.F. *Advanced Drilling and Well Technology*; Society of Petroleum Engineers: Richardson, TX, USA, 2009.
2. Irawan, S.; Kalwar, S.A.; Abbas, G.; Kumar, S. An Innovative Solution for Preventing Gas Migration through Cement Slurry in the Lower Indus Basin of Pakistan. In Proceedings of the Offshore Technology Conference Asia, Kuala Lumpur, Malaysia, 20–23 March 2018.
3. Kremieniewski, M.; Rzepka, M.; Stryczek, S.; Wiśniowski, R. Comparison of gas migration test and building structural strength of slurry in the aspect of limiting gas invasion. *AGH Drill. Oil Gas* **2016**, *33*, 595–608. Available online: http://journals.bg.agh.edu.pl/DRILLING/2016.33.3/drill.2016.33.3.595.pdf (accessed on 13 July 2021). [CrossRef]
4. Stryczek, S.; Gonet, A. Kierunki ograniczania migracji gazu z przestrzeni pierścieniowej otworu wiertniczego. *WUG* **2005**, *3*, 10–13. (In Polish)
5. Tao, C.; Rosenbaum, E.; Kutchko, B.G.; Massoudi, M.A. Brief Review of Gas Migration in Oilwell Cement Slurries. *Energies* **2021**, *14*, 2369. [CrossRef]
6. Kremieniewski, M. O konieczności prowadzenia serwisowych badań parametrów technologicznych zaczynów uszczelniających. *Nafta Gaz* **2019**, *1*, 48–55. [CrossRef]
7. Kremieniewski, M.; Rzepka, M. Poprawa szczelności płaszcza cementowego za pomocą innowacyjnych dodatków antymigracyjnych. *Nafta Gaz* **2018**, *6*, 8–15. [CrossRef]
8. Kremieniewski, M. Receptura zaczynu lekkiego do uszczelniania otworów w strefie niskich ciśnień złożowych. *Nafta Gaz* **2020**, *9*, 577–584. [CrossRef]
9. Kremieniewski, M. Recipe of Lightweight Slurry with High Early Strength of the Resultant Cement Sheath. *Energies* **2020**, *13*, 1583. [CrossRef]
10. Lavrov, A.; Torsaeter, M. *Physics and Mechanics of Primary Well Cementing*; Springer: Berlin/Heidelberg, Germany, 2016; pp. 9–62.
11. McLean, R.H.; Manry, C.W.; Whitaker, W.W. *Displacement Mechanics in Primary Cementing*; Society of Petroleum Engineers: Richardson, TX, USA, 1966; Volume SPE 1488.
12. Rogers, M.J.; Dillenbeck, R.L.; Eid, R.N. Transition time of cement slurries, definitions and misconceptions, related to annular fluid migration. In Proceedings of the SPE Annual Technical Conference and Exhibition, Houston, TX, USA, 26–29 September 2004; Society of Petroleum Engineers: Richardson, TX, USA, 2004.
13. Velayati, A. Gas migration through cement slurries analysis: A comparative laboratory study. *IJMGE Int. J. Min. Geo Eng.* **2015**, *49*, 281–288.
14. Kremieniewski, M. Ultra-Lightweight Cement Slurry to Seal Wellbore of Poor Wellbore Stability. *Energies* **2020**, *13*, 3124. [CrossRef]
15. Kremieniewski, M. Analiza parametrów technologicznych wybranych zaczynów lekkich stosowanych do uszczelniania płytkich otworów o temperaturze do 45 °C. *Nafta Gaz* **2020**, *10*, 710–718. [CrossRef]
16. Kotwica, Ł.; Chorembała, M.; Kapeluszna, E.; Stępień, P.; Deja, J.; Illikainen, M.; Gołek, Ł. Effect of metakaolinite on properties of alkali activated slag materials. *Key Eng. Mater.* **2018**, *761*, 69–72. Available online: https://www-1scientific-1net-1mc29xjjl31e7.wbg2.bg.agh.edu.pl/KEM.761.69.pdf (accessed on 13 July 2021). [CrossRef]
17. Kremieniewski, M.; Stryczek, S.; Wiśniowski, R.; Rzepka, M.; Gonet, A. Influence of bentonite addition on parameters of fresh and hardened cement slurry. *AGH Drill. Oil Gas* **2017**, *34*, 335–348. Available online: http://journals.bg.agh.edu.pl/DRILLING/2017.34.2/drill.2017.34.2.335.pdf (accessed on 13 July 2021). [CrossRef]
18. Mitchell, R.F.; Miska, S.Z. *Fundamentals of Drilling Engineering*; SPE 2011 Textbook Series; Society of Petroleum Engineers: Richardson, TX, USA, 2011; Volume 12.
19. Nelson, E.B.; Guillot, D. *Well Cementing*; Schlumberger Educational Service: Houston, TX, USA, 1990.
20. Stryczek, S.; Wiśniowski, R.; Gonet, A.; Złotkowski, A. The influence of time of rheological parameters of fresh cement slurries. *AGH Drill. Oil Gas* **2014**, *31*, 123–133. [CrossRef]
21. Kremieniewski, M. Wpływ drobnoziarnistej krzemionki na parametr czasu oczekiwania na cement—WOC. *Nafta Gaz* **2019**, *11*, 683–690. [CrossRef]
22. Kremieniewski, M. Korelacja skuteczności działania środków dyspergujących o różnym mechanizmie upłynniania. *Nafta Gaz* **2020**, *11*, 816–826. [CrossRef]
23. Kremieniewski, M. Influence of Graphene Oxide on Rheological Parameters of Cement Slurries. *Energies* **2020**, *13*, 5441. [CrossRef]
24. Kremieniewski, M.; Stryczek, S. Zastosowanie cementu wysokoglinowego do sporządzania zaczynów uszczelniających w technologiach wiertniczych. *Cement. Wapno Beton* **2019**, *3*, 215–226.
25. Kremieniewski, M.; Rzepka, M.; Stryczek, S.; Wiśniowski, R. Microstructure of porous space in cement sheath used for sealing oil wells. *AGH Drill. Oil Gas* **2018**, *35*, 433–442. Available online: http://journals.bg.agh.edu.pl/DRILLING/2018.35.3/drill.2018.35.3.433.pdf (accessed on 13 July 2021).
26. SPE. ATW: Cementing the interface—Best practices and techniques, Moscow, 15–17 November 2004.

energies

MDPI

Article

Invert Drilling Fluids with High Internal Phase Content

Sławomir Błaż, Grzegorz Zima, Bartłomiej Jasiński and Marcin Kremieniewski *

Oil and Gas Institute–National Research Institute, 31-503 Krakow, Poland; blaz@inig.pl (S.B.); zima@inig.pl (G.Z.); jasinski@inig.pl (B.J.)
* Correspondence: kremieniewski@inig.pl

Abstract: One of the most important tasks when drilling a borehole is to select the appropriate type of drilling fluid and adjust its properties to the borehole's conditions. This ensures the safe and effective exploitation of the borehole. Many types of drilling fluids are used to drill holes for crude oil and natural gas. Most often, mainly due to cost and environmental constraints, water-based muds are used. On the other hand, invert drilling fluids are used for drilling holes in difficult geological conditions. The ratio of the oil phase to the water phase in invert drilling fluids the most common ratio being from 70/30 to 90/10. One of the disadvantages of invert drilling fluids is their cost (due to the oil content) and environmental problems related to waste and the management of oily cuttings. This article presents tests of invert drilling fluids with Oil-Water Ratio (OWR) 50/50 to 20/80 which can be used for drilling HPHT wells. The invert drilling fluids properties were examined and their resistance to temperature and pressure was assessed. Their effect on the permeability of reservoir rocks was also determined. The developed invert drilling fluids are characterized by high electrical stability ES above 300 V, and stable rheological parameters and low filtration. Due to the reduced content of the oil, the developed drilling fluid system is more economical and has limited toxicity.

Keywords: oil-based mud; invert drilling fluid; water phase; oil phase; emulsion stability; emulsifier

Citation: Błaż, S.; Zima, G.; Jasiński, B.; Kremieniewski, M. Invert Drilling Fluids with High Internal Phase Content. *Energies* **2021**, *14*, 4532. https://doi.org/10.3390/en14154532

Academic Editor: Min Wang

Received: 6 July 2021
Accepted: 24 July 2021
Published: 27 July 2021

Publisher's Note: MDPI stays neutral with regard to jurisdictional claims in published maps and institutional affiliations.

1. Introduction

Invert drilling fluids are one type of fluid that can be used to drill boreholes in almost any geological condition. The invert drilling fluid consists of two phases: the oil phase and the water phase. The oil phase consists most often of mineral or synthetic oils. The oil phase is the continuous phase of the emulsion, and very often it is also referred to as the dispersion medium. Sometimes the emulsion phases are called the inner and outer phases. In this case, the concept of the outer phase is synonymous with the continuous phase, while the inner phase is the dispersed phase. The water phase, on the other hand, is most often brine solution, which is the internal phase of the emulsion. In addition to the oil and water phases, weighting agents that regulate the drilling fluid are added to the fluid [1–5]. The first invert drilling fluids were made on the basis of crude oil, which was successively replaced with diesel oil, and then with mineral oil. Each of the replaced oils has a lower amount of polycyclic aromatic hydrocarbons, which makes it less toxic to the environment. The next step towards reducing the toxic impact of invert drilling fluids on the environment was the use of enhanced mineral oil (EMO) or linear paraffin (LP) in their composition. These oils are characterized by limited toxicity due to the low amount of aromatic hydrocarbons. Environmental legislation imposes specific restrictions on the use of invert drilling fluids for offshore drilling. A particular problem is related to oil-contaminated drilled material that must be disposed of on the offshore platform or transported to the shore. It should be disposed of in an environmentally safe manner [6–10]. The increase in these requirements gave rise to multidirectional research aimed at developing chemical compounds with properties similar to diesel fuel, harmless to the natural environment. Based on these experiments, it is concluded that these compounds can be oils of plant and animal origin. However, the lack of thermal stability at higher temperatures and the rapid biodegradation of these compounds makes it impossible to use them in practice. Studies show that it is possible to obtain synthetic compounds from vegetable oils characterized by greater thermal and oxidative stability. The conducted research contributes to the development of chemical compounds such as esters, polyalphaolefins, ethers and acetyls, linear alkylbenzene (LAB), linear alpha-olefine (LAO) and internal (isomeric) olefin. Among the presented chemical compounds, the esters and chemical compounds from the olefin group have the greatest application in drilling fluid technology [11–16].

Invert drilling fluids are recommended to be used when drilling clay-shale and salt rock formations. Due to their low density, they can also be used for drilling boreholes in areas with low formation pressure. The excellent lubricating properties of the invert drilling fluids allow the drilling of directional and horizontal holes. They make it possible to reduce frictional resistance and torque. Invert drilling fluids also contribute to increasing the drilling speed, extending the lifetime of the drill bit and reducing the adhesion force of the drill string to the filter cake on the borehole wall. These are just some of the advantages of invert drilling fluids that are taken into account when choosing the type of drilling fluid in difficult geological and technical conditions [17–19].

The ratio of the oil phase to the water phase in inversion drilling fluids ranges from 65/35 to 90/10, with the most common ratio being from 70/30 to 80/20. Several factors determine such ranges. These are, among others: the solid phase content in the drilling fluid, including the weighting agents used to ensure the appropriate density of the drilling fluid adapted to the geological and technical conditions. An increased content of the oil phase in the invert drilling fluid is also needed in order to render the weighting agents and cuttings sufficiently oil wettable. At such ratios of the oil phase to the water drilling fluid, they are characterized by high temperature resistance and appropriate rheological and structural parameters that allow drilling in HTHP conditions [20–23].

One of the disadvantages of invert drilling fluids is their cost (due to the oil content) and environmental problems related to waste and the management of oily cuttings. Sometimes high costs are reduced by using a drilling fluid after proper cleaning to drill a few or several boreholes. Invert drilling fluids are also difficult to service in comparison to water dispersion drilling fluids. Drilling fluid losses during drilling in areas with caverns and rocks of high porosity are also costly. When drilling with an invert drilling fluid, special precautions must be taken to avoid direct contact with the invert drilling fluid. It is important to avoid inhaling its vapors and pay special attention to the fire hazard resulting from the possible ignition of its vapors. The use of invert drilling fluids requires special cleaning of the borehole prior to cementing and the near-hole area prior to operation. It is necessary to use special washing liquids based on solvents and surfactants. They enable the dissolution and removal of oil deposits and the reversal of the wettability of the borehole wall before the cementing procedure and of the near-hole area before exploitation [24–26].

2. Related Works

Al-Ajmi et al. [27] presented the laboratory tests invert emulsion fluids to protect the reservoir core from drilling fluid. Invert drilling fluids contained no damaging materials, such as barite, asphaltic material, or organophilic clay. Invert drilling fluids are more resistant to weighting material sag than conventional invert fluids systems of similar rheology.

Askø et al. [28] present laboratory tests for invert drilling fluids designed for drilling long horizontal wells in extremely depleted chalk reservoirs within the limitations of a narrow mud weight window, high overbalance, high solids contamination when drilling a water injection well on the Valhall Flank North in the southern Norwegian Continental Shelf.

Wagle et al. [29] describe the formulation of medium density organoclay-free invert emulsion drilling fluids. These fluids were formulated with acid-soluble manganese tetroxide as a weighting agent and specially designed bridging agent package. The tests carried showed that invert emulsion drilling fluids formulated with nanoparticles and rheology modifiers were stable at 120 °C and 150 °C.

In the article by Elkatatny [30], he presented the effect of using a new copolymer (based on styrene and acrylic monomers) on the rheological properties and stability of invert emulsion drilling mud. The deflection test was performed under static conditions (vertical and 45 inclination) and dynamic conditions to evaluate the ability of the copolymer to strengthen the suspension properties of drilling mud.

Ramasamy and Amanullah [31] conducted a comparative study has out by formulating invert emulsion OBM using commercially available emulsifier and the emulsifier derived from used (cooking)/vegetable oil.

Sheer et al. [32] describe the formulation of laboratory tests to analyze the optimal approach to drill invert drilling fluids of ratio o/w 60:40. The designed system reduces the use of diesel consumption by 26% in total oil-based mud formulation, lowers the percentage of Low Gravity Solids (LGS) compared to the 80:20 OWR mud, and decreases the impact on the environment. The articles presented show the standard compositions of invert drilling fluids with an o/w ratio of 60/40 to 80/20. There are no examples of invert emulsion fluids with an o/w ratio below 50/50 in the analyzed articles.

One of the examples of such scrubbers with an o/w ratio below 50/50 is invert emulsion fluid [33], which is designed to carry a gravel packing a wellbore in a subterranean formation. The present emulsion contains oil as the external phase, clear brine as the internal phase and an effective emulsifier with an HLB value of 3 to 8. In addition, the emulsion contains a wetting agent with an HLB value of about 4. An emulsifier was used to prepare the emulsion which was selected from the group consisting of amides and imidazolines. The invert emulsion fluids contained an oil-to-brine ratio of 45/55.

The second example is an invert emulsion fluid [34] in which the oil phase ratio is less than 50:50. The role of emulsifier in this type of emulsion was played by alkoxylated ether acid. The presented emulsion compositions were characterized by the electric stability of the emulsion depending on the ratio of the oil phase to the water phase in the range from 25 to max. of 287 Volt.

The article presents one of the methods of reducing the cost of an invert drilling fluid and reducing its toxic impact on the environment by developing invert drilling fluids with a reduced oil phase content and an increased water phase content. Formulated Drilling fluids with an o/w ratio less than or equal to 50/50 differ from conventional invert drilling fluids in their composition and properties. Due to the higher concentration of the dispersed internal phase, these drilling fluids are characterized by reduced stability and high values of rheological and structural parameters. The appropriate values of the drilling fluid's parameters and high stability can be maintained by use of new emulsifiers adapted to a given emulsion system with a specific oil phase to water phase ratio.

3. Materials and Methods

3.1. Materials

Mineral oil made by Brenntag is used to prepare an invert drilling fluid with a high concentration of the water phase. It has a boiling point above 200 °C and a low aromatic content below 0.2%.

The primary emulsifier used is ENI, which is the reaction product of unsaturated fatty acids with diethylenetriamine and fatty alcohol polyoxyalkylene glycol ether. The second type of emulsifier is an agent called ENII. It is a mixture of sorbitan monooleate and ethoxylated coconut oil fatty acid amides. The aqueous phase is a 35% solution of calcium chloride. Organophilic clay purchased from Polski Serwis Płynów Wiertniczych Sp. z o.o. Polska is used to modify the rheological and structural parameters of the emulsion.

The composition of applied drilling muds is shown in Table 1.

Table 1. Composition of applied drilling muds.

Composition	Invert Drilling Mud on Ratio o/w						
	o/w 50/50	o/w 50/50	o/w 40/60	o/w 35/65	o/w 30/70	o/w 25/75	o/w 20/80
Mineral oil, cm^3	472	472	377	331	283	235	189
emulsifier ENI, cm^3	37.5	37.5	37.5	37.5	37.5	37.5	37.5
emulsifier ENII, cm^3	18.5	18.5	18.5	18.5	18.5	18.5	18.5
35% solution $CaCl_2$, cm^3	472	472	566	613	660	707	754

3.2. *Preparation of the Invert Drilling Fluid*

In order to compare the properties of drilling fluids with different concentrations of the oil phase to the water phase, invert drilling fluids with the following coefficients are prepared: o/w = 50/50, o/w = 40/60, o/w = 35/65, o/w = 30/70, o/w = 25/75, o/w = 20/80. The emulsions are prepared as follows: organophilic clay is added to a specific amount of oil and mixed for 5 min at a speed of 16,000 rpm. Then, selected emulsifiers are introduced into the prepared emulsion and mixed for 5 min. Next, the water phase, which is made up of 35% $CaCl_2$, is added to the mixture and stirring is continued for 30 min. Then the type of the resulting emulsion is checked, and its properties are determined.

3.3. *Experimental Procedures*

The following procedure is performed to evaluate the feasibility of developing an invert drilling fluid with reduced oil phase content and increased water phase content.

Invert drilling fluids with different oil phase to water phase ratios were prepared.

(1) Density, filtration, electric emulsion stability, rheological and structural properties and filtration at ambient temperature were determined.
(2) The influence of the water phase content in the invert drilling fluids on the emulsion stability was investigated by determining the TSI index.
(3) The rheological properties of selected invert drilling fluids were measured in a wide temperature range from 20 to 120 °C.
(4) The filtration properties of the developed invert drilling fluids were measured at 120 °C and a differential pressure of 500 psi.
(5) The influence of the developed invert drilling fluids on the permeability of reservoir rocks was determined.

3.3.1. Testing Procedures

The technological properties of drilling fluids, such as density, rheological parameters, filtration and electric emulsion stability were determined in accordance with the PN-EN ISO 10414-2 standard. Petroleum and natural gas industries–Field testing of drilling fluids—Part 2: Oil-based fluids (ISO 10414-2:2011).

3.3.2. Density Measurement

The density of the drilling fluids was determined by means of a Baroid type fluid balance at a temperature of 20 °C and at atmospheric pressure. The fluid balance is designed in such a way that the vessel for drilling fluid situated on one end of the arm is balanced by a fixed counterweight placed on the other end of the arm. The balance arm is equipped with a weight moved along the scale and a level, enabling precise weighing [35].

3.3.3. Measurements of Rheological Parameters

The rheological parameters of drilling fluids are measured at 20 °C using an Ofite 900 viscometer. The viscometer is used to directly determine the relationship between the shear rate of a liquid and the existing shear stress, enabling the calculation of plastic viscosity, apparent viscosity, the yield point, and the measurement of the gel strength [35,36]. The

•

Ofite 77 high-temperature and high-pressure viscometer was used to measure rheological parameters at higher temperatures. The drilling fluids are tested in both the heating and cooling cycles in the temperature range from 20 to 120 °C with shear rates from 5.1 to 1020 s^{-1}.

3.3.4. Filtration Measurement

The filtration of drilling fluids at ambient temperature is determined using a low-pressure filter press at a pressure of 0.7 MPa for a period of 30 min. On the other hand, at a temperature of 120 °C, filtration is measured with a static HTHP filter press at a differential pressure of 3.4 MPa [36].

3.3.5. Stability Tests of Invert Drilling Fluids

The stability of the resulting emulsion is determined by the Emulsion Stability Meter, which measures the ES value by applying a voltage between plate electrodes immersed in the liquid. The resulting current remains low until the threshold voltage (61 ± 5A) is reached. Then the voltage of the current is increased to the point where a so-called puncture occurs. The voltage value is given in volts. If the breakdown occurs at a low voltage level, then this indicates an o/w emulsion. In the event of an electric breakdown only at high voltage, it indicated a w/o emulsion. The higher the ES value, the more stable the emulsion is [36].

Stability studies of invert drilling fluids with different oil phase to water phase ratios were also tested using the Turbiscan Tower device. It allows to measure the stability of suspensions, emulsions and foams. The analyzer works by analyzing multiple scattered light (transmitted and reflected back). Thanks to the moving head and two detectors working synchronously, the device collects data on the intensity of the transmitted and reflected light at a specific frequency. Then all signal changes in the sample are summed up to give a unique TSI number reflecting the destabilization of a given sample. The greater the number of TSIs, the more unstable the sample.

3.3.6. Test of the Influence of the Developed Invert Drilling Fluids on the Permeability of Reservoir Rocks

In the first stage of the test, the initial gas permeability of the sandstone cores (nitrogen) is determined. The measurement is made by passing nitrogen at a pressure of 2.2 MPa (inlet) through the core to a pressure of 2MPa (outlet) at a temperature of 120 °C. Using Darcy's law and the measurement data, the initial baseline permeability of the cores is calculated.

The next stage of the test is saturation of the core with selected invert drilling fluids. The pressure of the drilling fluid at the inlet to the core (P_1) is 5MPa and the pressure at the outlet from the core (P_2) is 2MPa. The core saturation process at a differential pressure of ΔP = 3MPa is carried out for 30 min.

Nitrogen is then passed through the core from the formation side, in the opposite direction to the direction of saturation with the drilling fluid, resulting in a pressure depression value of 2 MPa. The next step is to measure the permeability, which allows us to determine the degree of damage to the deposit by a given drilling fluid.

3.3.7. Determination of Loss of Permeability of a Sample

The coefficient of loss of permeability, WU (1), is defined as the percentage decrease in the permeability of the initial sample after the solid phase and filtrate particles penetrate into its pore space, and is expressed by the formula:

$$\mathrm{WU} = \left(1 - \frac{k_d}{k_0}\right) \cdot 100\% \tag{1}$$

where: k_d—permeability of damaged porous rock (after exposure to drilling fluid), taking into account the Klingenberg correction, [mD]; k_0—initial permeability (undamaged porous rock), taking into account the Klingenberg correction, [mD].

4. Results and Discussion

4.1. Properties of the Invert Drilling Fluid Depending on the Ratio of the Oil Phase to the Water Phase

Table 2 shows the properties of invert drilling fluids with an oil phase to water phase ratio of 50/50 to 20/80, prepared with the use of specially selected emulsifiers ENI and ENII. The electric stability ES of the prepared drilling fluids ranged from 610 Volt with an o/w 50/50 to 630 Volt ratio in drilling fluids with an internal phase content of 80% (Table 2). The differences between the drilling fluids can be seen mainly in the rheological and structural parameters, TSI value, and density. The lowest drilling fluid density of 1.1 g/cm^3 was obtained with an o/w phase mixing ratio of 50/50. This drilling fluid is also characterized by the lowest stability defined by the TSI index of 6.9 and the lowest values of rheological parameters (PV = 15 mPa·s, YP = 2.4 Pa). The TSI instability of 3.6 is defined in a drilling fluid with an oil to water ratio of 40/60. Increasing the proportion of the water phase to 40% increases the plastic viscosity to 27mPas and the yield point to 4.8Pa. Higher TSI values and low values of the yield point of drilling fluids with an o/w ratio of 50/50 and 40/60 indicate the necessity to increase the rheological and structural parameters in these drilling fluids by introducing organophilic clay into their composition or modifying the oil fraction composition. Therefore, 2% organophilic clay is added to the drilling fluid composition with an oil phase to water phase ratio of 50/50 (Table 2, item 2). Due to the use of hydrophobic bentonite in the invert drilling fluid with an oil phase to water phase ratio of 50/50, the yield point of the drilling fluid is now 4.3 Pa, and the sample stability index decreased from 6.9 to 0.5. (Table 2). On the other hand, in a drilling fluid with an oil phase to water phase ratio of 40/60, the use of 1% organophilic clay increases the yield point to 5.7 Pa and reduces the TSI index to 0.7 (Table 2, Figure 1).

Table 2. Parameters of applied drilling muds.

Invert Drilling Mud on Ratio, o/w	Density, g/cm^3	Plastic Viscosity, mPa·s	Apparent Viscosity, mPa·s	Yield Point, Pa	Gel Strength, I/II, Pa	TSI	Electrical Stability, Volts
	±0.05	±1	±1	±0.75	±0.25	±0.1	±1
	20 °C						
o/w—50/50	1.1	15	17.5	2.4	0.48/0.96	6.9	610
o/w—50/50 +2% Viscosifier	1.1	21	25.5	4.3	0.96/1.4	0.5	580
o/w—40/60	1.14	27	32	4.8	1.9/2.4	3.6	640
o/w—40/60 +1% Viscosifier	1.14	27	33	5.7	1.9/2.4	0.7	610
o/w—35/65	1.16	42	53	10.5	3.3/4.3	1.5	670
o/w—30/70	1.19	50	64.5	13.9	3.8/4.3	1.7	680
o/w—25/75	1.21	61	89.5	27.2	6.2/7.6	2.0	690
o/w—20/80	1.23	85	116.5	30.1	9.1/10.5	0.9	630

The reduction of the oil phase to water phase ratio in invert drilling fluids causes an increase in rheological and structural parameters and slight changes in the electrical stability of the drilling fluids, ranging from 610 Volts to 690 Volts (Table 2). A properly selected composition of emulsifiers allows for obtaining invert drilling fluids with a water phase content of up to 75%. Above this value, the drilling fluid is characterized by high rheological and structural parameters, a plastic viscosity of 85 mPa·s and a yield point of 30.1 Pa (Table 2).

The increase in the water phase content in the invert drilling fluid makes it possible to increase the density of the drilling fluid to 1.23 g/cm^3 without the need to add any weighting materials. In this type of drilling fluid, the rheological and structural properties can be partially controlled by slightly changing the ratio of the oil phase to the water phase. Reducing the o/w ratio increases the rheological parameters while increasing the ratio reduces them.

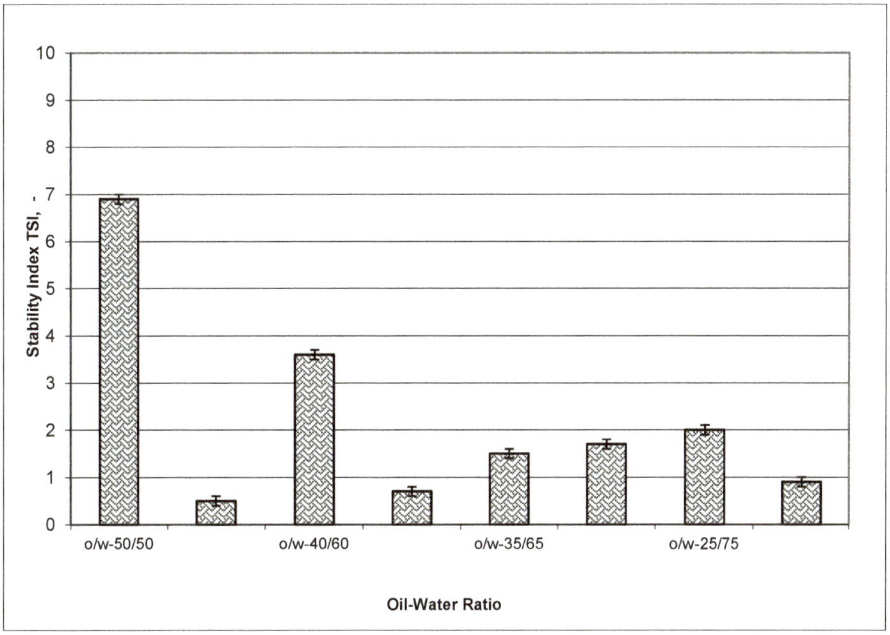

Figure 1. Destabilization kinetics of selected invert drilling fluids with a different ratio of oil to water phase.

4.2. Determining the Influence of Elevated Temperature on Rheological and Structural Properties and Filtration of the Developed Invert Drilling Fluids

Two invert drilling fluids with an oil phase to water phase ratio of 40/60 and 30/70 were selected for the research. For comparative analysis, the drilling fluids were weighted with barite to the same density of 1200 kg/m^3. The drilling fluids were tested in both the heating and cooling cycle in the temperature range from 20 to 120 °C. The measurement results are presented in Figures 2–5 in the form of a graph of changes in the plastic viscosity and the yield point versus temperature.

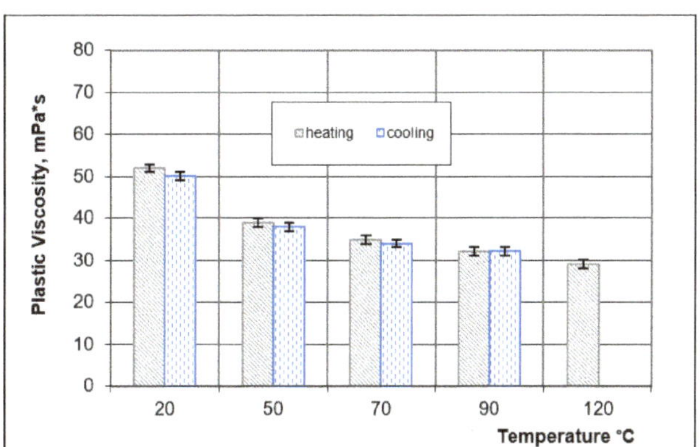

Figure 2. Changes in the plastic viscosity of invert drilling fluids with an oil to water ratio of 40/60 depending on temperature changes.

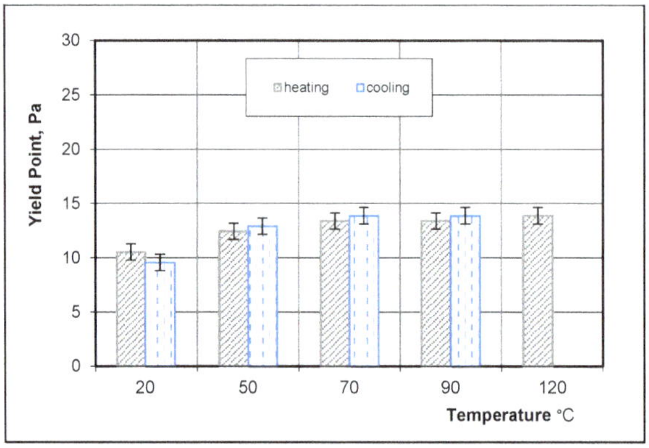

Figure 3. Changes in the yield point of invert drilling fluids with an oil phase to water phase ratio of 40/60 depending on temperature changes.

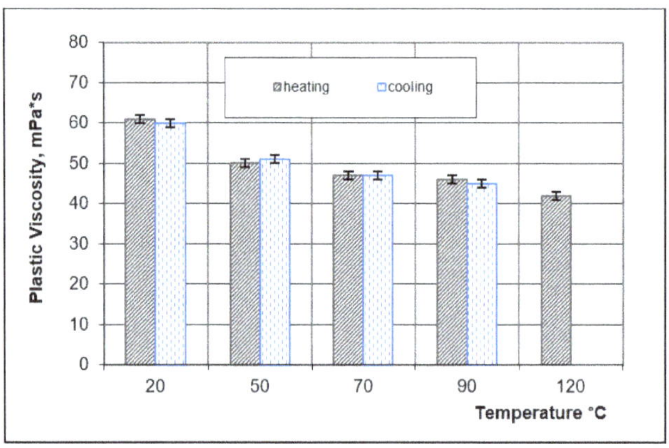

Figure 4. Changes in the plastic viscosity of invert drilling fluids with an oil to water ratio of 30/70 depending on temperature changes.

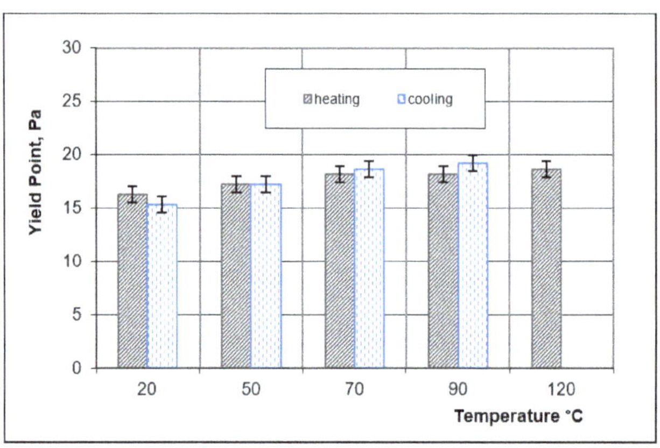

Figure 5. Changes in the yield point of invert drilling fluids with an oil phase to water phase ratio of 30/70 depending on temperature changes.

The invert drilling fluid with an oil phase to water phase ratio of 40/60 at 20 °C and atmospheric pressure is characterized by a plastic viscosity of 52 mPa·s and a yield point of 10.5 Pa (Figures 2 and 3). As a result of the temperature of up to 120 °C, the plastic viscosity of the drilling fluid is gradually reduced. At 50 °C it drops to 39 mPa·s, at 90 °C to 32 mPa·s, and at 120 °C it reaches the value of 29 mPa·s (Figure 2). Changes in the yield point due to the effect of temperature are similar only to the temperature of 50 °C. As a result of heating the drilling fluid to 50 °C, the yield point decreases to the value of 12.4 Pa, then when the temperature rises to 70 °C, the yield point slightly increases to the value of 13.4 Pa (Figure 3). Another increase in temperature to 90 and 120 °C does not change the yield point of the drilling fluid. In the drilling fluid cooling cycle to the temperature of 20 °C, the drilling fluid restores its rheological parameters to the initial values of the drilling fluid before annealing. Plastic viscosity is 2 mPa·s higher and the yield point is 1 Pa higher.

The invert drilling fluid with an oil phase to water phase ratio of 30/70 at 20 °C is characterized by a plastic viscosity of 61 mPa·s and a yield point of 16.3 Pa (Figures 4 and 5). As a result of the gradual increase in temperature, the plastic viscosity of the drilling fluid falls. At 50 °C, the plastic viscosity of the drilling fluid drops to 50 mPa·s, at 90 °C to 46 mPa·s, and at 120 °C to 42 mPa·s (Figure 4). In the cycle of cooling the drilling fluid to ambient conditions, the plastic viscosity is restored to the value of 60 mPa·s (Figure 4). When analyzing the diagram of changes in the yield point of the drilling fluid depending on the temperature, it is stated that the decrease in the yield point value takes place to the temperature of 50 °C. At 50 °C, the yield point has the lowest value of 17.2 Pa. A further increase in temperature to 120 °C increases the yield point to 18.6 Pa (Figure 5). In the cycle of cooling the drilling fluid to the ambient temperature, the yield point is reduced to 15.3 Pa (Figure 5). After cooling the drilling fluid to the ambient temperature, it was observed that the drilling fluid withstood the temperature of 120 °C and did not decompose into the water and oil phases, and the electric stability value of the ES invert drilling fluid was 660 Volts.

In the next stage of the research, the influence of temperature and pressure on the filtration of invert drilling fluids was observed. The filtration of the drilling fluids was tested on an HPHT filter press at a temperature of 120 °C with a differential pressure of 3.4 MPa. The influence of elevated temperature and pressure on the filtration value was investigated for invert drilling fluids with an oil to water phase ratio of 50/50, 40/60, 30/70 and 20/80. The test results are shown in Figure 6.

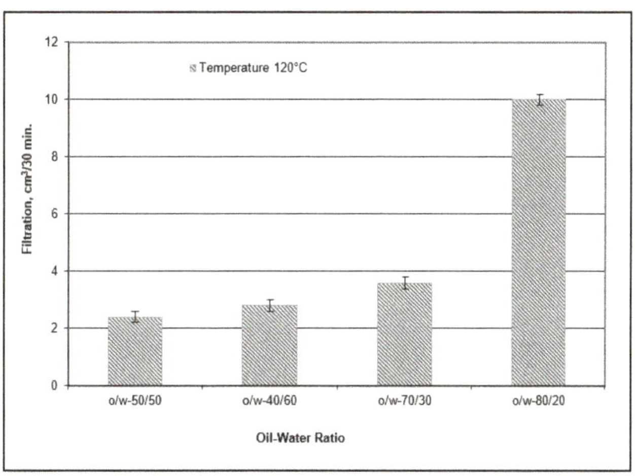

Figure 6. Filtration loss of invert drilling fluids depending on the ratio of the oil phase to the water phase at the temperature of 120 °C and a differential pressure of 3.4 MPa.

The filtration of an invert drilling fluid with a 50/50 ratio of the oil phase to the water phase at a temperature of 120 °C is 2.4 $cm^3/30$ min. Lowering the share of the oil phase in the invert drilling fluid to 40% increases filtration of the drilling fluid to 2.8 $cm^3/30$ min (Figure 6). As the concentration of the water phase in the invert drilling fluid increases, the filtration increases. In an invert drilling fluid with a 30/70 ratio of the oil phase to the water phase, filtration increased to 3.6 $cm^3/30$ min. The highest filtration, which is 10 $cm^3/30$ min, was recorded for the invert drilling fluid with an o/w ratio of 20/80 (Figure 6).

The performed tests confirm that the drilling fluid shows high temperature resistance while maintaining the electrical stability of ES of over 570 Volts under variable pressure and temperature.

4.3. Investigation of the Influence of Selected Invert Drilling Fluids on the Change of the Permeability Coefficient of Reservoir Rocks

The study of the reservoir rock permeability coefficient was carried out on sandstone cores with permeability from 28.57 to 34.55 mD. The results of the study of permeability changes of reservoir rock samples are presented in Figures 7 and 8. To determine the impact of changes in the permeability of reservoir rocks, the developed invert drilling fluids with an oil phase to water phase ratio of 50/50, 40/60, 30/70 and an invert drilling fluid with an oil phase to water phase ratio of 80/20 were selected for comparison.

On the basis of the analysis of the obtained results and the presented drawings (Figures 7 and 8), it is concluded that the tested drilling fluids, depending on the water phase content, have a different effect on the amount of damage to the permeability of the rocks in the near-hole area. The highest value of the sandstone reflux permeability coefficient is obtained by operating the drilling fluid with an oil phase to water phase ratio of 40/60. Measurements of changes in the hydraulic conductivity for gas show that the tested drilling fluid at a differential pressure of 2 MPa causes slight damage to the reservoir rock permeability. The final value of the permeability coefficient after the effect of the drilling fluid decreases from 34.55 to 31.78 mD (Figure 7). The calculated loss of sandstone hydraulic conductivity for gas is 8.02% (Figure 8).

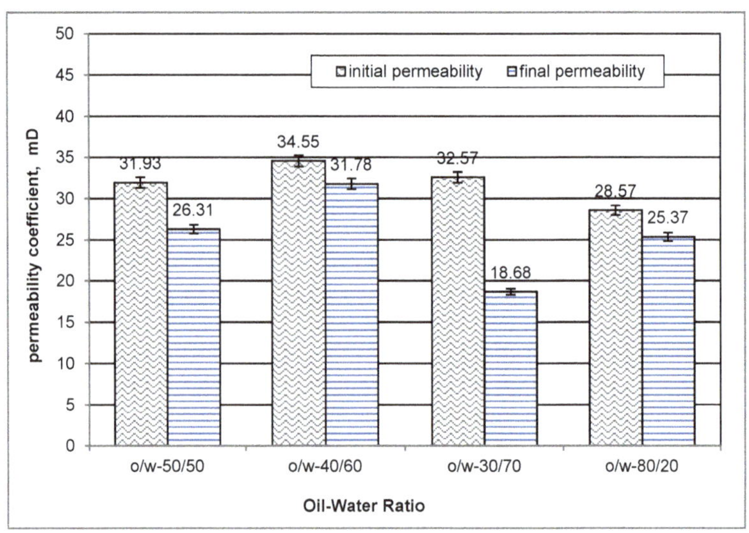

Figure 7. Influence of the type of invert drilling fluid on the change of sandstone gas permeability.

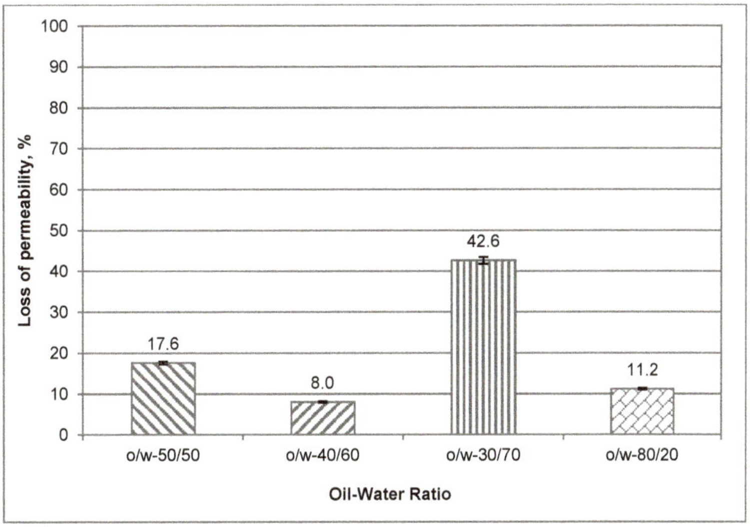

Figure 8. Loss of permeability of tested samples after the effect of invert drilling fluids for gas.

In comparison, the sandstone permeability coefficient was measured after treatment with an invert drilling fluid with an o/w ratio of 80/20. As a result of the contact of the drilling fluid with the reservoir rock, a decrease in the hydraulic conductivity of the reservoir rock sample is visible from 28.57 to 25.37 mD (Figure 7), which results in a loss of the sample's permeability by 11.20% (Figure 8).

Another permeation test was carried out for an invert drilling fluid with an oil phase to water phase ratio of 50/50. Balancing the amount of the oil phase in relation to the water phase reduces the sandstone permeability coefficient from 31.93 to 26.31 (Figure 7). The drilling fluid damages the sample permeability by 17.60% (Figure 8).

The greatest decrease in reservoir rock permeability among the invert drilling fluids selected for the tests was observed after using a drilling fluid with an oil phase to water phase ratio of 30/70. Permeation studies show that the drilling fluid causes a loss of permeability of the sandstone sample by 42.65% (Figure 8). The permeability coefficient after the effect of the drilling fluid decreased from the value of 32.57 to 18.68 mD (Figure 7).

5. Conclusions

(1) The conducted laboratory tests confirm that there is a real possibility of developing invert drilling fluid compositions with a low oil phase to water phase ratio below 50/50.
(2) Invert drilling fluids with a high internal phase concentration are obtained by using emulsifiers synthesized under laboratory conditions. These emulsifiers are the reaction product of unsaturated fatty acids with diethylenetriamine and fatty alcohol polyoxyalkylene glycol ether and an emulsifier which is a mixture of sorbitan monooleate and ethoxylated coconut fatty acid amides.
(3) The analysis of the quality, stability and resistance of the prepared emulsion systems under ambient conditions allows the selection of invert drilling fluid compositions characterized by high electrical stability ES ranging from 580 to 690 Volts and low TSI indexes from 0.5 to 6.9.
(4) The density of the developed invert drilling fluids can be partially controlled by the content of the internal phase, which may be salt solutions. This limits the solids content (weighting agents) in the drilling fluid. The density of the developed drilling fluids is determined by the content of the water phase, which is 35% r-r $CaCl_2$. The developed invert drilling fluids, depending on the ratio of the oil phase to the water phase, can be used without the use of any weighting agents to the density of:

- 1140 kg/m^3 invert drilling fluids with an o/w 40/60 phase ratio;
- 1190 kg/m^3 invert drilling fluids with an o/w 30/70 phase ratio;
- 1230 kg/m^3 invert drilling fluids with an o/w 20/80 phase ratio.

(5) The tests of the drilling fluids under HTHP conditions show that the developed invert drilling fluids with an oil phase to water phase ratio from 50/50 to 30/70 are characterized by high resistance to temperature and pressure. Invert drilling fluids maintain stable rheological parameters to a temperature of 120 °C and are characterized by low filtration—lower than 4 $cm^3/30$ min.

(6) Based on the obtained results, it is stated that the developed drilling fluids, depending on the ratio of the oil phase to the water phase, have a different effect on the permeability of reservoir rocks. The highest value of the sandstone reflux permeability coefficient is obtained after the treatment with a drilling fluid with an oil phase to water phase ratio of 40/60. The drilling fluid causes a loss of the sandstone's hydraulic conductivity for gas at the level of 8%. On the other hand, the impact of a drilling fluid with an oil to water ratio of 30/70 on the sandstone core reduces the formation damage index by approximately 43%. Based on the conducted studies of the permeability coefficient, it is concluded that the developed drilling fluids with an increased 50/50 and 40/60 water phase content affect the formation damage in a similar way as in the case of the conventional 80/20 invert drilling fluid, which reduced gas permeability by about 11%.

(7) The cost of preparing the developed invert drilling fluid with the ratio o/w—30/70 will be lower by approx. 60% in relation to the conventional invert drilling fluid with the ratio o/w 70/30. Due to the reduced content of the oil, its toxicity and its impact on the environment will be reduced.

(8) Further tests of the developed invert drilling fluids will be carried out to determine their thermal resistance to temperatures above 160 ° C and pressures above 34 MPa.

Author Contributions: Conceptualization, S.B., G.Z., B.J. and M.K.; methodology, S.B.; validation, S.B., G.Z., B.J. and M.K.; formal analysis, S.B. and G.Z.; investigation, S.B.; resources, G.Z., M.K. and B.J.; data curation, S.B.; writing—original draft preparation, S.B.; writing—review and editing, G.Z., B.J. and M.K.; visualization, S.B.; supervision, S.B.; project administration, S.B. All authors have read and agreed to the published version of the manuscript.

Funding: This research was funded by the Ministry of Science and Higher Education Warsaw (Internal order Oil and Gas Institute—National Research Institute Project No. 0044/KW/20).

Institutional Review Board Statement: Not applicable.

Informed Consent Statement: Not applicable.

Data Availability Statement: Not applicable.

Conflicts of Interest: The authors declare no conflict of interest.

Nomenclature

o/w	the ratio of the oil phase to the water phase in invert drilling fluids
HTHP	high temperature high pressure
MPa	Megapascal—pressure unit
mD	millidarcy—unit of permeability
Pa	pascal—unit of the yield point and structural strength
mPa·s	millipascal second—unit of plastic and apparent viscosity
ES	electric stability of the emulsion
TSI	Emulsion Stability Index
OBM	Oil based mud
HLB	Hydrophilic-Liophilic Balance

References

1. Błaż, S. Badania laboratoryjne nad opracowaniem składu płuczki inwersyjnej. *Nafta-Gaz* **2015**, *3*, 54–63.
2. Błaż, S. Odwracalne ciecze emulsyjne o wysokim stężeniu fazy wewnętrznej (HIPR). *Nafta-Gaz* **2021**, *3*, 175–186. [CrossRef]
3. Błaż, S. Analiza właściwości technologicznych płuczki inwersyjnej w warunkach HPHT. *Nafta-Gaz* **2016**, *6*, 403–412. [CrossRef]
4. Amani, M. The rheological properties of oil-based mud under high pressure and high temperature conditions. *Adv. Pet. Explor. Dev.* **2012**, *3*, 21–30.
5. Dearing, H.; Larry, H.; Flak, P.E. High Salinity, Low Activity Oil Mud Improves Performance in Near Salt Drilling. In Proceedings of the 2007 AADE National Technical Conference and Exhibition, Wyndam Greenspoint Hotel, Houston, TX, USA, 10–12 April 2007.
6. Fernandez, J.; Sharp, K.; Plummer, D. Enhanced Fluid Viscosity Using Novel Surfactant Chemistry Purposely Designed for Low-Aromatic Mineral and Synthetic Base Fluids. In Proceedings of the 2014 AADE Fluids Technical Conference and Exhibition, Hilton Houston North Hotel, Houston, TX, USA, 15–16 April 2014.
7. Amani, M.; Al-Jubouri, M.; Shadravan, A. Comparative study of using oil-based mud versus water-based mud in HPHT fields. *Adv. Pet. Explor. Dev.* **2012**, *4*, 18–27.
8. Elkatatny, S. Mitigation of barite sagging during the drilling of high-pressure high-temperature wells using an invert emulsion drilling fluid. *Powder Technol.* **2019**, *352*, 325–330. [CrossRef]
9. Alford, P.; Anderson, D.; Bishop, M.; Goldwood, D.; Stouffer, C.; Watson, E.; Karonka, M.; Moore, R. Novel oil based mud additive decreases HTHP fluid loss and enhances stability. In Proceedings of the 2014 AADE Fluids Technical Conference and Exhibition, Hilton Houston North Hotel, Houston, TX, USA, 15–16 April 2014.
10. Growcock, F.B.; Patel, A.D. The Revolution in Non-Aqueous Drilling Fluids. In Proceedings of the 2011 AADE National Technical Conference and Exhibition, Hilton Houston North Hotel, Houston, TX, USA, 12–14 April 2011.
11. Jasiński, B. Badania nad zastosowaniem emulsji olejowowodnych jako cieczy roboczych o obniżonej gęstości. *Nafta-Gaz* **2012**, *12*, 1155–1164.
12. Karimi, A.; Tahmasbi, K.; Arsanjani, N. The feasibility study of replacing oil based mud with more environmentally acceptable paraffin based system in Iranian oil fields. In Proceedings of the Asia Pacific Health, Safety, Security and Environment Conference, Jakarta, Indonesia, 4–6 August 2009. [CrossRef]
13. Geng, T.; Qiu, Z.; Zhao, C.; Zhang, L.; Zhao, X. Rheological study on the invert emulsion fluids with organoclay at high aged temperatures. *Colloids Surf. A Physicochem. Eng. Asp.* **2019**, *573*, 211–221. [CrossRef]
14. Paswan, B.K.; Jain, R.; Sharma, S.K.; Mahto, V.; Sharma, V.P. Development of Jatropha oil-in-water emulsion drilling mud system. *J. Pet. Sci. Eng.* **2016**, *144*, 10–18. [CrossRef]
15. Maghrabi, S.; Wagle, V.; Teke, K.; Kulkarni, D.; Kulkarni, K. Low plastic viscosity invert emulsion fluid system for HPHT wells. In Proceedings of the 2011 AADE National Technical Conference and Exhibition, Hilton Houston North Hotel, Houston, TX, USA, 12–14 April 2011.
16. Paswan, B.K.; Mahto, V. Development of environment-friendly oil-in-water emulsion based drilling fluid for shale gas formation using sunflower oil. *J. Pet. Sci. Eng.* **2020**, *191*, 107–129. [CrossRef]
17. Wagle, V.; Maghrabi, S.; Kulkarni, D. Formulating Sag-Resistant, Low-Gravity Solids-Free Invert Emulsion Fluids. In Proceedings of the SPE Middle East Oil and Gas Show and Conference, Manama, Bahrain, 18–21 March 2013. [CrossRef]
18. Whitby, C.P.; Fornasiero, D.; Ralston, J. Effect of oil soluble surfactant in emulsions stabilised by clay particles. *J. Colloid Interface Sci.* **2008**, *323*, 410–419. [CrossRef] [PubMed]
19. Young, S.; De Stefano, G.; Lee, J. Advances in invert emulsion performance through novel emulsifier chemistry. In Proceedings of the SPE Oil and Gas India Conference and Exhibition, Mumbai, India, 28–30 March 2012. [CrossRef]
20. Tiwari, R.; Kumar, S.; Husein, M.M.; Rane, P.M.; Kumar, N. Environmentally benign invert emulsion mud with optimized performance for shale drilling. *J. Pet. Sci. Eng.* **2020**, *186*. [CrossRef]
21. Hajiabadi, S.H.; Aghaei, H.; Ghabdian, M.; Kalateh-Aghamohammadi, M.; Esmaeilnezhad, E.; JinChoi, H. On the attributes of invert-emulsion drilling fluids modified with graphene oxide/inorganic complexes. *J. Ind. Eng. Chem.* **2020**, *93*, 290–301. [CrossRef]
22. Kulkarni, S.D.; Jamison, D.E. Determining association of particles and emulsion in invert emulsion drilling fluids: Experiments and modeling. In Proceedings of the 2015 AADE National Technical Conference and Exhibition, Henry B. Gonzalez Convention Center, San Antonio, TX, USA, 8–9 April 2015.
23. Li, Q.; Li, S. Ultra-High Density Oil-Based Drilling Fluids and its Applications in Ultra-Deep Petroleum Systems. In Proceedings of the SPE Latin American and Caribbean Petroleum Engineering Conference, Virtual, 27–31 July 2020. [CrossRef]
24. McMillan, D.N.; Lunde, O.H.; Mikalsen, R.; Maeland, Y.; Wróblewski, T.; Vatne, A.; Dillner, B. Development and Field Application of an Innovative, Minimally Damaging, Low-ECD Invert Emulsion Fluid for Enhanced Drilling and Completion of HP/HT Wells. In Proceedings of the SPE European Formation Damage Conference and Exhibition, Budapest, Hungary, 3–5 June 2015. [CrossRef]
25. Miller, M.; Kulkarni, S.D.; King, D.; Valenziano, R. Gellant for oil-based drilling fluid behind casing. In Proceedings of the 2014 AADE Fluids Technical Conference and Exhibition, Hilton Houston North Hotel, Houston, TX, USA, 15–16 April 2014.

26. Rojas, J.C.; Daugherty, B.; Renfrow, D.; Bern, P.; Greene, B.; Irby, R.; Gusler, B.; Grover, P.; Trotter, N.; Dye, B. Increased Deepwater Drilling Performance Using Constant Rheology Synthetic-based Mud. In Proceedings of the 2007 AADE National Technical Conference and Exhibition, Wyndam Greenspoint Hotel, Houston, TX, USA, 10–12 April 2007.
27. Al-Ajmi, A.; Al-Rushoud, A.; Gohain, A.; Khatib, F.; Al-Haj, H.; Al-Naqa, F.; Al-Mutawa, F.; Al-Gharib, M.; Shinde, H.; Arora, S.; et al. Successful Field Application of Organophilic Clay-Free Invert Emulsion Fluid to Protect the Reservoir Core from Drilling Fluid Damage: Case Study from a Kuwait Field. In Proceedings of the SPE Middle East Oil and Gas Show and Conference, Manama, Bahrain, 18–21 March 2019. [CrossRef]
28. Askø, A.; Alsvik, E.T.; Danielsen, T.H.; Haga, M.A. Low-Density Invert Emulsion Drilling Fluid Enables Recovery of Oil Reserves in Extremely Depleted Reservoirs: A Case History from Valhall, Norway. In Proceedings of the IADC/SPE International Drilling Conference and Exhibition, Galveston, TX, USA, 3–5 March 2020. [CrossRef]
29. Wagle, V.; Yami, A.; Onoriode, M.; Butcher, J. Design, Qualification and Field Deployment of Low ECD Organoclay-Free Invert Emulsion Drilling Fluids. In Proceedings of the SPE Russian Petroleum Technology Conference, Virtual, 26–29 October 2020. [CrossRef]
30. Elkatatny, S. Enhancing the Stability of Invert emulsion drilling fluid for Drilling in High-Pressure High-Temperature Conditions. *Energies* **2018**, *11*, 2393. [CrossRef]
31. Ramasamy, J.; Amanullah, M. Converting Waste Vegetable Oil to Emulsifier for Invert-Emulsion Oil Based Mud. In Proceedings of the International Petroleum Technology Conference, Beijing, China, 26–28 March 2019. [CrossRef]
32. Sheer, S.; Alotaibi, A.; Gadhiya, K.; Al-Khaldy, M.; Abhijit, D.; Al-Failkawi, K.; Al-Saadi, D.; Al-Saeedi, A.; Hamed, A.; Al-Azmi, F.; et al. The Dynamics of Drilling with Oil-Based Mud, 60:40 Oil-Water Ratio–Case History in South East Kuwait Fields. In Proceedings of the SPE Gas & Oil Technology Showcase and Conference, Dubai, United Arab Emirates, 21–23 October 2019. [CrossRef]
33. Jones, T.A.; Mckellar, A.J.; Quintero, L. Invert carrier fluid and oil-wetting agent and metod of using same. U.S. Patent 7,803,743B2, 28 September 2010.
34. Luster, M.; Patel, A.D.; Lim, S.K. Methods of using invert emulsion fluids with high internal phase concentration. U.S. Patent 9,004,167 B2, 14 April 2015.
35. Standard PN-EN ISO 10414-1. Petroleum and Natural gas Industries. Field Testing of Drilling Fluids. Part 1: Water-Based Fluids. Available online: https://www.iso.org/standard/41939.html (accessed on 25 May 2021).
36. Standard PN-EN ISO 10414-2. Petroleum and Natural Gas Industries. Field Testing of Drilling Fluids. Part 2: Oil-Based Fluids. Available online: https://www.iso.org/standard/41192.html (accessed on 25 May 2021).

energies

Article

Study of the Long Term Acid Gas Sequestration Process in the Borzęcin Structure: Measurements Insight

Marcin Warnecki *, Mirosław Wojnicki, Jerzy Kuśnierczyk and Sławomir Szuflita

Oil and Gas Institute—National Research Institute, 25 A Lubicz Str., 31-503 Krakow, Poland;
wojnicki@inig.pl (M.W.); kusnierczyk@inig.pl (J.K.); szuflita@inig.pl (S.S.)
* Correspondence: warnecki@inig.pl

Abstract: Geological sequestration of acid gases, including CO_2, is now a growing solution to prevent progressive Earth climate change. Disposal of environmentally harmful greenhouse gases must be performed safely and securely to minimise leakage risk and possible uncontrolled emissions of injected gases outside the sequestration structure. The paper describes a series of research activities at the Borzęcin sequestration site located in western Poland, which were designed to study the migration paths of injected acid gases (mainly mixture of CO_2 and H_2S) into the water-bearing layers underlying natural gas reservoir. Along with understanding the nature and dynamics of acid gases migration within the sequestration structure, the research was also addressed to assess its leak-tightness and the long-term safety of the entire reinjection facility. As a part of the research works, two downhole sampling campaigns were completed in 2018–2019, where samples of water underlying the Borzęcin reservoir were taken and subsequently studied to determine their physicochemical parameters that were never before examined. Compositions of gas dissolved in downhole brine samples were compared with produced and injected gas. Relevant studies of reservoir water from selected wells were performed, including isotopic analyses. Finally, four series of soil gas analyses were performed on the area surrounding the selected well, which are important for the hazardous gas sequestration safety analysis in the Borzęcin facility. All the above mentioned research activities aimed to acquire additional knowledge, which is valuable for risk assessment of the acid gas sequestration process taking place on the specific example of the Borzęcin site operating continuously since 1996.

Keywords: CO_2 and H_2S geological sequestration; acid gas reinjection; acid gas migration; bottom-hole sampling; isotopic composition; gas and water chemical analysis; soil gas analysis; leakage risk analysis

Citation: Warnecki, M.; Wojnicki, M.; Kuśnierczyk, J.; Szuflita, S. Study of the Long Term Acid Gas Sequestration Process in the Borzęcin Structure: Measurements Insight. *Energies* **2021**, *14*, 5301. https://doi.org/10.3390/en14175301

Academic Editors: Dino Musmarra and Attilio Converti

Received: 18 June 2021
Accepted: 22 August 2021
Published: 27 August 2021

1. Introduction

The safety of carbon dioxide and other acid gases underground storage is an obvious and critical issue for this kind of activity. Geological storage must ensure leak-tightness for the hazardous substances deposited there for many hundreds of years. The paper describes selected research activities conducted by the Oil and Gas Institute—National Research Institute (OGI—NRI) within the SECURe project focused on the acquisition of measurement data confirming (or not) safety of long-term acid gas sequestration process in the Borzęcin site located in Western Poland.

First acid gas (containing significant concentration of acid components e.g., CO_2, H_2S) reinjection into a depleted oil reservoir was started in 1989 in Alberta, Canada. By the end of 2003, about 4.5 Mt of acid gas (2.5 Mt of CO_2, and 2.0 Mt of H_2S) was injected into deep saline aquifers and depleted hydrocarbon reservoirs at 48 sites in Canada [1–3]. It should be emphasised that the Borzęcin injection project was the first full scale, acid gas reinjection process of practical value into the producing gas reservoir, put into operation in 1996 [4,5]. In 2004, a similar process at a larger scale has been executed in the Krechba field in Algeria by BP and Statoil [6,7], and the reinjection of produced CO_2 started at the K12-B in the Netherlands operated by GDF-SUEZ [8,9]. In the Borzęcin project, acid gas has been reinjected into the water-bearing zone underlying the gas cap, from which natural gas has been constantly produced.

Slightly different projects were then implemented in Sleipner gas field and Weyburn oil field. In the case of gas produced from Sleipner, the recovered CO_2 is reinjected into the separate Utsira aquifer, whereas Weyburn is a typical EOR-CO_2 project [10]. At that time, the sequestration of acid gas into the same reservoir was in the early stages of development. Together with K12-B, the Borzęcin project was then innovative on a global scale, also in the domain of reservoir engineering. It required the performance of detailed simulations with the assumption that it was possible to rationally reconcile the simultaneous recovery of gas and reinjection of acid gas into the water-bearing zone, and that it would be in accordance with applicable laws and the principles of rational exploitation. The challenge also was to not inject sour natural gas, which was already done in Canada and the Netherlands, but a reactive mixture of H_2S and CO_2.

Borzęcin is considered to be a unique experimental plant which allows us to investigate the acid gas sequestration process for 25 years of its operation and possibly for a few more years to come. Contrarily to other large reservoirs, the small capacity of the onshore Borzęcin structure enable to effectively control and analyse the detailed mechanisms taking part in the sequestration process at a relatively short term scale [4].

The described unique site of acid gas injection to depleted natural gas reservoir was examined. The goal of the research was to identify the intensity and propagation of acid gas migration within the structure and potential pathways towards the ground surface. A number of tests and measurements were conducted, that directly and indirectly were connected with the assessment of containment, and thus the analysis of safety in the long-term perspective. Most of the tests and measurements described in the paper were never performed at the Borzęcin sequestration structure before, that is why the data obtained are highly valuable.

2. Bottomhole Sampling of Reservoir Water Saturated with Gas

The reinjected acid gas is a waste product of the amine sweetening plant used for processing produced sour natural gas. The injection of acid gas into the B-28 well is continuous, and injection rate is dependent on the current production rates of four producing wells. The acid components that have been separated from produced sour natural gas are reinjected into the sequestration structure in a closed loop. The current injection rate of acid gas is 0.9 t/day. The acid gas reinjected through the B-28 well partly dissolved in formation water adjacent to the well. Because of full hydrodynamic connectivity, the water- and gas-bearing zones remain in equilibrium. Water underlying the gas reservoir remains naturally saturated with natural gas. The undissolved gas occurs in the aquifer in the form of immobile dispersed bubbles. Having exceeded the critical gas saturation of the brine, the acid gas will gravitationally migrate upwards in the reservoir, and will penetrate the gas zone leading to a gradual mixing with the reservoir gas. Compositional changes of gas produced by the individual wells, and actually the occurrence of increased CO_2 or H_2S concentrations provides information on the propagation of acid gases plume migration within the gas zone of the Borzęcin sequestration structure. From the beginning of functioning of the acid gases reinjection facility, there were attempts to monitor migration of fluids in the reservoir. In quarterly intervals the composition of natural gas produced from different wells was analysed.

To obtain a more consistent picture of acid gas plume migration in the Borzęcin structure it was decided to obtain bottomhole samples (underground sampling—recovery of reservoir samples from the well) of reservoir water, i.e., brine situated immediately below the gas reservoir. In the history of the Borzęcin sequestration operation, such samples were not obtained before. Small volumes of liquid are sampled using a bottomhole samplers driven into the well (Figure 1) [11]. After lowering to a pre-set depth, the sampler is hermetically closed and the fluid (gas/oil/water) remains isolated in its chamber preserving the pressure and temperature at the moment of sampling, hence in situ reservoir conditions [12–15]. Taking such samples allows the determination of the composition of gas dissolved or dispersed in the reservoir water before these gases will penetrate to the gas zone of the structure. After pulling out the sampler from the well, the sample of water saturated with gas was subject to separation, which resulted in obtaining the gas, entirely dissolved or dispersed in the water phase at in situ reservoir conditions, and degassed reservoir water.

(a) (b)

Figure 1. Sampling of bottomhole samples; (**a**) assembling lubricator at B-24 well; (**b**) pulling out the bottomhole samplers from the B-22 well lubricator.

Within the SECURe project, 10 operations of bottomhole sampling of reservoir water underlying the Borzęcin gas reservoir were carried out within two campaigns in 2018–2019. The majority of bottomhole sampling operations were successful. Samples of reservoir water saturated with gas were obtained from wells B-4, B-6, B-22, and B-24. In wells B-4, B-6, and B-24 it was possible to obtain the research material twice—year after year—which is an additional advantage for comparative analyses. Despite attempts, no bottomhole samples from wells B-21, B-27, and B-30 were obtained. The main reason was the lack of water at the pre-set depth of samplers driving—albeit there were attempts to drive them to the maximum safe depth determined by the drift mandrel with an overflow bailer driven down to the well.

In the two producing wells (B-4 and B-22), the gas saturation with water was increased, approximately 0.6 Nm^3/m^3 (gas dispersed and dissolved), in the others closed, the gas saturation was at dissolved gas level, i.e., 0.3 Nm^3/m^3. Figure 2 presents the location of wells on the structural map of Borzęcin reservoir. The B-28 injection well is marked with a black square. The wells, where sampling of reservoir water was successful, are marked with green triangles, while those, from which despite efforts made, were not possible, with red ones.

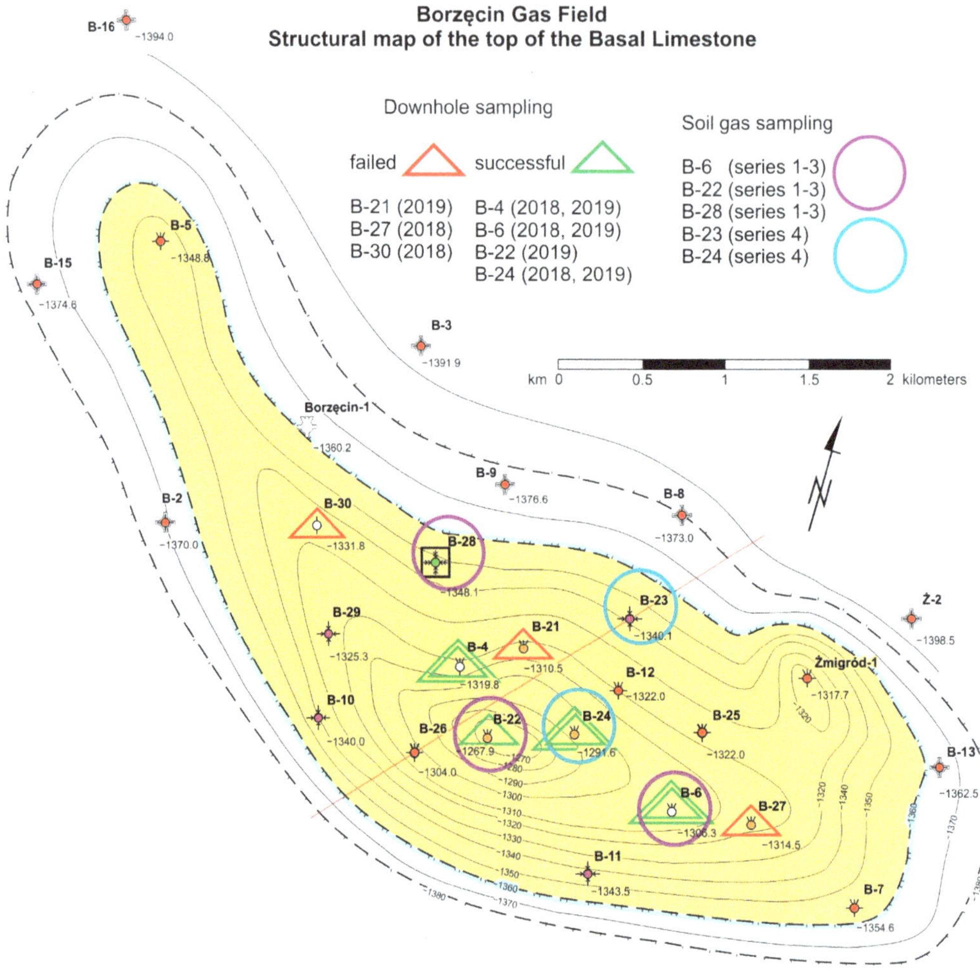

Figure 2. Map of Borzęcin reservoir—sampling activities [16].

3. Gas Analyses

As previously described, after the separation of water and gas phases, gas and water samples for further tests were obtained. Their volume was measured, they were then sealed for transport to the laboratory. Water samples were subject to detailed physical and chemical analyses. The samples of gas released from the reservoir water were subject to chromatographic analyses and isotope composition determinations (abundance of each isotope in atom %). To perform chromatographic analyses of gas, a two-channel gas chromatograph was used—Agilent 7890A. A Nickel Catalyst Tube G2747A connected with a flame detector was used to determine trace amounts of CO and CO_2. Argon and nitrogen were used as the carrier gas in gas chromatography. The uncertainty of individual components measurement is 3%. The isotopic composition of stable carbon, hydrogen, and nitrogen isotopes was performed using Trace GC Ultra chromatograph (30-m HP-PLOT/Q capillary column with a diameter of 0.32 mm) coupled with Delta V Advantage isotope ratio mass spectrometer.

3.1. Gas Released from the Reservoir Water

The recording, in the tested gas composition, of increased (in respect to the natural) concentrations of components of the gas injected to the reservoir (such as CO_2 and H_2S) is an important piece of information on directions and intensity of acid gases migration within the structure. The simplified results of gas analyses are presented in Figure 3.

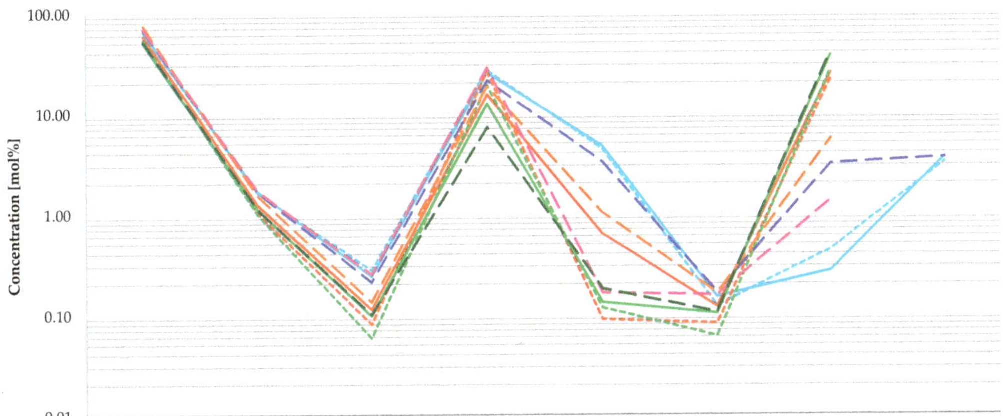

	CH4	C2H6	C3H8	N2	CO2	He	H2	H2S
B-4, Lab 1, 2018	64.57	1.68	0.24	24.82	4.61	0.15	0.27	3.55
B-4, Lab 2, 2018	63.46	1.62	0.27	26.43	4.33	0.12	0.43	3.18
B-6, Lab 1, 2018	58.17	1.23	0.11	15.01	0.62	0.12	24.48	0.00
B-6, Lab 2, 2018	51.92	1.02	0.08	25.46	0.09	0.08	21.34	0.00
B-24, Lab 1, 2018	49.49	1.11	0.10	12.23	0.13	0.10	36.63	0.00
B-24, Lab 2, 2018	53.20	0.99	0.06	18.84	0.12	0.06	26.73	0.00
B-4, Lab 1, 2019	67.14	1.65	0.21	20.92	3.21	0.16	3.08	3.53
B-6, Lab 1, 2019	73.47	1.53	0.13	18.13	1.01	0.16	5.44	0.00
B-22, Lab 1, 2019	68.39	1.70	0.25	27.91	0.16	0.15	1.34	0.00
B-24, Lab 1, 2019	51.71	1.12	0.10	7.16	0.18	0.10	39.58	0.00

Figure 3. Composition of the gas dissolved in the Borzęcin reservoir water.

It should be noted that from certain wells (B-4, B-6, B-24) samples were taken twice (in 2018 and 2019), and, in addition, a part of those samples were analysed in separate laboratories. Methane is the main hydrocarbon component of the gas released from the reservoir water, with an average content of approximately 60 mol% based on 10 analyses. The next are: ethane (1.4 mol%) and propane (0.16 mol%)—the other hydrocarbon components are negligible (below the significance level to the studied phenomena), and they were not included in the presented specification. The non-hydrocarbon components are dominated by nitrogen (approximately 20 mol%), followed by carbon dioxide (on average 1.5 mol%, but in samples originating from the B-4 well its amount is highest, reaching 4.6 mol%). Small amounts of helium (on average slightly more than 0.1 mol%) were also identified in each sample. Additionally, strongly varying hydrogen concentrations were determined in the studied gases. For two productive wells (B-4 and B-22), the values are relatively low, i.e., on average, approximately 1.3 mol%, while for the other wells, turned out from gas production, the hydrogen concentrations were much higher—on average 30 mol%. The highest hydrogen concentrations (reaching almost 40 mol%) were recorded in the B-24 well, which will be discussed hereafter. Hydrogen sulphide, on an average level of 3.4 mol%, was identified only in the gas samples from the B-4 well.

To estimate the direction and intensity of acid gases migration, the CO_2 and H_2S concentration in relation to the distance from the injection well was also analysed. Studies showed that water originating from the Borzęcin-4 well is saturated to the largest extent with carbon dioxide and hydrogen sulphide, hence the acid gases reinjected to water-bearing layers. In each of the three analyses of gas released from the B-4 reservoir water, the determined groups exceeded many times the levels observed in the other wells from which bottomhole samples were taken (B-6, B-22, B-24). The analysis of arrangement and mutual distances of the wells mentioned above to the injecting well helps understand the concentration distribution. As shown in Figure 2, the B-4 well is situated closest to the well injecting acid gases. Moreover, it is located south of B-28, that is in the direction of the expected migration of acid gases upwards the structure, directed by gravity forces, but also by a depression of pressure caused by continuous gas production from the reservoir—now by four wells, i.e., B-4, B-21, B-22 (most productive), and B-27. It also should be noted that the gas containing hydrogen sulphide at an average level of 3.4 mol% was detected only in samples originating from the B-4 well (along a straight line 630 m away from the B-28 well). This is a very high concentration, considering that in samples from the other wells hydrogen sulphide was not identified at all.

Considering the chromatographic results, attention should be paid to the unusually high concentration of hydrogen originating from the B-24 well, which has been out of operation since 2013. During the workover carried out in 2014, the set of tubing section in the well was replaced. Unfortunately, part the of tubing section was lost and left in the well. The last pulled out production tubing section was entirely damaged by corrosion. Despite reconstruction works, no satisfactory gas production was achieved, and the well remains out of operation. Some information on the well history was provided here to explain that in the gas originating from the B-24 well, record hydrogen concentrations of 26 ÷ 40 mol% were registered. Unexpectedly high hydrogen concentrations were the reason for transferring the backup gas samples to an additional laboratory (Lab-2) to verify the analysis. The confirmed very high hydrogen concentration is related to progressing corrosion of the lost part of the tubing. The bottomhole brine was sampled approximately 13 m above the tubing failure. It features a very low degree of saturation with gas, estimated on average at 0.2 Nm^3/m^3, therefore, already small amounts of hydrogen can substantially increase its percentage concentration. The well remained out of operation for years, resulting in minimal water exchange between the well annulus and the aquifer. This provides favourable conditions for progressive corrosion, which leads to continuous water saturation with hydrogen. The methane, nitrogen, and other gas components contents are necessarily proportionally lower at very high hydrogen concentrations. No hydrogen sulphide was found in any of the analysed samples, which could be partially related to the H_2S consumption during the corrosion reaction. The carbon dioxide concentration, on average, is 0.14 mol%, so it maintains a very low level. It should be concluded that the acid gases migration did not also reach the out-of-operation B-24 well.

3.2. Produced Gas

The gas from the Borzęcin reservoir is currently produced with four wells, i.e., B-4, B-21, B-22, and B-27. Samples of the gas, produced by individual wells and of water produced with gas, were taken from the surface installation in May 2019. The samples for testing were taken from the reduction-measurement lines, separate for each productive well [17,18]. The analyses showed that methane prevails among hydrocarbon gas components, with an average content of 66.77 mol%; the next ones are ethane (1.91 mol%) and propane (0.31 mol%)—the other hydrocarbon components due to their negligible concentrations are omitted. The non-hydrocarbon components are dominated by nitrogen (on average 30.16 mol%), carbon dioxide is next (on average 0.41 mol%, but in samples originating from the B-4 well its amount is highest—0.68 mol%). The analysis of gas released from bottomhole water samples revealed increased CO_2 and H_2S concentrations only in the B-4. As shown in Figure 2, the B-4 well is located closest to the well injecting the acid gas. Therefore, there is a phenomenon of increasingly rising penetration of the injected gas to the produced gas. Additionally, small amounts of helium (on average 0.18 mol%, at relatively small concentration differences between the wells) were identified in each of the four producing wells. In addition, small hydrogen concentrations (approximately 0.01 mol%, but in the B-4 again, the highest, 0.03 mol%) were determined in the tested gases. Hydrogen sulphide was identified in the gas from each well. Its amount is highest in the gas from the B-4 (0.18 mol%) and from the B-27 (0.15 mol%), albeit the analysis of historical data shows that H_2S concentrations exceeding 0.2 mol% were determined in the past. The results are summarized in Figure 4.

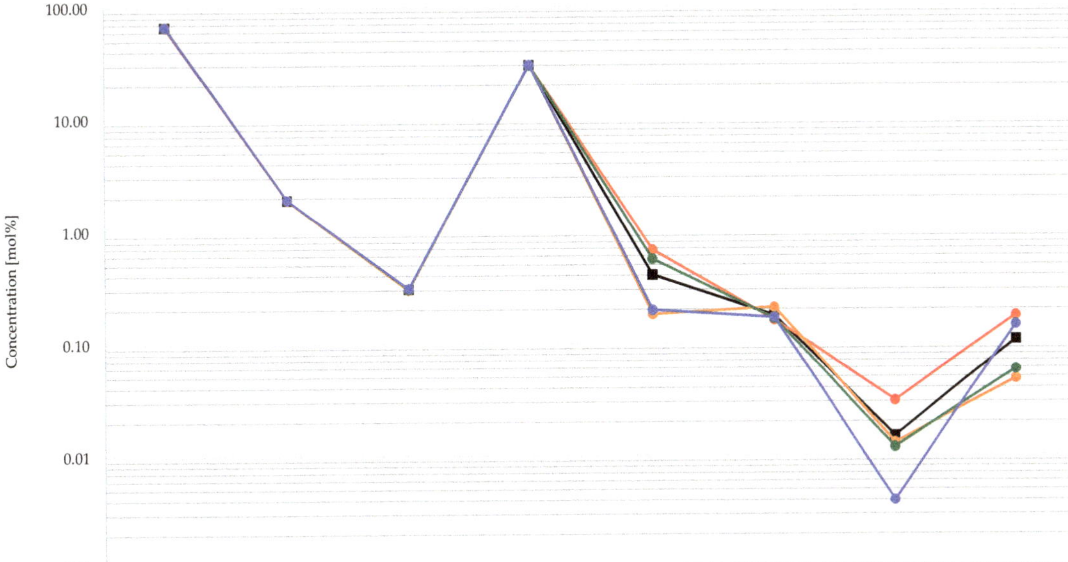

	CH4	C2H6	C3H8	N2	CO2	He	H2	H2S
Average	66.77	1.91	0.31	30.16	0.41	0.18	0.01	0.11
B-4	66.53	1.91	0.31	30.07	0.68	0.16	0.03	0.18
B-21	67.09	1.91	0.30	30.11	0.18	0.21	0.01	0.05
B-22	66.77	1.92	0.31	30.06	0.56	0.17	0.01	0.06
B-27	66.70	1.91	0.31	30.40	0.20	0.17	0.00	0.15

Figure 4. Composition of the produced gas.

3.3. Injected Gas

In February 2020, a pressurized gas sample was taken from the acid gases injection installation. Currently reinjected to the Borzęcin reservoir gas composition, revealed using chromatographic analysis is shown Figure 5. Carbon dioxide prevails among the main gas components (with an average concentration of 79.9 mol%), hydrogen sulphide is next (on average 19.0 mol%). In general, the acid components (CO_2 + H_2S) constitute 99% of the injected gas composition. Minor amounts of hydrocarbon components are the remainder. Methane was identified with an average content of 0.93 mol%, the next ones are ethane (0.04 mol%), and propane (0.01 mol%). The other hydrocarbon components were omitted in the presented specification due to their negligible concentration. Trace amounts of helium and hydrogen appear occasionally; in most cases, both gases are below the threshold of method detectability (<0.01 mol% for He and <0.001 mol% for H_2).

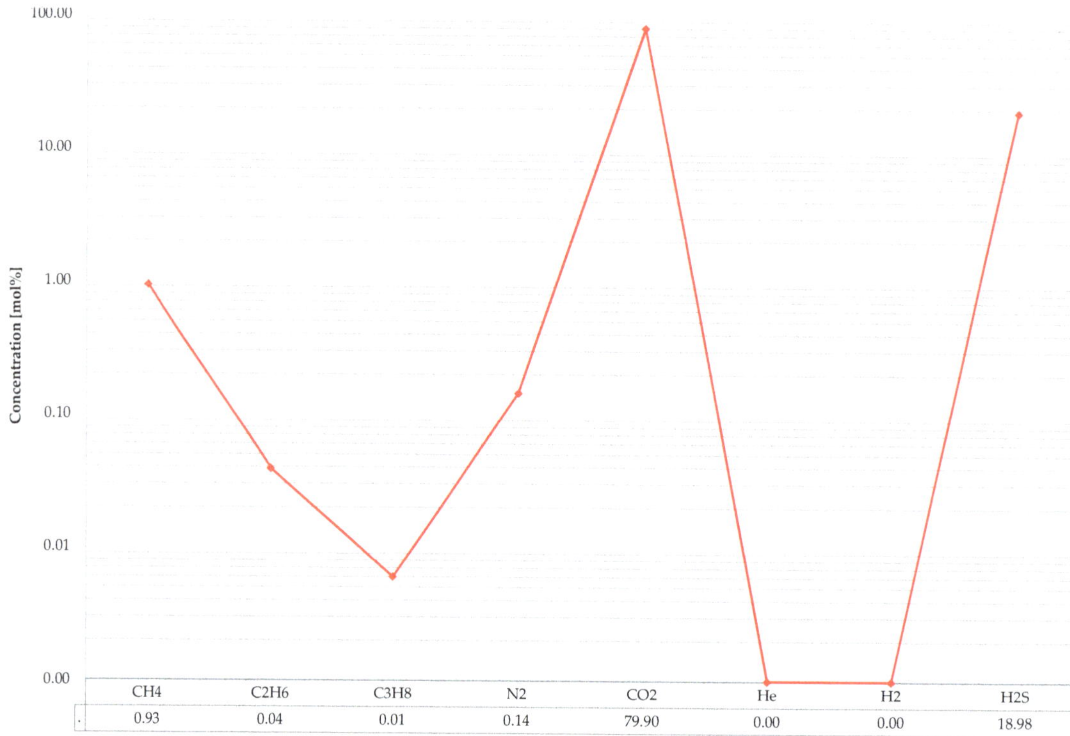

Figure 5. Current composition of the injected gas.

3.4. Isotopic Composition

Molecular and isotopic compositions of gases can provide information on their origin, maturity, and crucially on compositional change due to migration [19]. Their applications for development, production, and operation issues are increasing [20,21]. The determination of isotope composition of stable carbon, hydrogen, and nitrogen isotopes in gas dissolved in brine and in the produced (free) gas was performed. It allowed to investigate the possible temporal and spatial variability and to make an attempt to assess the migration of the injected gas components based on analyses from individual wells. In 2018, isotope determination was carried out for gas from degassing of downhole reservoir brine samples from the B-4, B-6, and B-24. A year later, in addition to repeated isotope determination in the above wells, an analysis for the B-22 well was also successfully completed. These are the first studies of such type at that site. As part of the isotope analyses, the values of carbon ($\delta^{13}C$) in methane, ethane, propane, and carbon dioxide, deuterium (δD) in methane, and nitrogen ($\delta^{15}N$) in molecular nitrogen were determined. In total, in the framework of the SECURe project, the above isotope composition was determined for 11 samples, including 3 brine degassing gas samples drawn in 2018, 4 brine gas degassing samples, and 4 produced gas samples drawn in 2019.

The isotope composition of carbon in methane ranged from −35.9‰ to −34.4‰ (average −35.1‰), in ethane from −30.57‰ to −30.1‰ (average −30.2‰), in propane from −26.4‰ to −26.0‰ (average −26.2‰), and in carbon dioxide from −17.58‰ to −3.71‰ (average −10.4). The isotope composition of hydrogen in methane was in the range of: −133.5‰ ÷ −106.9‰ (average −118.4‰), while the isotope composition of nitrogen ranged from −0.4‰ to 5.2‰ (average 3.4‰).

A relationship of the carbon stable isotopes values in methane and carbon dioxide is used to determine the genetic type of the gas (Figure 6).

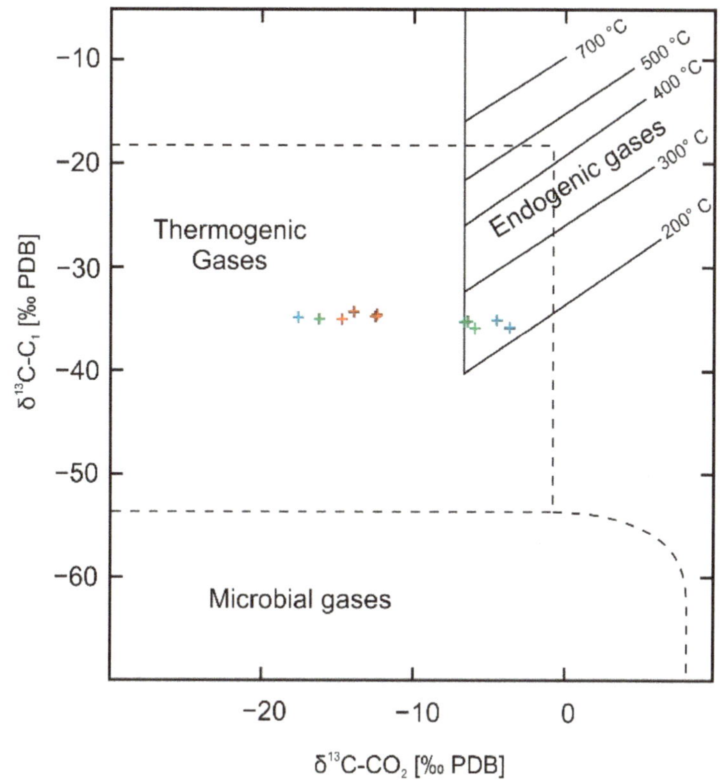

Figure 6. Carbon isotopic composition of methane versus carbon dioxide [22].

The location of points corresponding to the analysed samples on the graph of the $\delta^{13}C$-C_1 and $\delta^{13}C$-CO_2 relationships indicates mainly the thermogenic origin of the gas, although a shift of some points towards the area typical for endogenous gases is also noticeable (Figure 6). The shift of those results towards the area of endogenic gases can be attributed to the enrichment of carbon dioxide in the heavy ^{13}C isotope and it applies to some of the samples of gas dissolved in brine, drawn both, in 2018 and 2019. In the samples of gas dissolved in brine carbon isotopic composition of CO_2 could be shifted towards heavier values as a result of easier dissolving of isotopically lighter CO_2 in brine. Such dissolution results in enrichment of CO_2 with ^{13}C in natural gas. The enrichment of carbon dioxide in a heavier isotope may result from pressure changes in the reservoir caused by annual downtime and release of gas from the blind parts of the reservoir (ventilation/homogenisation). Such an enrichment in the heavier ^{13}C isotope can also result from the reduction of carbon dioxide.

The isotope composition of carbon in methane, ethane, and propane appears to be unchanged and independent of the type of gas sample drawn and the time of sampling. The differences in the extreme values of carbon isotope composition in methane are about 1.5‰ (approximately 4%) and, in ethane and propane, ca. 0.4‰ (approximately 1.5%).

However, some changes in the isotope composition of hydrogen in methane are noticeable. The isotope composition of hydrogen in the extracted gas samples is distinguished by enrichment in a heavier isotope, the deuterium, relative to gas samples from brine degassing drawn in 2019 (Figure 7).

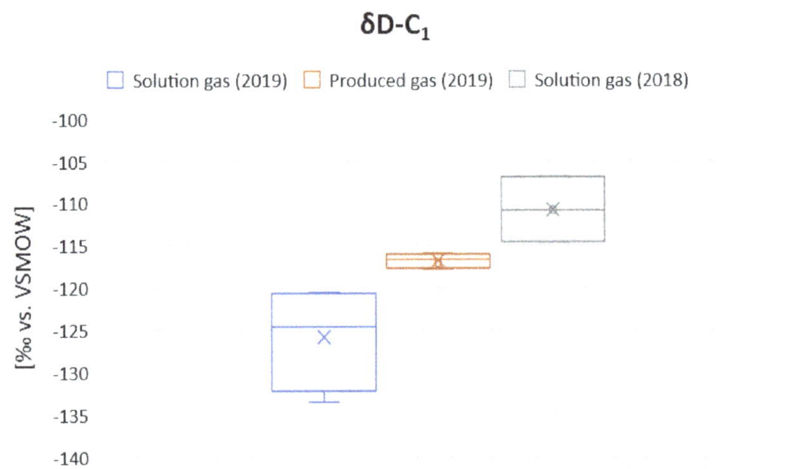

Figure 7. Variability of the isotope composition of hydrogen in methane.

4. Reservoir Water Analysis

The analyses of physiochemical parameters of bottom water underlying the Borzęcin reservoir were carried out by testing downhole samples drawn from the productive wells B-4, B-6, B-24, and B-22 and surface samples of water drawn from the wells B-21, B-22, and B-27. The tests of the downhole samples in wells B-4, B-6 and B-24 were repeated at approximately one-year interval.

In most cases, the analysed water is in the form of reservoir brine. Only the samples drawn from the B-22 (downhole sample) and B-22 (surface sample) wells bear the marks of condensation water and will not be a part of further considerations.

The liquid was from colourless to yellowish green. In most cases, the liquid had an opalescent suspension of sulphides visible to the naked eye. The odour of hydrogen sulphide was clearly perceptible.

The amount of substances dissolved in the analysed samples falls within the range 228 to 251 g/L, with 243 g/L on average. The density takes values in the range 1.153 to 1.166 g/cm^3, 1.156 g/cm^3 on average. The conductivity is on average 205 mS/cm. The pH of the samples analysed ranges from 4.5 to 6.5 with an average of 5.4. The chemical composition of the brine analysed is dominated by chloride and sodium ions. The percentage of chloride ion among anions is greater than 99%. While that of sodium ions among cations is ~78%. The averaged percentage of anions and cations (%mval) is shown in Table 1.

Table 1. Average ionic composition of the reservoir brine.

Cation (%mval)						Anion (%mval)					
Na^+	Fe + Al	K^+	Mg^{2+}	Ca^{2+}	NH_4^+	Cl^-	Br^-	SiO_3^{2-}	HCO_3^-	SO_4^{2-}	S^{2-}
2.03	0.38	0.074	4.59	15.75	0.14	99.37	0.26	0.01	0.03	0.28	0.08

Differences in the concentration of individual components between samples drawn from individual wells are insignificant and, in the case of anions, even negligible (on average below 0.1 percentage point). For cations, the differences observed, mainly concerning Ca^{2+} and K^+, are slightly higher and amount to a maximum of ~3 percentage points.

The chemical analyses repeated for the same wells at an annual interval did not show significant changes in the concentration of individual components over time either. A slight increase in the percentage of Na^+. NH_4^+ and Mg^{2+} cations can be observed at the cost of the Ca^{2+} and K^+ ones. However, those changes are slight, below 1 percentage point.

In the water samples analysed, the presence of dissolved hydrogen sulphide was found in the range of 0.015 to 0.065 g/L (0.039 g/L on average) and sulphides in the range of 0.0016 to 0.0056 (0.0033 g/L on average).

Salinity and alkalinity of brine calculated by the Palmer method [4], take the following averaged values:

- First order salinity: 79.45;
- Second order salinity: 20.49;
- Second order alkalinity: 0.05.

Table 2 presents a set of hydrochemical indicators calculated on the basis of the ionic content of individual components

Table 2. Average hydrochemical indicators.

Ion Ratio										
$\frac{Cl^-}{HCO_3^-}$	$\frac{Cl^-}{HCO_3^- + CO_3^{2-}}$	$\frac{Cl^-}{Br^-}$	$\frac{Cl^-}{SO_4^{2-}}$	$\frac{Na^+}{Cl^-}$	$\frac{Na^+}{Ca^{2+}}$	$\frac{Na^+}{Ca^{2+} + Mg^{2+}}$	$\frac{Na^+}{SO_4^{2-}}$	$\frac{Ca^{2+}}{Mg^{2+}}$	$\frac{\%Cl - \%Na}{\%Mg}$	$\frac{\%SO_4}{\%Cl}$
2859.39	2859.38	368.86	349.26	0.81	4.81	3.76	283.09	3.59	4.73	0.31

Based on the physiochemical parameters and hydrochemical indicators presented above, it can be stated that the bottom water underlying the Borzęcin reservoir, to which acid gas is injected, is composed of approximately 24% Cl-Na binary brine. According to Sulin classification [23], it is a calcium chloride type. The values of hydrochemical coefficients indicate a high degree of metamorphosis of the brine analysed. It is a quality typical of highly metamorphosed fossil brines originating from formations with very high hydrogeological tightness. Local variability in the chemical composition of brine is highly related to the performance of the wells injecting waste brine. Reservoir water with diverse chemistry account for the majority of waste brine injected, originating from the nearby oil and gas reservoirs.

5. Solubility Study

Determination of the solubility (Rs) both of the produced gas and the acid gas reinjected to the reservoir brine was a significant test for considerations of migration of the injected gases in the aquifer underlying the Borzęcin reservoir. Solubility tests were carried out at the reservoir temperature and in five pressure steps, based on which the solubility curve was drawn. Similar studies are presented in the number of papers [24–26], however it was crucial to determine the solubility properties of particular fluids originating from the Borzęcin site. The injected acid gas features definitely much higher solubility in the reservoir brine, even 23 times more—depending on the pressure, than the produced gas. Figure 8 present results of solubility study—solubility coefficient versus pressure.

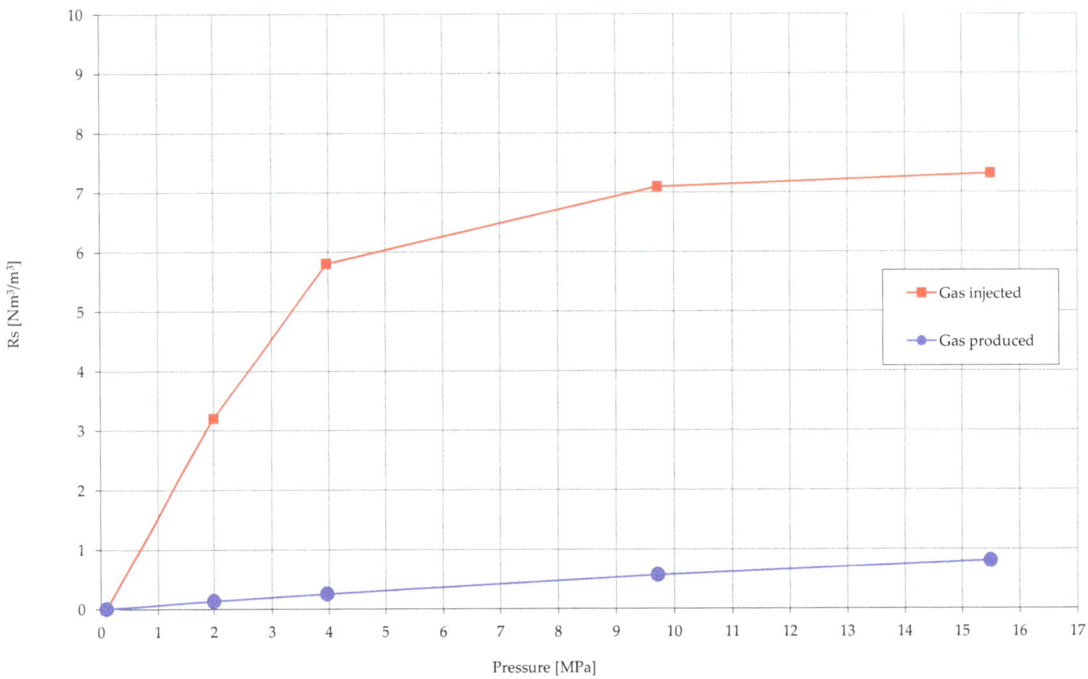

Figure 8. Solubility of gases in Borzęcin reservoir water.

6. Soil Gas Analysis

For containment evaluation of the Borzęcin sequestration structure and the safety of acid gas storage, it was necessary to perform environmental studies, including soil gas analysis. This task aimed to investigate changes in the soil gas composition of the ground located directly above the reservoir [27,28]. Currently, these areas are mainly agricultural (cultivated fields, and partly forested). The study was conducted in four cycles between October 2019 and March 2020.

At selected sampling points, mainly around the wellheads, drive-in probes were installed, or appropriately designed monitoring wells were dug in to facilitate the inflow and collection of soil gas from a given area. Then, the chromatographic analysis of the collected gas was conducted to reveal its chemical composition. The composition of soil gas was analysed along with its variation potentially resulting from different locations of sampling points, different age, and purpose of the wells around which the probes were placed or different time of sampling (e.g., autumn vs. winter) [29].

6.1. Well Selection and Sampling—Measurement Series 1–3

In order to determine the potential influence of sour gases injection into the Borzęcin reservoir on soil air composition, the measurements of its composition were performed. The air specimens were sampled from monitoring wells, located at three boreholes, that is B-6—currently inactive, observation borehole, although most of gas from this reservoir has been produced by this borehole; B-22—currently the best producer, and B-28—the borehole injecting sour gases into an aquifer, underlying the natural gas reservoir—Figure 2.

The Borzęcin-28 borehole was drilled in 1987. Despite several intensification jobs, including two-stage acid-treatment, an industrial production of gas was not achieved from it, so it has been intended for liquidation. In 1994, workover jobs have been completed to adapt the well for sour gas reinjection. The B-28 well is the crucial element of sour gas reinjection on the Borzęcin sequestration facility, and its proper technical condition is principally decisive of whole undertaking safety.

The monitoring wells were installed on October 2019. They were located directly at the wells (at the wellhead within the near well zone) and at a distance of approximately 35 m from the wellhead. An example of the location of the measuring points at the B-6 well is shown in Figure 9.

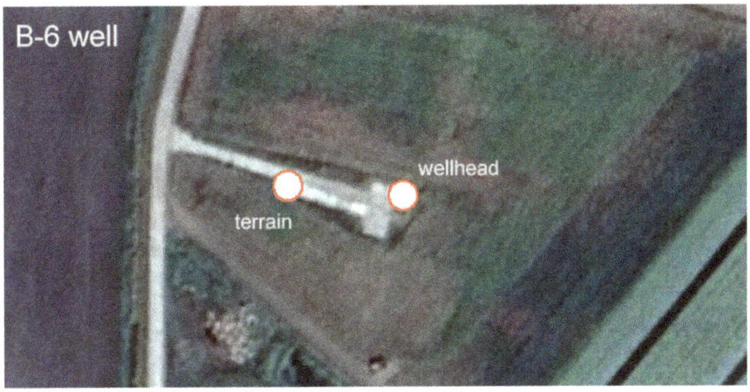

Figure 9. Location of measuring points at the B-6 well.

The gas was sampled from six 2-metre-deep monitoring wells. A monitoring well was made of PE pipe embedded in gravel backfill. In the lower part of the pipe, along an 80 cm section, there was a perforation made enabling the free inflow of soil gas. The sampling was carried out using glass pipettes and a vacuum pump—Figure 10.

(**a**) (**b**)

Figure 10. Soil gas sampling from a monitoring well located within the B-28 well pad: (**a**) from a monitoring well located within the B-28 well pad; (**b**) from a monitoring well located within 35 m of the B-28 wellhead.

6.2. Determination of the Sampled Gas Composition—Measurement Series 1–3

The composition of gas sampled in the first three measurement series was examined. Chromatographic analyses of the sampled gas were performed using a two-channel valve Clarus 680 GC Perkin Elmer-Arnel chromatograph with thermal conductivity (TCD) and flame ionisation (FID) detectors. The apparatus was coupled with TotalChrom Navigator software. A total of three measurement series were completed. No hydrogen sulfide was detected (threshold of 0.0001 mol%) in any of the samples, while carbon dioxide concentrations did not exceeded acceptable natural levels associated with agricultural areas and ranged 0.1 to 0.7 mol%. Simplified chart with hydrocarbons content in samples taken at the B-6, B-22, and B-28 wells (series 1–3) is presented in Figure 11.

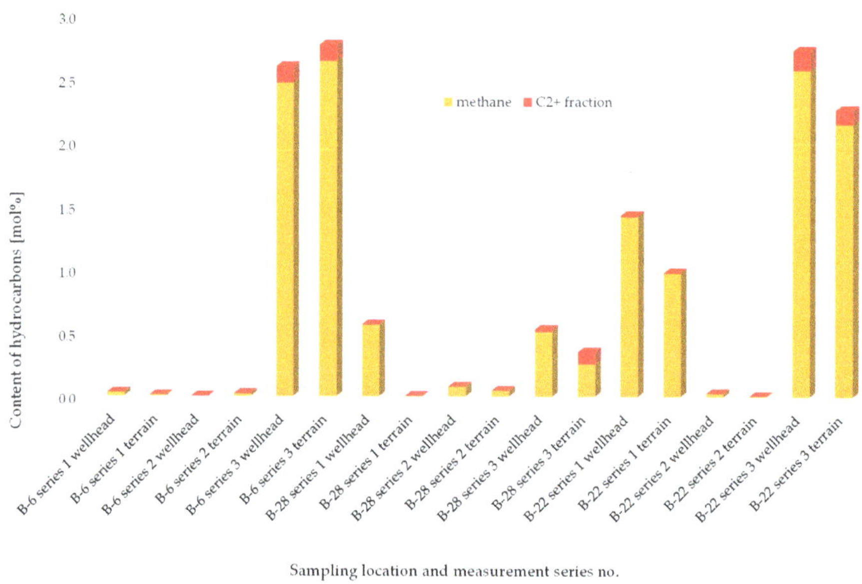

Figure 11. Hydrocarbons content in samples taken at the B-6, B-22 and B-28 wells.

Among the results from the first three measurement series, no values were observed which could suggest uncontrolled emission (leakage) of injected acid gases into the soil air. However, due to the presence of higher hydrocarbons in some samples, it was decided to perform another series of measurements, taking into account an increased number of measuring points around the indicated productive wells.

6.3. Wells Selection and Sampling—Measurement Series 4

Two wells were selected for further examination: B-23—the well injecting waste brine into the reservoir and B-24—the well in which progressing corrosion has led to a partial loss of production tubing. The B-23 well has been producing gas from the reservoir since 1988. The well has been inactive since December 1997 due to water encroachment. In 2004, the well was reconstructed to adapt it for waste brine injection (mainly water produced from other reservoirs). In 2005 and 2007, injectivity tests were performed on the well. For two years it has been one of the four wells through which waste brine to the Borzęcin reservoir is injected. The B-24 was productive well since 1985 till 2013, when during workover a section of tubing was lost, and the well remains out of production. Within the SECURe project, the downhole reservoir water sampling was performed twice in the B-24 well (in September 2018 and July 2019). The sampling depth was 1330 m below ground level. Chemical analyses of the gas separated from the brine demonstrated an increased concentration of hydrogen—over 30 mol%, which, given the above producing tubing issue, can prove the occurrence of electrochemical corrosion and hydrogen release in the well. Therefore, the B-24 well was selected for the extended examination of the soil air composition.

Around the two wells selected for measurements, 18 driven-in monitoring probes were deployed, each with a 1 m depth, 9 units per well. The measuring points were centrally located within approximately 25 m and 50 m distances from the wells and at the wells themselves—Figure 12. A hammer drill was used to make the holes for the probes. Then, a perforated PE pipe with a socket with a built-in measuring connector was placed in the holes. Perforation made on a part of the pipe allowed free flow of soil gas into the probe. The probes were installed in such a manner that the measuring connections were located at the ground surface. After 9 days, soil air samples from the installed probes were drawn into glass bulbs and handed over for further analysis. Samples were taken from 16 holes. Sampling soil air from two probes located beside the B-24 well was unsuccessful. That was presumably caused by the groundwater level being too high, preventing free flow of air into the probes.

Figure 12. Location of measuring points around the B-23 well.

6.4. Determination of the Sampled Gas Composition—Measurement Series 4

Examination of gas composition from 16 samples contained in glass bulbs within the fourth measurement series was performed. As in the measurement series 1–3, no hydrogen sulphide was detected (threshold of 0.0001 mol%) in any of the samples, while carbon dioxide concentrations did not exceeded acceptable natural levels associated with agricultural areas and ranged 0.1–2.1 mol%. Simplified chart with hydrocarbons content in samples taken at the B-23 and B-24 wells (series 4) is presented in Figure 13.

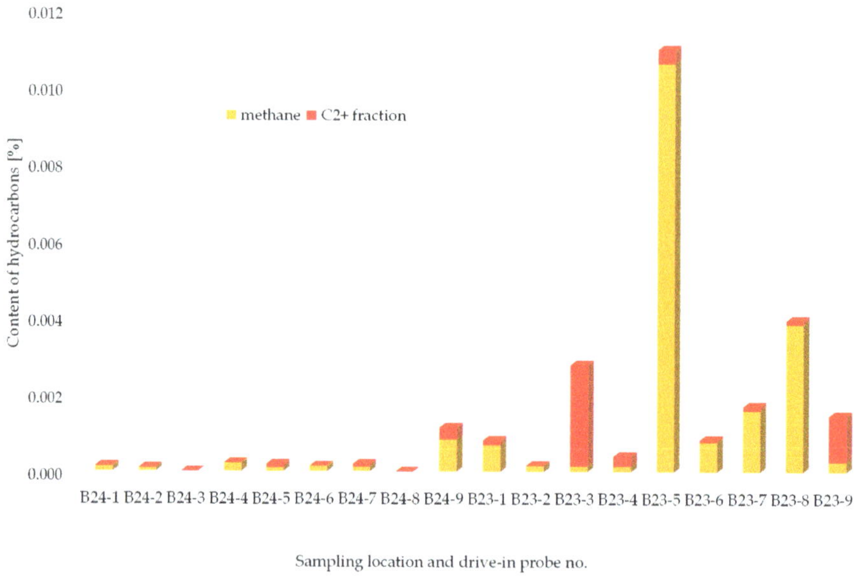

Figure 13. Hydrocarbons content in samples taken at the B-23 and B-24 wells.

The potential negative impact of acid gases injection into the Borzęcin reservoir should be considered primarily in terms of the possibility of direct uncontrolled leakage of acid gases into the soil air along the production well structure. That process could then be manifested by an increased content of, in particular, CO_2 as the dominant component, and the presence of H_2S in the soil air. In order to determine the potential of such hazards, a total of 24 monitoring wells/probes were deployed, and 34 soil air samples were drawn for examination.

Among all the results of analyses carried out in series 1–4, including the B-28 well used for injecting acid gases into the reservoir, no values were observed that could suggest uncontrolled emission of injected gases into the soil air. Analysing the results, it should be kept in mind that carbon dioxide is a gas that is part of typical soil air, and the main factors affecting its level include the activity of microorganisms and the roots of higher plants, temperature, humidity, depth, soil aeration degree, and atmospheric pressure. Furthermore, the area where the research was conducted should be taken into account. In the case of the four wells, it was an agricultural area, and forest for the B-6 well. Organic substances occurring naturally in the soil or penetrating into it as a result of human activity (natural fertilizers, crops residues) are subject to digestion and fermentation processes during which biogenic gas is generated, which consists mainly of carbon dioxide (CO_2) and methane (CH_4). It can, therefore, be concluded that the low methane concentrations found in some samples are also a consequence of natural processes occurring in the soil. The CO_2 levels observed should, therefore, be assumed to be absolutely normal for the type of area [29]. Moreover, the lack of even trace amounts of hydrogen sulphide (H_2S) in the samples examined proves the absence of uncontrolled emissions into the soil.

Out of the first three measurement series, completed at the B-6, B-22, and B-28 wells, the results of the third measurement series at the B-6 well and the first and third series at the B-22 well depart from the typical soil air composition that can be found in agricultural areas. The high level of methane (approx. 2.5%) and, in particular, the presence of higher hydrocarbons is likely to result from local gas exhalations from the reservoir. Those exhalations may take place along casing and cement, which may be indicated by higher levels of contaminants in the samples drawn directly at the wells. Increased methane and ethane values in samples drawn from the B-28 well, in the first and third series, due to the trace content of C_3 and higher hydrocarbons, are most likely caused by natural processes occurring in the soil. Those compounds may also have an anthropogenic origin, e.g., soil pollution by ongoing farming and the use of heavy equipment. The remaining samples from series 1–3 contained insignificant amounts of methane, which can be considered as the regional geochemical background.

The fourth measurement series was completed at the B-22 and B-24 wells. The presence of small amounts of methane in soil air samples (up to 0.01%) is a consequence of digestion and fermentation processes (e.g., natural fertilisers), during which biogenic gas is produced. In none of the samples was the content of higher hydrocarbons found in quantities which could imply significant gas exhalations from the reservoir.

7. Summary and Conclusions

The extensive sampling and testing campaign results, including reservoir fluids and soil gas analyses, are vital sources of knowledge concerning the phenomena occurring in the sequestration structure due to acid gas injection. All the activities conducted are summarised in Table 3.

Table 3. Sampling and testing activities.

Well Number	Well Status	Sampling Activity	Fluid Analyzed	Test Performed
B-4	producing	bottomhole sampling *, surface sampling	reservoir water, dissolved gas produced gas	reservoir water physico-chemical analysis, dissolved gas chemical composition, dissolved gas isotopic composition, produced gas composition
B-6	shut-in	bottomhole sampling *	reservoir water, dissolved gas	reservoir water physico-chemical analysis, dissolved gas chemical composition, dissolved gas isotopic composition
B-21	producing	bottomhole sampling—unsuccessful surface sampling	produced gas produced water	produced water physico-chemical analysis, produced gas composition
B-22	producing	bottomhole sampling, surface sampling	reservoir water, dissolved gas produced gas produced water	reservoir water physico-chemical analysis, dissolved gas chemical composition, dissolved gas isotopic composition, produced gas composition, produced water physico-chemical analysis
B-23	injecting waste brine	surface sampling	soil gas	soil gas composition
B-24	shut-in	bottomhole sampling *	reservoir water, dissolved gas	reservoir water physico-chemical analysis, dissolved gas chemical composition, dissolved gas isotopic composition
B-27	producing	bottomhole sampling—unsuccessful	produced gas produced water	produced gas composition, produced water physico-chemical analysis
B-28	injecting acid gases	surface sampling	soil gas injected gas	soil gas composition, injected gas composition
B-30	shut-in	bottomhole sampling—unsuccessful		

* Repeated twice.

Downhole samples of water saturated with gas are very valuable material for research, and their proper utilisation gives a lot of useful information for better understanding of the processes occurring within the Borzęcin reservoir. So far, such samples have never been obtained before.

Conducted actions and measurements allowed to reach the main goal of the research, which was the identification of acid gas migration characteristics within the structure and potential leakage pathways. The results of solution gas analysis revealed that the acid gas injected into the aquifer through the B-28 injector well migrate southwards towards the B-4 well. The investigations proved that this is the main direction of the injected acid gas plume propagation. Water taken from the B-4 well is most intensely saturated with reinjected carbon dioxide and hydrogen sulphide. In each of the three analyses of gas separated from B-4 reservoir water, the concentration of CO_2 and H_2S exceeded many times the levels observed in the other downhole-sampled wells. The B-4 well is closest to the B-28. Moreover, it is located in the direction consistent with the expected migration of acid gas upward the structure, driven by gravity forces as well as by pressure depletion caused by continuous gas production from the reservoir (especially by the B-4 and B-22 wells, where B-22 accounts for 80% of current production).

Research actions were also related to the surface sampling of gas and associated water. As in the case of gas separated from the reservoir water, the chemical analysis of produced gas confirmed the inflow of the injected acid gas into the gas produced (especially) by the B-4 well. Furthermore, an increased concentration of CO_2 was also recorded in the gas produced by the B-22 well, which may indicate that the migration zone of the injected acid gas is going to affect the major production well.

Samples of the acid gas were also taken from the reinjection facility. The results of the chemical analysis confirmed stability in the composition of the gas over time. Currently, after the modernisation of the acid gas sweetening facility, the reinjected gas is composed mainly (in 99 mol%) of carbon dioxide and hydrogen sulphide, while the hydrocarbon components represent less than 1 mol%.

The high dynamics of individual phase systems occurring in the structure of the Borzęcin reservoir implies temporal and spatial variability in the isotope composition of the analysed gas from brine degassing and of the produced gas. The system is subject to periodic disturbances related to the process of reservoir exploitation, repeated injection of acid gases and waste brine coming from outside the reservoir. Moreover, the solubility of CO_2 in brine and isotope fractionation (hydrogen between methane and brine) may also affect the isotope composition of carbon in CO_2 and hydrogen in methane for the brine degassing samples. Additionally, other processes, as microbial or abiotic oxidation of C_1 might play role in the d2H (C_1) variations.

Based on the physiochemical parameters and hydrochemical indicators, it can be stated that the water underlying the Borzęcin reservoir, to which acid gas is injected, is composed of approximately 24% Cl-Na binary brine. According to the Sulin classification [23], it is a calcium chloride type brine. Local variability in the chemical composition of brine is highly related to the performance of the wells injecting waste brine. Reservoir water with diverse chemistry account for the majority of waste brine injected, originating from the nearby oil and gas production facilities.

Among others, the ability to dissolve the produced and reinjected gases in the reservoir water was also examined. The studies showed that acid gas has excellent solubility in Borzęcin reservoir water. In the current thermobaric reservoir conditions, the solubility of acid gas is about 20 times higher than of the produced gas. This feature indicates that a significant volume of acid gas injected directly into the aquifer dissolves in water and migrate within the structure, remaining in the water zone. This is a positive phenomenon when concerning acid gas sequestration within an active gas reservoir. The dissolution of CO_2 and H_2S in the reservoir water certainly delays the breakthrough of these components into the natural gas zone.

The soil air analyses carried out so far did not reveal surface leaks of acid gases injected into the Borzęcin structure. Periodically elevated hydrocarbon concentrations at the B-6 and B-22 wells are likely to occur due to natural gas exhalation. To ascertain what their origin is, it is recommended to perform long-term monitoring tests on the identified wells that could be the future venues of research.

Author Contributions: Conceptualisation, M.W. (Marcin Warnecki) and M.W. (Mirosław Wojnicki); methodology, M.W. (Marcin Warnecki) and M.W. (Mirosław Wojnicki); investigation, M.W. (Marcin Warnecki); J.K., S.S. and M.W. (Mirosław Wojnicki); writing—original draft preparation, M.W. (Marcin Warnecki); writing—review and editing, M.W. (Mirosław Wojnicki). All authors have read and agreed to the published version of the manuscript.

Funding: This report is part of a project that has received funding by the European Union's Horizon 2020 research and innovation programme under grant agreement number 764531. Project acronym and title: SECURe—Subsurface Evaluation of Carbon capture and storage and Unconventional Risks. D2.2 Report on effects of long-term sequestration process in the Borzęcin structure—observation evidence of the injected gas migration and possible leakage.

Institutional Review Board Statement: Not applicable.

Informed Consent Statement: Not applicable.

Conflicts of Interest: The authors declare no conflict of interest.

References

1. Bachu, S.; Gunter, W.D. Acid-gas injection in the Alberta basin, Canada: A CO_2-storage experience. *Geol. Soc. Lon. Spéc. Publ.* **2004**, *233*, 225–234. [CrossRef]
2. Bachu, S.; Gunter, W. Overview of acid-gas injection operations in Western Canada. In *Greenhouse Gas Control Technologies 7*; Elsevier Inc.: Amsterdam, The Netherlands, 2005; pp. 443–448. [CrossRef]
3. Wong, S.; Keith, D.; Wichert, E.; Gunter, B.; Mccann, T. Economics of Acid Gas Reinjection: An Innovative CO_2 Storage Opportunity. In *Greenhouse Gas Control Technologies—6th International Conference*; Elsevier Inc.: Amsterdam, The Netherlands, 2003; pp. 1661–1664. [CrossRef]
4. Lubaś, J.; Szott, W. 15-year experience of acid gas storage in the natural gas structure of Borzęcin—Poland. *Nafta-Gaz* **2010**, *66*, 333–338.
5. Lubaś, J.; Szott, W.; Jakubowicz, P. Effects of Acid Gas Reinjection on CO_2 Concentration in Natural Gas Produced from Borzęcin Reservoir. *Nafta-Gaz* **2012**, *68*, 405–410.
6. Baroni, A.; Estublier, E.; Deflandre, J.P.; Daniel, J.M. Modelling surface displacements associated with CO_2 reinjection at Krechba. In Proceedings of the 45th US Rock Mechanics/Geomechanics Symposium, San Francisco, CA, USA, 26–29 June 2011.
7. Ringrose, P.; Mathieson, A.; Wright, I.; Selama, F.; Hansen, O.; Bissell, R.; Saoula, N.; Midgley, J. The In Salah CO_2 Storage Project: Lessons Learned and Knowledge Transfer. *Energy Procedia* **2013**, *37*, 6226–6236. [CrossRef]
8. Van der Meer, L.G.H. The K12-B CO_2 injection project in the Netherlands. In *Geological Storage of Carbon Dioxide (CO_2): Geoscience, Technologies, Environmental Aspects and Legal Frameworks*; Elsevier Inc.: Amsterdam, The Netherlands, 2013. [CrossRef]
9. Vandeweijer, V.; Hofstee, C.; Graven, H. 13 years of safe CO_2 injection at K12-B. In Proceedings of the 5th CO_2 Geological Storage Workshop, Utrecht, The Netherlands, 21–23 November 2018; European Association of Geoscientists and Engineers, EAGE: Houten, The Netherlands, 2018. [CrossRef]
10. Chadwick, R.A.; Marchant, B.; Williams, G.A. CO_2 storage monitoring: Leakage detection and measurement in subsurface volumes from 3D seismic data at Sleipner. *Energy Procedia* **2014**, *63*, 4224–4239. [CrossRef]
11. Williams, J. Getting the Best Out of Fluid Samples. *J. Pet. Technol.* **1994**, *46*, 752. [CrossRef]
12. Yonebayashi, H.; Miyagawa, Y.; Ikarashi, M.; Watanabe, T.; Maeda, H.; Yazawa, N. How Many Back Up Prepared For Asphaltene Onset Pressure Measurement Using Costly Collected Single Phase Bottomhole Fluid Samples? In Proceedings of the SPE Annual Technical Conference and Exhibition, Dubai, United Arab Emirates, 26–28 September 2016. [CrossRef]
13. Jamaluddin, A.; Ross, B.; Ross, D.; Hashem, M. Single-Phase Bottomhole Sampling Technology. *J. Can. Pet. Technol.* **2002**, *41*. [CrossRef]
14. Piazza, R.; Vieira, A.; Sacorague, L.A.; Jones, C.; Dai, B.; Pearl, M.; Aguiar, H. Innovative Formation Tester Sampling Procedures for Carbon Dioxide and Other Reactive Components. *Petrophys.—SPWLA J. Form. Eval. Reserv. Descr.* **2021**, *62*, 65–72. [CrossRef]
15. Gutman, R. Formation Fluid Sampling Technique Allowing to Control Physical Properties of Sampling Medium. In Proceedings of the SPE Annual Technical Conference and Exhibition, San Antonio, TX, USA, 9–11 October 2017. [CrossRef]
16. Lubaś, J.; Szott, W.; Łętkowski, P.; Gołąbek, A.; Miłek, K.; Warnecki, M.; Wojnicki, M.; Kuśnierczyk, J.; Szuflita, S. *Long-Term Sequestration Process in the Borzęcin Structure—Observation Evidence of the Injected Acid Gas Migration and Possible Leakage*; Instytut Nafty i Gazu—Państwowy Instytut Badawczy: Kraków, Poland, 2020. [CrossRef]
17. Haji, A.; Veerakumar, U. Overcoming Hazards in Sour Gas Sampling by Controlled Measures. In Proceedings of the Abu Dhabi International Petroleum Exhibition & Conference, Abu Dhabi, United Arab Emirates, 7–10 November 2016. [CrossRef]
18. Balthar, A.R.; Oliveira, V.M.; Souza, J.R.; Bayer, M.M.; Macedo, V.C. The influence on gas chromatography analysis caused by the Joule-Thomson cooling effect during the natural gas sampling. In Proceedings of the 17th World Petroleum Congress, Rio de Janeiro, Brazil, 1–5 September 2002.
19. Waseda, A.; Iwano, H.; Kato, S. Geochemical Evaluation of Reservoir Compartment, Effective Pay Zone and Origin of Gas Seepage in Production Site Using Gas Molecular and Isotopic Composition. In Proceedings of the International Petroleum Technology Conference, Kuala Lumpur, Malaysia, 3–5 December 2008. [CrossRef]

20. Lu, F.; Harthi, O.; Dubaissi, J. Isotopic Characterization of Formation Waters: Applications and Challenges. In Proceedings of the SPE Middle East Oil and Gas Show and Conference, Manama, Bahrain, 18–21 March 2019. [CrossRef]
21. Schloemer, S.; Krooß, B.M. Molecular transport of methane, ethane and nitrogen and the influence of diffusion on the chemical and isotopic composition of natural gas accumulations. *Geofluids* **2004**, *4*, 81–108. [CrossRef]
22. Kotarba, M.J.; Nagao, K. Composition and origin of natural gases accumulated in the Polish and Ukrainian parts of the Carpathian region: Gaseous hydrocarbons, noble gases, carbon dioxide and nitrogen. *Chem. Geol.* **2008**, *255*, 426–438. [CrossRef]
23. Tikhomirov, V.V. *Hydrogeochemistry Fundamentals and Advances: Groundwater Composition and Chemistry*; John Wiley & Sons, Inc.: Hoboken, NJ, USA, 2016; Volume 1. [CrossRef]
24. Ratnakar, R.R.; Venkatraman, A.; Kalra, A.; Dindoruk, B. On the Prediction of Gas Solubility in Brine Solutions for Applications of CO_2 Capture and Sequestration. In Proceedings of the SPE Annual Technical Conference and Exhibition, Dallas, TX, USA, 24–26 September 2018. [CrossRef]
25. Kowta, R.; Erickson, D.; Barker, R.; Neville, A.; Hua, Y. Models for Calculating Corrosion Rates in Water-Saturated and Under-Saturated CO_2 Systems & Water Solubility in CO_2 Systems at Supercritical Conditions. In Proceedings of the Carbon Management Technology Conference, Houston, TX, USA, 15–18 July 2019. [CrossRef]
26. Agarwal, R.; Li, Y.-K.; Nghiem, L. An Efficient Method for Modelling Gas Solubility in the Aqueous Phase for Compositional Simulators. *J. Can. Pet. Technol.* **1993**, *32*. [CrossRef]
27. Erno, B.; Schmitz, R. Measurements of Soil Gas Migration Around Oil And Gas Wells In the Lloydminster Area. *J. Can. Pet. Technol.* **1996**, *35*. [CrossRef]
28. Barker, G.W.; Brown, D.R.; Corgan, J.M.; Fisher, J.B.; Raterman, K.T.; Trent, G.L. Soil Gas Surveys: A Cost-Effective Site Assessment Technique. In Proceedings of the SPE/EPA Exploration and Production Environmental Conference, Houston, TX, USA, 27–29 March 1995. [CrossRef]
29. Jóźwiak, K. Variability of concentrations of gases in the air of the vadose zone in the natural and agriculturally converted environments. *Przegląd Geol.* **2017**, *65*, 1075–1079.

energies

Article

An Experimental Investigation of WAG Injection in a Carbonate Reservoir and Prediction of the Recovery Factor Using Genetic Programming

Mirosław Wojnicki [1,*], Jan Lubaś [1], Mateusz Gawroński [2], Sławomir Szuflita [1], Jerzy Kuśnierczyk [1] and Marcin Warnecki [1]

[1] Oil and Gas Institute—National Research Institute, 31-503 Krakow, Poland; lubas@inig.pl (J.L.); szuflita@inig.pl (S.S.); kusnierczyk@inig.pl (J.K.); warnecki@inig.pl (M.W.)

[2] Department of Robotics and Mechatronics, AGH University of Science and Technology, 30-059 Krakow, Poland; matgawr@agh.edu.pl

* Correspondence: wojnicki@inig.pl

Citation: Wojnicki, M.; Lubaś, J.; Gawroński, M.; Szuflita, S.; Kuśnierczyk, J.; Warnecki, M. An Experimental Investigation of WAG Injection in a Carbonate Reservoir and Prediction of the Recovery Factor Using Genetic Programming. *Energies* **2022**, *15*, 2127. https://doi.org/10.3390/en15062127

Academic Editor: Reza Rezaee

Received: 20 January 2022
Accepted: 13 March 2022
Published: 14 March 2022

Abstract: Production from mature oil fields is gradually declining, and new discoveries are not sufficient to meet the growing demand for oil products. Hence, enhanced oil recovery is emerging as an essential link in the global oil industry. This paper aims to recognize the possibility of increasing oil recovery from Polish carbonate reservoirs by the water alternating gas injection process (WAG) using various types of gases, including CO_2, acid gas (a mixture of CO_2 and H_2S of 70/30% vol/vol) and high-nitrogen natural gases occurring in the Polish Lowlands. A series of 17 core flooding experiments were performed under the temperature of 126 °C, and at pressures of 270 and 170 bar on composite carbonate cores consisting of four dolomite core plugs. Original reservoir rock and fluids were used. A set of slim tube tests was conducted to determine the miscibility conditions of the injected fluids with reservoir oil. The WAG process was compared to continuous gas injection (CGI) and continuous water injection (CWI) and was proven to be more effective. CO_2 WAG injection resulted in a recovery factor (RF) of up to 82%, where the high nitrogen natural gas WAG injection was less effective with the highest recovery of 70%. Based on the core flooding results and through implementing a genetic programming algorithm, a mathematical model was developed to estimate recovery factors using variables specific to a given WAG scheme.

Keywords: enhanced oil recovery; WAG; carbonate reservoir; CO_2; acid gas; high-nitrogen natural gas; water alternating gas; EOR; recovery factor; genetic programming

1. Introduction

The oil recovery factor in conventional reservoirs varies from field to field since it depends on many different variables. The worldwide average is about 30% IOIP (initial oil in place) which means that there is great potential to recover more [1]. That is why enhanced oil recovery (EOR) has been one of the most investigated areas in the petroleum industry in the last decades. Tremendous work has been done so far which has resulted in a vast range of published papers concerning lab-scale research, reservoir modelling, and the outcome of field applications of different EOR methods [2,3]. Despite that, the ultimate profit from EOR applications is below expectation (<10% of total production), and recent studies show that EOR is still in the top priority research and innovation areas in the energy industry [4]. The need to further explore EOR concepts comes from the fact that every EOR process is strongly case-specific and it is difficult to make an analogy to another case, thus requiring specific research, optimization, expertise, and trial field demonstration.

Carbonate reservoirs are of special interest, because they contain more than 60% of the world's remaining conventional oil reserves and account for over 30% of the world's daily oil production [5–8]. Due to the complex oil recovery process in carbonates caused by their unfavorable reservoir characteristics, recovery factors are even lower with an average of 20% [9–11]. The above include high heterogeneity, mixed- to oil-wet conditions, and dual permeability—poor in rock matrices and high in fractures [12,13].

Most of the EOR projects in carbonates are gas injections (nearly 60%), and the vast majority use CO_2 (continuously or alternately with water). Currently, a majority of CO_2-EOR projects utilize CO_2 from natural sources, but as global discussion on climate change and worldwide efforts on carbon emission reduction intensifies, it is expected that anthropogenic CO_2 sources will be more frequently used. Hydrocarbon gas injection projects have a significantly lower contribution for EOR in carbonates. Others such as nitrogen or acid gas (mixture of H_2S and CO_2) are even less common [1,3,14].

However, EOR gas injection poses significant challenges connected with the high mobility ratio (caused by the significantly lower dynamic viscosity of the injected gas compared to reservoir oil), including viscous fingering and early breakthrough of the injected fluid into production wells [15,16].

To counteract that, a Water Alternating Gas (WAG) injection method was initially designed to control gas mobility and stabilize the gas displacement front during continuous gas injection (CGI) and finally improve sweep efficiency. The method, which combines CGI and waterflooding (continuous water injection—CWI) methods, was first implemented in 1956 in the North Pembina field (Alberta, Canada), and since then has been effectively used worldwide [16–19]. The combination of improved microscopic displacement of CGI with an improved macroscopic sweep of CWI led, in most cases, to enhanced oil recovery. Water slugs stabilize the displacement front and help to sweep crude oil from the lower part of the reservoir [20]. A further essential benefit of WAG is that less gas is required for injection, in favor of the usually cheaper water. In the WAG method, both water and gas are injected to the same well. There are different WAG injection schemes where, e.g., water and gas are injected simultaneously (SWAG) [21,22], a huge slug of gas is followed by a number of conventional WAG cycles—HWAG [23], or the volume of injected gas is gradually reduced over time—TWAG [24].

Gas can be injected under miscible (MWAG) or immiscible (IWAG) conditions that are differentiated by the Minimum Miscibility Pressure (MMP). When the injection pressure is slightly lower than MMP, it is hard to distinguish between miscible and immiscible types because of the subsequent mass transfer mechanisms involved (swelling/stripping), and such conditions should be referred to as "near-miscible" regime (nMWAG) [21,25]. Both MWAG and nMWAG are considered more effective than IWAG [17,26], but many studies revealed that IWAG is also efficient in enhancing oil recovery [27,28]. Miscibility development is not always required for successful WAG implementation, but helps to achieve better results in most cases [10]. The general scheme of WAG injection is shown in Figure 1.

WAG injection has been comprehensively studied with a particular consideration of factors affecting its performance such as reservoir parameters including wettability [29,30] heterogeneity [31–33], and fractures [34], injected fluid parameters including water salinity [35–37] and gas type [15,38], the WAG parameters including the WAG ratio [39,40], the number of cycles, slug sizes [41,42], the timing of injection [43], and finally the injection rates of the gas and water phases [11].

The cyclic nature of WAG leads to an increase in water saturation during the water injection half cycle and a decrease of water saturation during the gas injection half cycle involving associated hysteresis phenomena resulting in complex three phase-flow which makes the prediction of WAG performance very difficult. This was extensively studied by Fatemi, Sohrabi, and Shahverdi [44,45].

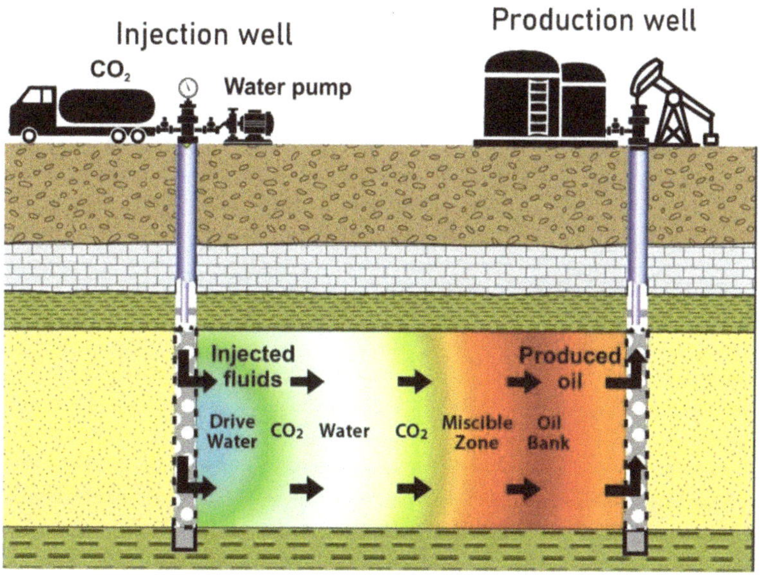

Figure 1. General scheme of WAG injection (miscible).

Physical simulation of the WAG process in the laboratory is generally done using the core flooding test, which is also used in the presented paper. Several important works were also conducted using micromodel visualization [46–49]. Over the years, a number of experimental studies has been performed in mixed wet carbonates, revealing different process issues and enabling the more effective use of WAG in complex reservoir conditions.

However, considering that WAG process efficiency is strongly site-specific, there is a gap concerning its suitability in the specific conditions of mixed-wet and fractured carbonate sour crude oil reservoirs of the Polish Lowlands. The novelty of the research is defined by the type of crude oil, reservoir conditions, and injected gases used. There is a lack of published WAG experimental data for such kind of settings. The previous study of the authors focused on evaluating high-nitrogen sour natural gas WAG injection and the impact of fractures on its efficiency [50]. Whereas the current work aims to evaluate the efficiency of four different types of gases in WAG injection and focus on empirical modelling of the oil recovery factor based on experimental data using evolutionary algorithms. Genetic programming was used to generate a correlation to predict the oil recovery factor as a function of variables defining core flooding experiments, i.e., injected gas composition, pressure, and the gas to water ratio in the injected fluid stream.

The estimated oil recovery factor is one of the most significant parameters for an operator when selecting the proper EOR method. The recovery factor is affected by several engineering and geological aspects, that make the estimation of the RF complex. RF estimation based on experimental data could be applied strictly in the tested conditions (or very similar) characterized by reservoir fluids, rock type, temperature, or flow conditions in porous media (e.g., natural fractures).

Implementation of evolutionary algorithms in the petroleum industry is widespread and concentrates on parameter estimation, correlation generation, and predictive analytics. They are particularly useful in solving problems where the relationships between variables are unknown or poorly understood [51,52]. Examples of the application of evolutionary algorithms in reservoir engineering include modelling and production optimization [53–55], estimation of effectiveness and optimization of EOR methods (including WAG) [56–59], estimating values of parameters such as MMP [60], the formation volume factor [61] or the emulsion viscosity [62], and issues related to the reservoir development [63–67].

2. Materials and Methods

Core flooding experiments were performed using original reservoir rock saturated with original reservoir fluids (brine, live oil) at thermobaric conditions of one of the major Polish oil reservoirs located in the Polish Lowlands. The reservoir has been developed in naturally fractured Late Permian Zechstein carbonates (mainly dolomites) of the Main Dolomite formation. The carbonates are both the source and the reservoir of the rocks and are sealed above and below by evaporites (Werra–Strassfurt cyclothemes) creating a closed petroleum system. This results in the presence of residual organic matter in reservoir rock, strongly affect the rocks' wettability leading to mixed-wet conditions. Experimental studies of WAG injection efficiency using very high-nitrogen natural gas (KG) for the same conditions were performed in a previous study [50]. In the referenced work, issues concerning reservoir characteristic rock material core flooding and the experimental process are described in detail. In the current study 3 new gas types, that are likely to be used in the WAG process, were tested. These include carbon dioxide, acid gas—a mixture of CO_2 and H_2S of 70/30% vol/vol corresponding to post-process gas from an amine sweetening plant (AG)—and nitrogen natural gas (MG).

2.1. Reservoir and Injected Fluids

Reservoir fluid was prepared individually prior to each core flooding and slim tube experiment by physical recombination from separator oil and gas samples. To ensure accurate recombination of reservoir fluid and determine the parameters required to correctly design, perform, and balance the experiments a full spectrum of PVT analyses was performed. Initial and current PVT data were used to develop a reservoir fluid model. The phase diagram of the considered reservoir fluid is shown in Figure 2.

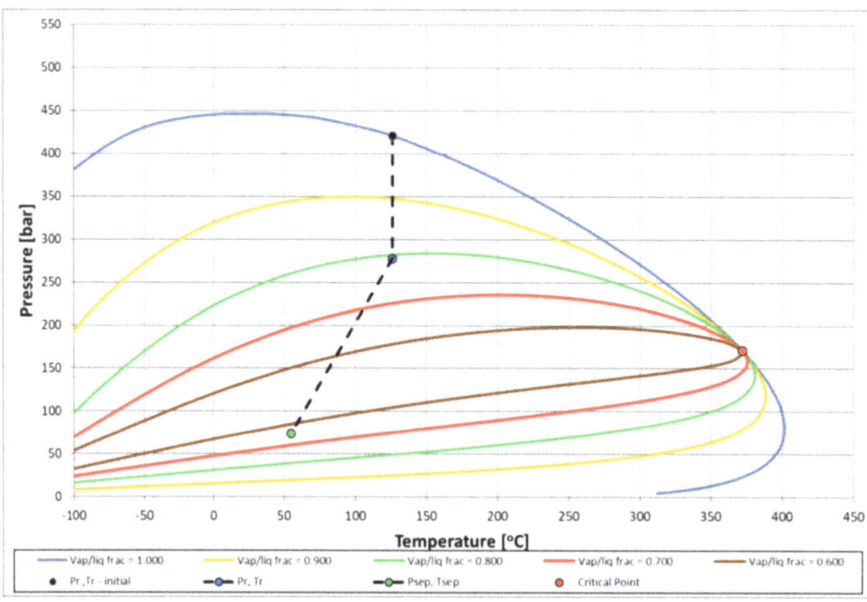

Figure 2. Phase diagram of reservoir fluid.

As can be seen from the PT diagram the initial reservoir pressure lies in the bubble point curve that indicates saturated oil conditions, along with an isothermal pressure depletion caused by the ongoing production reservoir fluid which reached the two-phase region. The dashed line indicates the path from the initial reservoir, through the current reservoir to the separator conditions.

The oil is light and sour crude oil with dynamic viscosity of 0.5 cP, and density of 0.65 g/cm^3 at reservoir conditions (P = 270 bar, T = 126 °C). The simplified composition of reservoir fluid is presented in Table 1.

Table 1. Composition of reservoir (live) oil.

Component	N_2	CO_2	H_2S	C_1	C_2	C_3	C_4	C_5	C_6	C_7	C_8	C_9	C_{10}−	C_{11}	C_{12}	C_{13}	C_{14}	C_{15+}
Mol %	30.4	0.5	4.8	19.3	3.4	2.6	2.3	3.2	2.4	2.8	3.1	2.9	2.7	2.0	1.7	1.6	1.4	13.1

The water phase used for core saturation, as well as the injection fluid in the WAG water cycle was sampled from the separator of the same well as the hydrocarbon fluids and proved to have pH of 7.9 and density of 1.006 g/cm^3. Its dynamic viscosity at test conditions (P = 270 bar, T = 126 °C) was 0.411 cP. The simplified composition of formation water is tabulated in Table 2.

Two types of nitrogen rich natural gases were used. The first one abbreviated KG is characterized by a very high nitrogen (~87%) content and the presence of hydrogen sulfide (2.7%) and carbon dioxide (1.2%). Its dynamic viscosity at the test conditions (P = 270 bar, T = 126 °C) was 0.0273 cP. The simplified compositional analysis of injected gas is shown in Table 3. It was sampled from the separator during a production test in the undeveloped gas field.

The second one abbreviated MG has lower nitrogen and carbon dioxide content but much higher methane content. Its density is 1.164 kg/m^3, and the dynamic viscosity at test conditions (P = 270 bar, T = 126 °C) is 0.0260 cP. It was sampled from the producing gas field located in the vicinity of the considered oil field. The simplified compositional analysis of injected gas is shown Table 4.

Table 2. Composition of formation water.

Total Salinity (g/L)	Cation (g/L)					Anion (g/L)					
	Na^+	K^+	Mg^{2+}	Ca^{2+}	NH_4^+	Cl^-	Br^-	SiO_3^{2-}	HCO_3^-	SO_4^{2-}	S^{2-}
8.932	2.03	0.462	0.062	0.389	0.259	5.265	0.037	0.013	0.177	0.147	0.505

Table 3. Simplified composition of KG natural gas.

Density (kg/m^3)	Component Concentration (%mol)							
	N_2	H_2S	CO_2	H_2	C_1	C_2	C_3	C_{4+}
1.2507	86.9	2.7	1.2	1.1	5.6	0.8	0.7	1

Table 4. Simplified composition of MG natural gas.

Density (kg/m^3)	Component Concentration (%mol)						
	N_2	H_2S	CO_2	C_1	C_2	C_3	C_{4+}
1.164	58.6	3.3	0.3	28.7	4.7	2.6	0.7

2.2. Rock Material

In the core flooding experiments original reservoir rock from the Upper Permian Main Dolomite Formation of the Zechstein Basin in western Poland was used. Reservoir rock samples were taken from the pay zone cored interval of one of the producing wells at a depth of around 3000 m. Mineral composition was quite uniform and consisted mainly of dolomite (~83%) with ankerite (~16%), anhydrite (~1%), and quartz (~0.5%) [50]. From the whole drilling of core samples, core plugs with 2.54 in diameter, and length of ~5–6 cm, were drilled horizontally. Then they were end faced, polished, cleaned, and dried. Most of the core plug preparation procedures were conducted following the guideline from API RP40 [68]. After that, their parameters such as absolute permeability (using a steady-state nitrogen permeameter), effective porosity (using a helium porosimeter) pore volume, bulk density, as well as grain density were determined. The samples with similar parameters were selected and grouped into composite cores consisting of four core plugs. Core plugs were arranged with Langaas criterion (decreasing permeability in the flow direction), such that the core with the highest permeability was placed at the inlet and the core with the lowest permeability at the outlet [69]. A set of parameters characterizing exemplary composite cores is presented in Table 5.

Six composite cores were assembled with the average porosity in the range of 23–30% and permeability of 70–80 mD. The basic parameters of the composite cores used in the core flooding tests are presented in Table 6. As the availability of the original reservoir rock samples from drilling cores is very limited, the composite cores were reused. Routine cleaning in a Soxhlet apparatus was replaced by dynamic mild cleaning using kerosene/heptane, DI water, and nitrogen to reduce the cleaning impact on the core properties. This involves injecting (in PT conditions) about 5 PV of kerosene, followed by about 5 PV of heptane, totaling 10 PV of solvent flooding. Then DI water was injected to remove dissolvable salts, followed by nitrogen which helps remove residues and dry out the sample. Directly after, cores were dried in the oven until they achieved a constant weight. Such an approach helps maintain and restore the original reservoir wettability in the carbonates [70].

Table 5. Individual core parameters of composite core no 1.

Core ID	Permeability	Effective Porosity	Core Volume	Grain Density	Length	Diameter
	[mD]	[%]	[cm^3]	[g/cm^3]	[cm]	[cm]
46	96.4	19.01	26.916	2.823	5.35	2.54
85	83.1	31.64	24.925	2.818	4.98	2.54
49	59.1	28.25	26.511	2.817	5.29	2.54
25	43.5	29.54	28.006	2.817	5.59	2.54
	Composite core parameters					
	Average permeability [mD]			70.5		
	Average permeability [%]			27.1		
	Composite core volume [cm^3]			106.4		
	Composite core pore volume [cm^3]			28.8		
	Composite core length			21.21		

Table 6. Properties of composite cores.

Composite Core No.	Length [cm]	Average Porosity [%]	Average Permeability [mD]	Pore Volume [cm^3]
1	21.21	27.1	70.5	28.8
2	21.73	30.4	77.3	33.0
3	21.95	24.6	72.1	26.4
4	22.30	22.7	80.7	25.7
5	21.40	24.6	80.4	26.7
6	22.15	28.7	77.0	31.8

2.3. Minimum Miscibility Pressure

Oil displacement through WAG gas injection is most effective when the injected gas is completely or near miscible with the oil in the reservoir. The main factor responsible for the increased oil displacement during miscible gas injection is the mass transfer of components between the flowing gas phase and the oil phase present in the reservoir. The efficiency of this process increases along with the miscibility of both phases. Immiscible WAG recovery mechanisms (also present in miscible injection) include oil volume expansion (oil swelling), oil viscosity reduction, 3-phase relative permeability, and oil film flow [71].

Fluids are considered miscible when they mix in all proportions to form a single homogeneous phase, so the miscibility is a physical condition between two (or more) fluids that permits them to mix in all proportions without the existence of any interface [72,73]. Under reservoir conditions of constant temperature and quasi-constant composition, a factor determining miscibility is pressure, and the lowest pressure at which the first or multiple-contact miscibility can be achieved is called the minimum miscibility pressure (MMP). It needs to be determined for each specific pair of fluids (reservoir oil and injected gas).

The slim tube method, which is the primary and most commonly used method for laboratory determination of MMP in industry, was used to determine MMP for the injected gas and reservoir oil. The slim tube is a one-dimensional model of the reservoir in the form of narrow and long stainless-steel tube packed with a porous material (typically with sand). This simple design allows multiple contact conditions between flowing fluids in a porous medium and provides dimensional dispersion free displacement of oil. Gravity override caused by the gravity effect is negligible because the tube is coiled, so that flow is basically horizontal. Tests are conducted at a constant reservoir temperature controlled by a thermostatic bath, where the tube is placed. The observation of the ongoing phenomena is possible through a mounted sight glass. Schematic diagram of the slim tube apparatus is shown in Figure 3.

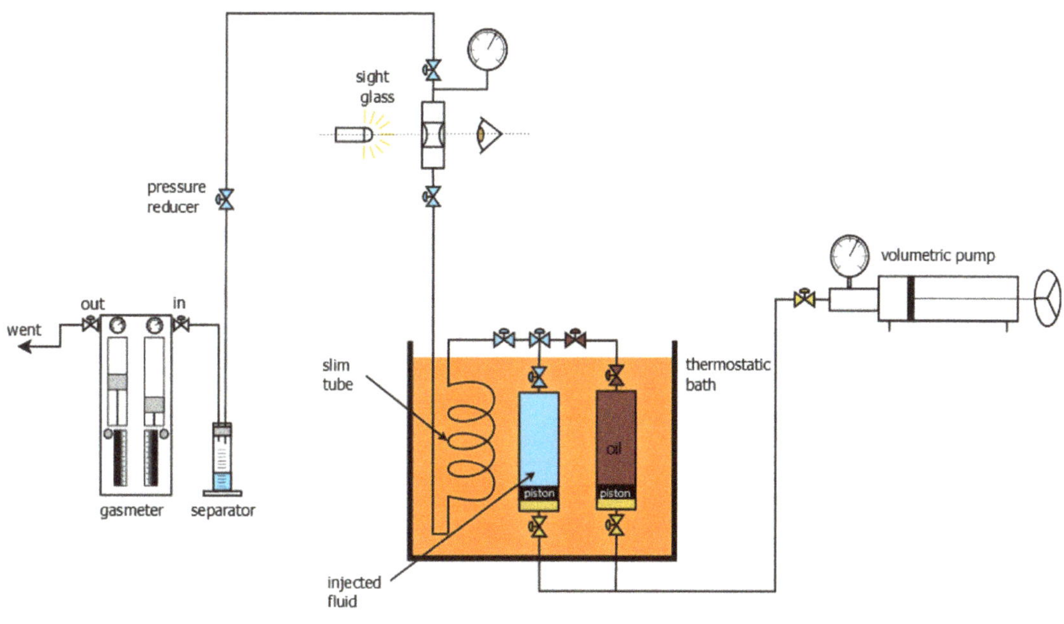

Figure 3. Schematic diagram of the slim tube set-up.

The test starts by saturating the porous media with reservoir oil. Then the oil is displaced by injecting gas at a constant rate and inlet pressure. The differential pressure between the inlet and outlet of the tube is so small compared to the pressure in the system that the displacement pressure is considered constant. The injected and produced fluid volumes are precisely monitored and measured during the test, which ends after injecting 1.2 pore volume (PV) of gas. Miscibility conditions are determined by conducting the displacement at various pressures and plotting the oil recoveries as a function of displacement pressure. The study is normally performed at between 4 to 6 test pressure. After every test the slim tube apparatus must be carefully cleaned with solvents and dried prior to subsequent saturation. Based on the plot observation, MMP could be identified as the pressure break in the curve (the recovery-pressure curve starts to flatten when the displacement becomes near miscible). Some researchers use specific recovery factor, e.g., 90% at 1.2 PV of injected gas as MMP [74]. The specification of the slim tube apparatus setup used in the study is presented in Table 7.

Experimental MMP studies were performed for CO_2, AG, and MG. The tests for KG were not undertaken because miscibility conditions would be unlikely to occur, in the considered pressure range, due to the very high nitrogen content in KG. The basic parameters of the performed slim tube tests are presented in Table 8. The MMP was also double-checked via simulations in PVTsim software using the original reservoir fluid model.

Table 7. Slim tube setup properties.

Parameter	Setting
Length	25 m
Internal diameter	5 mm
Pore volume	174.659 cm^3
Porosity	15 D
Permeability	35%
Grain type	Quartz sand
Grain size	0.15–0.20 mm
Injection rate	
Temperature	126 °C

Table 8. Basic properties of Slim tube tests.

Test No.	Injected Gas	Oil Saturation Pressure [bar]	Gas Injection Pressure [bar]
1	CO_2	130	140
2	CO_2	160	170
3	CO_2	200	210
4	CO_2	280	290
5	CO_2	330	340
6	CO_2	420	430
7	AG	130	140
8	AG	160	170
9	AG	200	210
10	AG	280	290
11	AG	330	340
12	AG	420	430
13	MG	240	250
14	MG	280	290
15	MG	330	340
16	MG	420	430
17	MG	460	470

2.4. Coreflooding Process

In the core flooding experiments a customized and properly adapted PVT apparatus upgraded with an additional core-holder cell was used. The core holder was designed based on the analysis of the available technical solutions with a special focus on the specification of the experiments and conditions where it would be used. A radial core holder can accommodate cores with a diameter of 2.54 cm (1 inch) and length up to 25 cm. Composite cores were placed in a rubber sleeve and then in the core holder (Figure 4). Tightness protection between the sleeve and composite core was maintained by a pressurized water system.

Figure 4. The core holder with rubber sleeve and core plugs forming the composite core.

A confining pressure of 100 bar higher than the test pressure was held during the injection. Stable temperature conditions (126 °C ± 0.5 °C) for the horizontally placed core holder and fluid bearing pressure cells were maintained using a thermostatic air bath. The core holder has one inlet and one outlet port connected to the pressure and temperature transducers. The fluid flow was controlled through the set of precise HTHP valves. Produced liquids were measured using a graduated cylinder, while the produced gas was measured using a gas meter. The proper configuration of an experimental set-up and its features such as dead volumes, location of PT transducers, location and type of pressure connections are essential for the accuracy of the volume measurements of injected and withdrawn fluids, and thus for the reliability and consistency of the obtained results. Therefore, special attention was paid to the correct design and then the verification and testing of the solutions used. Figure 5 shows a simplified scheme of the core flooding experimental setup.

Initially, the composite core was saturated with reservoir water to reach the preset pressure, and the pore volume (PV) was determined. Then, the composite core was flooded with live oil under the given test pressure, with constant flow rate q = 0.3 cm^3/min to determine the irreducible water saturation, and subsequently the hydrocarbon saturation—hydrocarbon pore volume (PV_{HC}). The total amount of injected fluids during coreflooding experiments was 1.2 PV_{HC}. The injection flow rate was fixed at 0.07 cm^3/min, which in the composite core resulted in velocity within the range 2.5 ÷ 3.3 cm/h. In the WAG process, the injection cycle started with gas in most of the experiments, and the slug size was 0.2 PV_{HC}. A total of 17 coreflooding experiments was conducted, including 5 variants difering in injection scheme/WAG ratio (CGI; WAG 1:1, 1:2, 2:1) and injection pressure (270 and 170 bar) for each newly tested gas (MG, AG and CO_2), and two supplementary tests to complement previously conducted studies for KG gas [50]. To compare WAG process efficiency, tests with continuous gas injection were preformed. The properities of the core flooding experiments are summarized in Table 9.

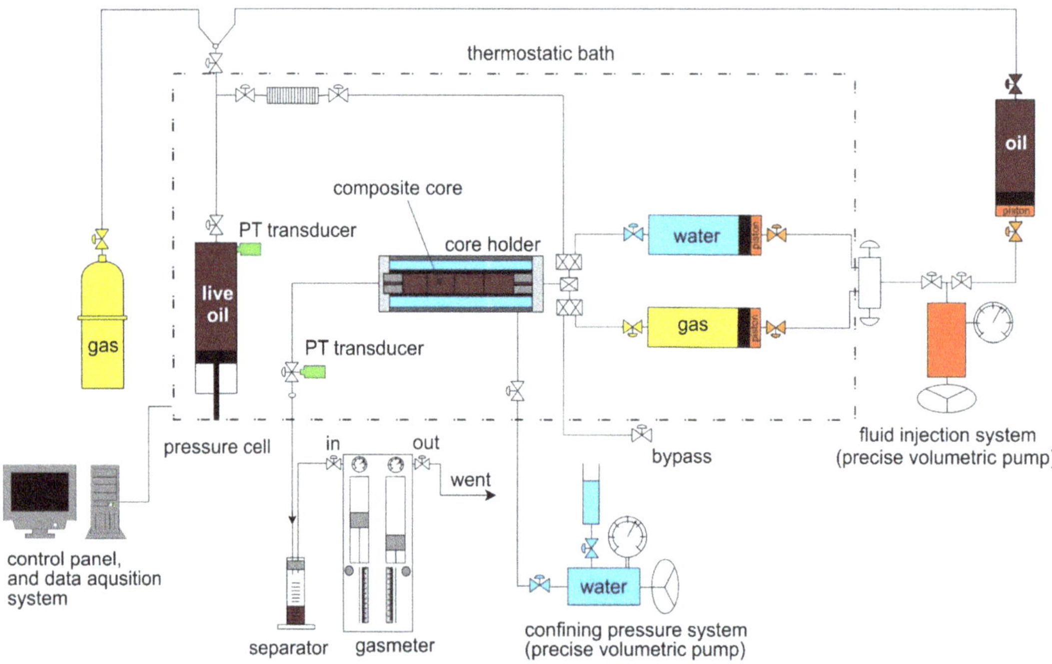

Figure 5. Simplified scheme of the core flooding setup.

Table 9. Core flooding experiments properties.

Test No	Composite Core No	Injection Type	Injected Fluid	WAG Ratio	Number of WAG Cycles	P_i [bar]	S_{wi} [%]	S_{oi} [%]	TWI [1] $[PV_{HC}]$	TGI [2] $[PV_{HC}]$
1	6	CGI	CO_2	0:1	-	270	45.6	54.4	0	1.2
2	2	WAG	CO_2/Water	1:1	3	270	53	47	0.6	0.6
3	3	WAG	CO_2/Water	2:1	2	270	59.3	40.7	0.8	0.4
4	4	WAG	CO_2/Water	1:2	2	270	54.2	45.8	0.4	0.8
5	5	WAG	CO_2/Water	1:1	3	170	54.8	45.2	0.6	0.6
6	6	CGI	MG	0:1	-	270	42.1	57.9	0	1.2
7	5	WAG	MG/Water	1:1	3	270	37.1	62.9	0.6	0.6
8	1	WAG	MG/Water	2:1	2	270	43.5	56.5	0.8	0.4
9	2	WAG	MG/Water	1:2	2	270	42.2	57.8	0.4	0.8
10	2	WAG	MG/Water	1:1	3	170	59.5	40.5	0.6	0.6
11	1	CGI	AG	0:1	-	270	40.6	59.4	0	1.2
12	6	WAG	AG/Water	1:1	3	270	41.7	58.3	0.6	0.6
13	3	WAG	AG/Water	2:1	2	270	42.6	57.4	0.8	0.4
14	5	WAG	AG/Water	1:2	2	270	33.4	66.6	0.4	0.8
15	4	WAG	AG/Water	1:1	3	170	38.3	61.7	0.6	0.6
16	5	WAG	KG	1:2	2	270	34.9	65.1	0.4	0.8
17	6	WAG	KG	1:1	3	170	31.7	68.3	0.6	0.6

[1] TWI—total water injected; [2] TGI—total gas injected.

2.5. Empirical Correlation for Recovery Factor

Empirical modelling is a process that allows an experimental input–output data set to be transformed into a functional relationship that can be used to estimate results. Genetic programming (GP) uses an evolutionary computation paradigm to generate computer programs that automatically solve a specific problem. Its application involves a transformation of computer programs into a new generation of programs by applying naturally occurring genetic operations [75–77]. Initially, LISP was chosen as the main language for GP in which the program structure is expressed as a parse tree. However, recently many other modern languages such as Python, Java, C++, and the languages associated with several scientific programming tools (e.g., MATLAB and Mathematica) have been used to develop tree-based GP applications. The variables and constants in the program are leaves of the tree called terminal nodes (terminals), where the arithmetic operations are internal nodes called functions [78]. GP programs can be composed of multiple components (set of trees) grouped under a root node.

Since genetic programming cannot be applied directly to identify nonlinear input–output models, the way to address this problem is to extend GP operators with a tool that uses the Orthogonal Least Squares (OLS) algorithm to create an equation with linear structure described in detail in [79]. Generally, the GP algorithm generates many potential solutions in the form of binary tree structures. These contain terms (subtrees) that affect the accuracy of the model to a greater or lesser extent. The OLS implementation in the GP algorithm involves decomposing tree structures (individual members of population) into subtrees—function terms of the linear in-parameter models. In the next step, the calculation of error reduction ratios of these functions is followed by eliminating the less significant terms. This method, called "tree pruning", is utilized before the calculation of the fitness values of the trees and conducted in every fitness evaluation. The approach is used to simplify the trees by keeping their accuracy close to the original ones. It is essential to preserve the tree structure because GP works with it. The proposed approach is implemented in the freeware GP-OLS Toolbox available for MATLAB software. This method results in more robust and interpretable models than the classical GP method. To develop a mathematical model, the RF results from core flooding experiments were used. There were seven variables selected, characterizing a given injection scenario, i.e., concentration of components such as CO_2, H_2S, N_2, C_{3+}, C_1 in the injected gas, the total share of gas compared to water in the injected fluid stream (Cg), and injection pressure (P). General GP parameters settings are presented in Table 10.

Table 10. GP parameters used in development of the RF correlation.

Parameter	Setting
Population size	1500
Max. tree depth	5
Number of generations (iterations)	500
Generation gap	0.8
Probability of mutation	0.3
Probability of crossover	0.7
Type of selection	Tournament
Type of crossover	One-point
Type of mutation	Randomly selected node
Type of replacement	Least fitness score
Input variables	7

3. Results

3.1. MMP Determination

The slim tube recovery factor for CO_2 obtained from the tests conducted in subsequent pressure steps (Table 8) was in the range of 59.1–97.1%. An example of the slim tube test set in given pressure steps is shown in Figure 6. The results of RF were plotted against pressure, and the MMP was determined using the plot (Figure 7). The pressure determined from the curve break point was 195 bar, while the value corresponding to the RF of 90% was 188 bar. The average, i.e., 192 bar was taken as MMP. The MMP obtained in the PVTsim simulator using the reservoir fluid model was 194 bar.

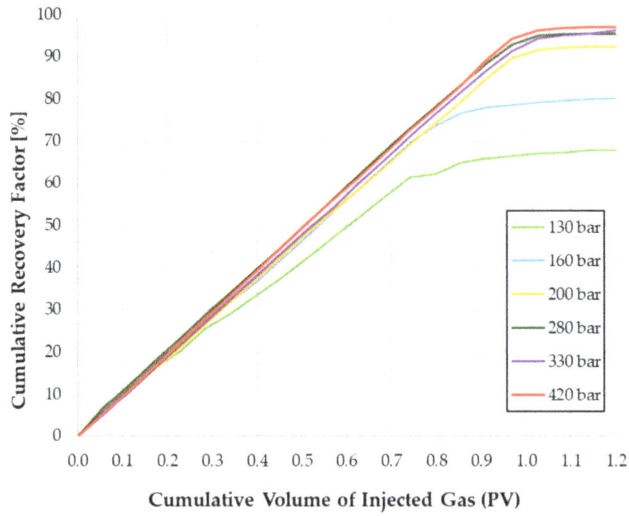

Figure 6. Overview of slim tube recovery factor curves for live oil and CO_2 injected.

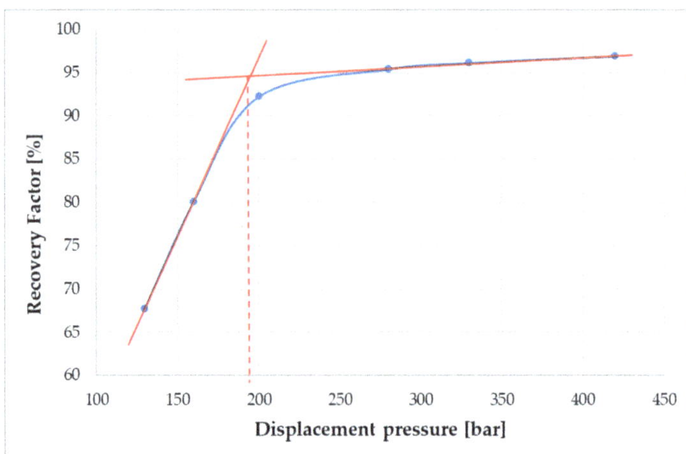

Figure 7. Slim tube recovery factors for CO_2 plotted against pressure—MMP determination.

Similarly, the MMP for acid gas was also determined using the slim tube RF (Figure 8) vs. the displacement pressure plot (Figure 9), and it turned out to be lower than for pure CO_2. The pressure determined from the curve break point was 171 bar, while the value corresponding to the RF of 90% was 173 bar, so the average of 172 bar was taken as MMP. The MMP obtained from simulation, as for pure CO_2 showed very good compliance with the experimental results (174 bar).

Determination of MMP for MG in the considered pressure range based on the slim tube results was impossible. The recovery factor plotted versus pressure follows an almost linear pattern reaching a maximum value of 63% at the highest pressure of 460 bar (Figure 10). For the KG measurement it was omitted because of the much higher nitrogen content of the gas which suggests an even higher MMP value (beyond the testing range). The MMP for KG and MG determined using PVTsim with values of 836 and 1245 bar, respectively, are surprisingly high and undermine the reliability of the simulations when considering the published data for pure nitrogen, where MMP is generally in the 350–650 bar range [80–83]. Based on the above considerations, it should be concluded that the injection of CO_2 and AG during core flooding will occur in miscible (or near miscible conditions—in the case of experiments conducted at pressure of 170 bar), while high nitrogen gases (MG and KG) are immiscible with reservoir oil in the considered pressure range.

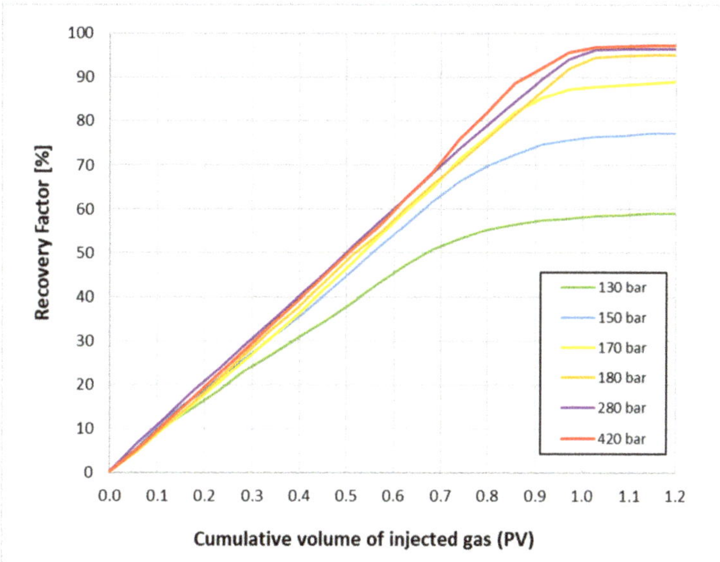

Figure 8. Overview of slim tube recovery factor curves for live oil and AG injected.

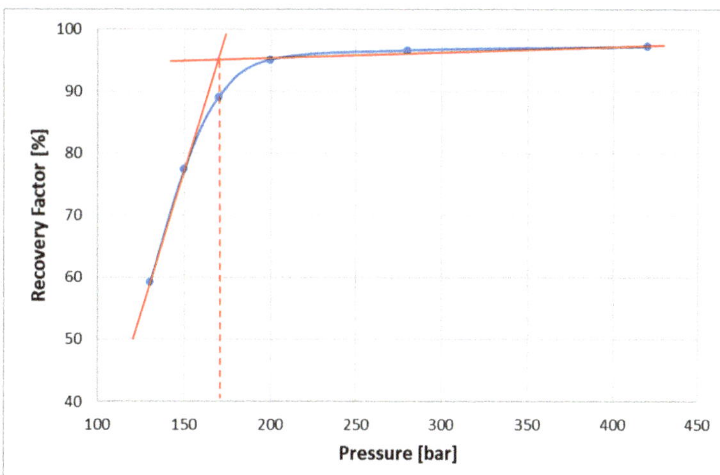

Figure 9. Slim tube recovery factors for H_2S plotted against pressure—MMP determination.

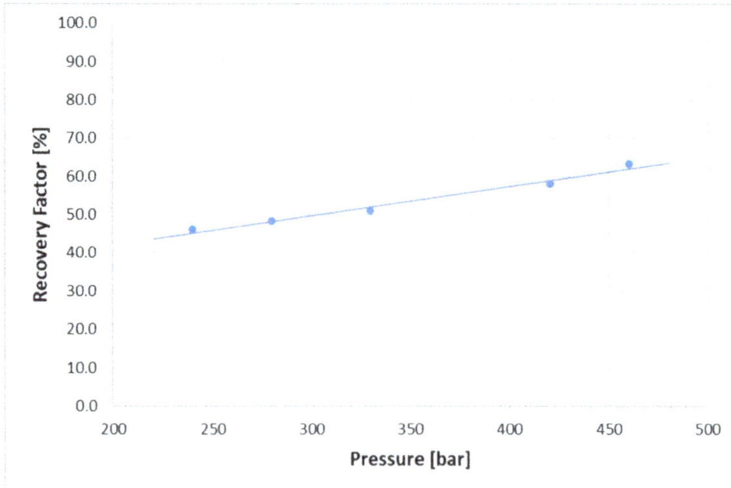

Figure 10. Slim tube recovery factors for MG plotted against pressure.

3.2. *Coreflooding*

The best efficiency of the WAG process among all the experiments performed was recorded for CO_2, where the recovery factor was in the range of 65.1–82.9% (Figure 11). The application of the WAG process increased the recovery factor by 11.1–28.5 pp when compared to the continuous water injection (CWI). CGI with RF of 79.8% outperformed CWI, and even two WAG schemes (WAG 2:1 at 270 bar, and WAG 1:1 at 170 bar). The highest RF was observed in the WAG 1:2 scheme, where the volume of gas injected within the WAG cycle was two times greater than the water volume. The lowest RF was recorded for the WAG 1:1 scheme conducted at lower pressure (170 bar) accounting for the immiscible nature of the injection process.

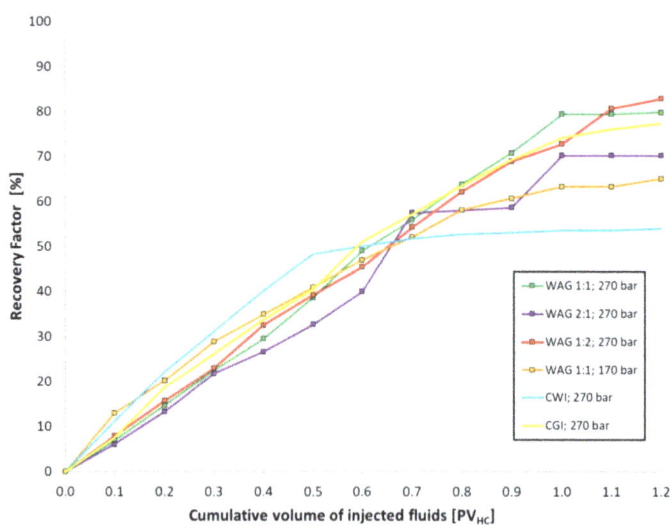

Figure 11. Oil recovery efficiency using different CO_2-WAG schemes compared to CWI and CGI injection.

MG WAG efficiency was considerably lower than that of CO_2-WAG with the RF in the range of 57.1–69.3% (Figure 12). The weakest WAG efficiency was observed in WAG 2:1; 270 bar, with an increased amount of water in the WAG cycle and WAG 1:1; 170 bar, with decreased test pressure. The RF for those injection schemes is only slightly higher than of CWI. The remaining WAG schemes appeared significantly more efficient and resulted in increased RF up to 15 pp. Continuous MG injection turned out to be the least effective injection scheme with RF 10 pp lower than with CWI.

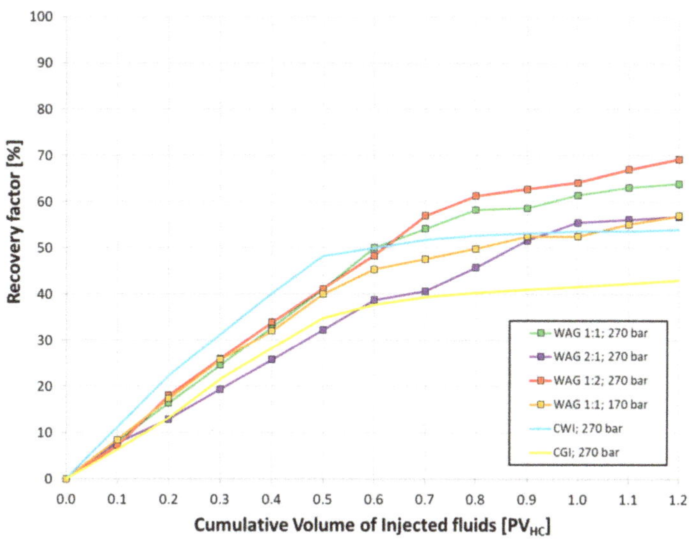

Figure 12. Oil recovery efficiency using MG with different WAG schemes compared to CWI and CGI injection.

Utilization of acid gas in the WAG injection resulted in high recovery factors that are mostly slightly lower than derived by pure CO_2, and definitely higher than derived by KG and MG. The RF of the WAG process was in the range of 60.7–72.6%, indicating an increase vs. CWI of up to 18.6 pp. The highest AG WAG efficiency was observed similar to CO_2 and MG in the WAG 1:2; 270 bar scheme. The lowest WAG efficiency was recorded at reduced test pressure (170 bar). The highest recovery efficiency among the tested schemes was obtained using continuous AG injection with RF of 81.2%. A comparison of different AG injection schemes in relation to CWI expressed in total RF is presented in Figure 13.

The efficiency of KG in the considered WAG injection schemes was in the range of 58.1–72% RF and was much more effective than CGI (RF higher even by 35 pp.) and CWI (RF higher up to 18). The best WAG efficiency was observed for the WAG ratio 1:1, where equal volumes of water and gas were injected in each cycle. The efficiency of WAG with an increased volume of gas compared to water (1:2), and injected at lower pressure was reduced, but still higher than CWI and CGI.

A comparison of different KG injection schemes in relation to CWI expressed in total RF is presented in Figure 14.

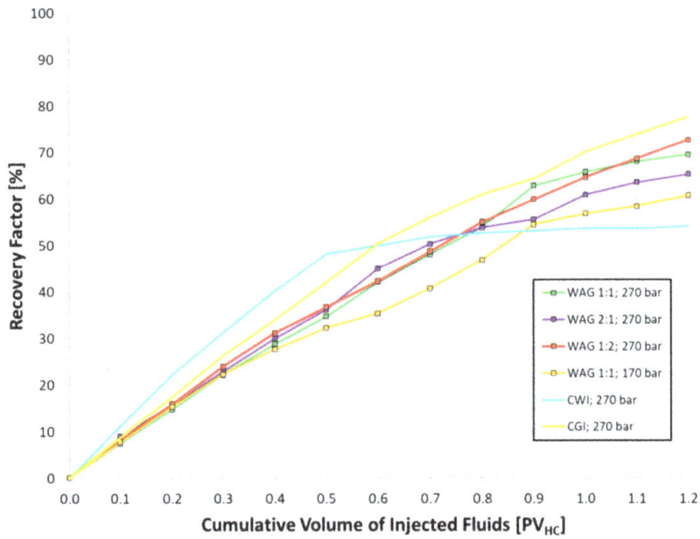

Figure 13. Oil recovery efficiency using AG with different WAG schemes compared to CWI and CGI injection.

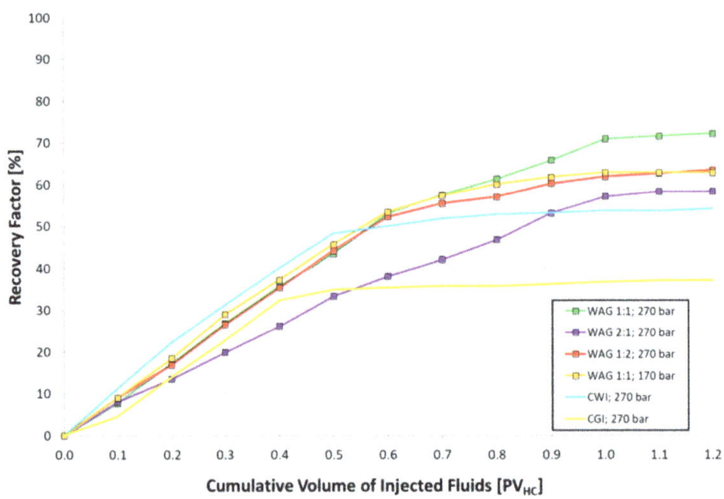

Figure 14. Oil recovery efficiency using KG with different WAG schemes compared to CWI and CGI injection.

3.3. Empirical Modelling

Empirical modelling of RF using GP was based on the experimentally derived RF and other variables characterizing the conducted experiments as gas composition, WAG ratio (reflected as injected gas contribution in the total injected fluids volume), and injection pressure (Table 11).

Table 11. Input and output data used in GP.

Exp No.	Gas	RF from Coreflooding [%]	Variables						
			CO_2	H_2S	N_2	C_{3+}	C_1	C_g [%]	P [bar]
1	-	54	0	0	0	0	0	0	270
2	CO_2	79.8	100	0	0	0	0	50.0	270
3		70.2	100	0	0	0	0	33.3	270
4		82.9	100	0	0	0	0	66.7	270
5		65.1	100	0	0	0	0	50.0	170
6		82.5	100	0	0	0	0	100.0	270
7	GM	43	0.28	3.26	58.5	4.57	28.66	100.0	270
8		64	0.28	3.26	58.5	4.57	28.66	50.0	270
9		56.8	0.28	3.26	58.5	4.57	28.66	33.3	270
10		69.3	0.28	3.26	58.5	4.57	28.66	66.7	270
11		56.9	0.28	3.26	58.5	4.57	28.66	50.0	170
12	Acid gas (CO_2+H_2S)	81.2	70	30	0	0	0	100.0	270
13		69.5	70	30	0	0	0	50.0	270
14		65.3	70	30	0	0	0	33.3	270
15		72.6	70	30	0	0	0	66.7	270
16		60.7	70	30	0	0	0	50.0	170
17	GK	37	1.2	2.73	86.85	1.73	5.63	100.0	270
18		72.1	1.2	2.73	86.85	1.73	5.63	50.0	270
19		58.1	1.2	2.73	86.85	1.73	5.63	33.3	270
20		63.4	1.2	2.73	86.85	1.73	5.63	66.7	270
21		62.8	1.2	2.73	86.85	1.73	5.63	50.0	170

Despite the initiation of the algorithm with seven variables, the best fit to the experimental values (95%) was obtained using only five of them. The equation relating the relationship of the RF and the variables characterizing a particular injection scheme takes the following formula:

$$\begin{aligned} RF = -0.000223 \quad & \cdot\left[\left(C_{3+} + C_g\right)\cdot\left(C_g\cdot N_2\right)\right] + 0.088866\cdot P + 0.189312\cdot C_g \\ & + 0.02397\cdot\left(C_g\cdot N_2\right) - 0.534878\cdot N_2 + 0.130390\cdot CO_2 \\ & + 28.526032 \end{aligned} \tag{1}$$

where:

C_{3+}—C_{3+} hydrocarbon fraction content in the injected gas [%]
C_g—Injected gas contribution in the total injected fluid volume [%]
N_2—N_2 content in the injected gas [%]
P—Test pressure [bar]
CO_2—CO_2 content in the injected gas [%]

The fitting of the results derived with a mathematical model presented above to the RF experimental values is presented in Figure 15.

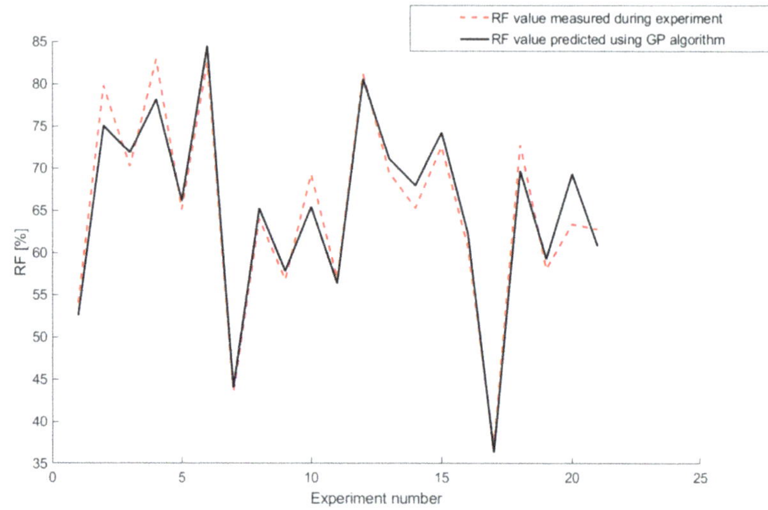

Figure 15. RF model fitting to the experimental results.

The minimum value of the absolute error expressed as the modulus of the difference between calculated and experimentally derived RF is about 0.5 p.p., the maximum value is about 6 p.p. when the average is about 2 p.p. (Figure 16).

A relative error, expressed as the modulus of the absolute error divided by the magnitude of the experimentally derived RF, was also used for the fitting evaluation. The mean relative error was about 3% (Figure 17).

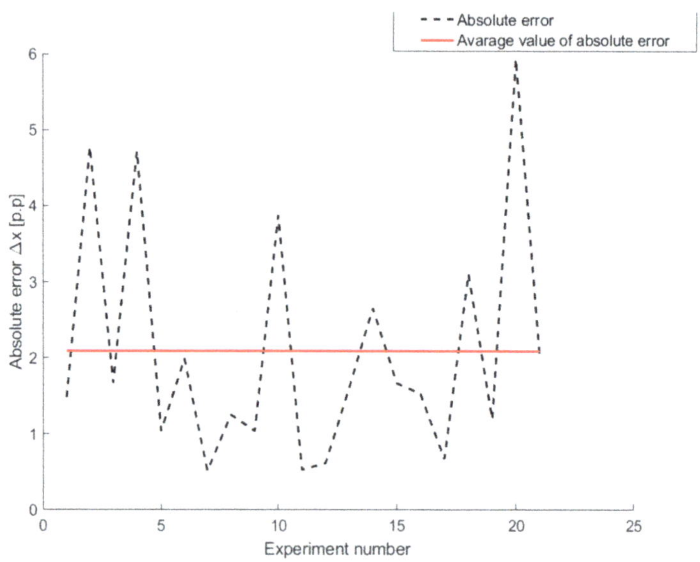

Figure 16. Absolute error.

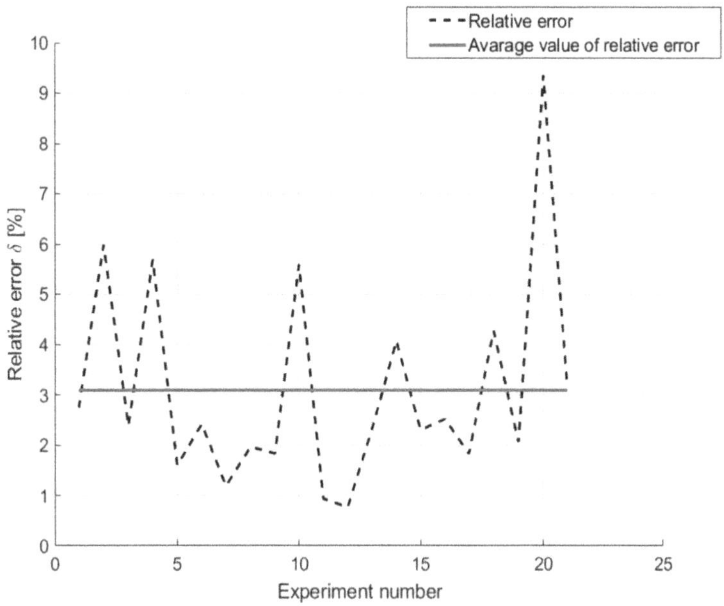

Figure 17. Relative error.

4. Conclusions

An extensive experimental investigation enabled the identification of the efficiency of the WAG method in conditions related to one of most important carbonate oil reservoirs in Poland, and moreover in the determination of the influence of factors such as gas type (composition), WAG ratio, injection pressure, and miscibility of the oil recovery.

The slim tube tests and PVTsim simulations showed that injection of CO_2 and AG during core flooding experiments conducted at 270 bar occurred in miscible conditions. At a lower test pressure of 170 bar, injection of these gases occurred more likely in near-miscible conditions. Injection of the two other gases (MG and KG) occurred in immiscible conditions.

Core flooding studies showed increased efficiency of the WAG process compared to CWI and CGI (except acid gas injection), which were taken as a baseline for evaluating WAG efficiency. As expected, oil recovery efficiency was strongly dependent on the injected gas type and injection parameters (i.e., test pressure, WAG ratio).

The highest WAG efficiencies were observed in scenarios where injection occurred under miscible conditions, i.e., CO_2 and AG. In the most effective injection scheme (CO_2 WAG 1:2, 270 bar), the recovery factor was over 82%, which, compared to CWI, allows the RF to be increased by nearly 30 pp.

The WAG efficiency using high nitrogen natural gases (MG and KG) injected at immiscible conditions was considerably lower, where the most effective schemes resulted in RF of about 70%. The oil recovery was noticeably lower than that obtained with miscible displacement but still significantly higher compared to CWI.

Based on the core flooding results, a mathematical model was constructed for estimating the RF using parameters such as: composition of injected gas, gas contribution in the injected fluid stream, and injection pressure. The equation developed using genetic programming features had a good fit with the experimental result (95%) and can be applied to estimate the RF in conditions specific to domestic oil reservoirs located in the Main Dolomite.

Author Contributions: Conceptualization, M.W. (Mirosław Wojnicki); methodology, M.W. (Mirosław Wojnicki) and J.L.; software, M.W. (Mirosław Wojnicki) and M.G.; validation, J.K.; investigation, M.W. (Mirosław Wojnicki), M.G., J.K., S.S. and M.W. (Marcin Warnecki); resources, S.S.; writing—original draft preparation, M.W. (Mirosław Wojnicki); writing—review and editing, M.W (Mirosław Wojnicki) and J.L.; supervision, J.L.; project administration, M.W. (Mirosław Wojnicki). All authors have read and agreed to the published version of the manuscript.

Funding: The work was financially supported by the Polish Ministry of Science and Higher Education (Oil and Gas Institute—National Research Institute Internal Order No. 0033/KB/19 and 0064/KB/20).

Institutional Review Board Statement: Not applicable.

Informed Consent Statement: Not applicable.

Data Availability Statement: Not applicable.

Conflicts of Interest: The authors declare no conflict of interest.

Nomenclature

AG	Acid gas
C_g	Total share of gas in injected fluid stream
CGI	Continuous Gas Injection
CWI	Continuous Water Injection
DI	Deionized water
EOR	Enhanced Oil Recovery
GP	Genetic Programming
IWAG	Immiscible WAG
KG	Very high-nitrogen natural gas
MG	High-nitrogen natural gas
MG	Nitrogen natural gas
MMP	Minimum Miscibility Pressure
nMWAG	Near miscible WAG
OLS	Orthogonal Least Squares
P_i	Injection pressure
pp	Percentage point
PV	Pore Volume
PV_{HC}	Hydrocarbon Pore Volume
RF	Recovery factor
TWI	Total water injected
TGI	Total gas injected
WAG	Water Alternating Gas

References

1. McGlade, C.; Sondak, G.; Han, M. Whatever Happened to Enhanced Oil Recovery? International Energy Agency. 2018. Available online: https://www.iea.org/commentaries/whatever-happened-to-enhanced-oil-recovery (accessed on 10 January 2022).
2. Sheng, J.J. *Enhanced Oil Recovery Field Case Studies*, 1st ed.; Gulf Professional Publishing: Oxford, UK, 2013. [CrossRef]
3. Manrique, E.J.; Muci, V.E.; Gurfinkel, M.E. EOR Field Experiences in Carbonate Reservoirs in the United States. *SPE Reserv. Eval. Eng.* **2007**, *10*, 667–686. [CrossRef]
4. Alnuaim, S. The Role of R&D in Energy Sustainability: Global Oil and Gas Supply Security and Future Energy Mix. *J. Pet. Technol.* **2019**, *71*, 10–11.
5. Kargarpour, M.A. Carbonate reservoir characterization: An integrated approach. *J. Pet. Explor. Prod. Technol.* **2020**, *10*, 2655–2667. [CrossRef]
6. Akbar, M.; Vissapragada, B.; Alghamdi, A.H.; Allen, D.; Herron, M.; Carnegie, A.; Dutta, D.; Olesen, J.R.; Chourasiya, R.D.; Logan, D.; et al. A Snapshot of Carbonate Reservoir Evaluation. *Oilfield Rev.* **2000**, *12*, 20–41.
7. Jia, C. Petroleum Geology of Carbonate Reservoir. *Adv. Top. Sci. Technol. China* **2012**, 495–532. [CrossRef]
8. Huang, Z.; Xing, H.; Zhou, X.; You, H. Numerical study of vug effects on acid-rock reactive flow in carbonate reservoirs. *Adv. Geo-Energy Res.* **2020**, *4*, 448–459. [CrossRef]
9. Masalmeh, S.K.; Wei, L.; Blom, C.; Jing, X. EOR Options for Heterogeneous Carbonate Reservoirs Currently under Waterflooding. In Proceedings of the Abu Dhabi International Petroleum Exhibition and Conference, Abu Dhabi, UAE, 10–13 November 2014; Society of Petroleum Engineers: Calgary, AB, Canada, 2014. [CrossRef]

10. Mogensen, K.; Masalmeh, S. A review of EOR techniques for carbonate reservoirs in challenging geological settings. *J. Pet. Sci. Eng.* **2020**, *195*, 107889. [CrossRef]
11. Jafari, M. Laboratory Study for Water, Gas and Wag Injection in Lab Scale and Core Condition. *Pet. Coal* **2014**, *56*, 175–181.
12. Martyushev, D.A.; Yurikov, A. Evaluation of opening of fractures in the Logovskoye carbonate reservoir, Perm Krai, Russia. *Pet. Res.* **2021**, *6*, 137–143. [CrossRef]
13. Chichinina, T.I.; Martyushev, D.A. Specific anisotropy properties of fractured reservoirs: Research on Thomsen's anisotropy parameter delta. In Proceedings of the Geomodel 2021—23th Conference on Oil and Gas Geological Exploration and Development, Gelendzhik, Russia, 6–10 September 2021; Volume 2021, pp. 1–5. [CrossRef]
14. Manrique, E.J.; Thomas, C.P.; Ravikiran, R.; Izadi, M.; Lantz, M.; Romero, J.L.; Alvaradov, V. EOR: Current status and opportunities. In Proceedings of the SPE Symposium on Improved Oil Recovery, Tulsa, OK, USA, 24–28 April 2010; Society of Petroleum Engineers (SPE): Calgary, AB, Canada, 2010; Volume 2. [CrossRef]
15. Kulkarni, M.M.; Rao, D.N. Experimental investigation of miscible and immiscible Water-Alternating-Gas (WAG) process performance. *J. Pet. Sci. Eng.* **2005**, *48*, 1–20. [CrossRef]
16. Christensen, J.R.; Stenby, E.H.; Skauge, A. Review of WAG Field Experience. *SPE Reserv. Eval. Eng.* **2001**, *4*, 97–106. [CrossRef]
17. Afzali, S.; Rezaei, N.; Zendehboudi, S. A comprehensive review on Enhanced Oil Recovery by Water Alternating Gas (WAG) injection. *Fuel* **2018**, *227*, 218–246. [CrossRef]
18. Christensen, J.R.; Stenby, E.H.; Skauge, A. Review of WAG Field Experience. In Proceedings of the International Petroleum Conference and Exhibition of Mexico, Villahermosa, Mexico, 3–5 March 1998; Society of Petroleum Engineers: Calgary, AB, Canada, 1998. [CrossRef]
19. Teigland, R.; Kleppe, J. EOR Survey in the North Sea. In Proceedings of the SPE/DOE Symposium on Improved Oil Recovery, Tulsa, OK, USA, 22–26 April 2006; Society of Petroleum Engineers: Calgary, AB, Canada, 2006. [CrossRef]
20. Surguchev, L.M.; Korbel, R.; Haugen, S.; Krakstad, O.S. Screening of WAG Injection Strategies for Heterogeneous Reservoirs. In Proceedings of the European Petroleum Conference, Cannes, France, 16–18 November 1992. [CrossRef]
21. Skauge, A.; Serbie, K. Status of fluid flow mechanisms for miscible and immiscible WAG. In Proceedings of the Society of Petroleum Engineers—SPE EOR Conference at Oil and Gas West Asia 2014: Driving Integrated and Innovative EOR, Muscat, Oman, 31 March–2 April 2014; Society of Petroleum Engineers: Calgary, AB, Canada, 2014. [CrossRef]
22. Caudle, B.H.; Dyes, A.B. Improving Miscible Displacement by Gas-Water Injection. *Trans. AIME* **1958**, *213*, 281–283. [CrossRef]
23. Lin, E.C.; Poole, E.S. Numerical evaluation of single-slug, WAG, and hybrid CO2 injection processes, Dollarhide Devonian Unit, Andrews County, Texas. *SPE Reserv. Eng.* **1991**, *6*, 415–420. [CrossRef]
24. Khan, M.Y.; Kohata, A.; Patel, H.; Syed, F.I.; Al Sowaidi, A.K. Water alternating gas WAG optimization using tapered WAG technique for a giant offshore middle east oil field. In Proceedings of the Society of Petroleum Engineers—Abu Dhabi International Petroleum Exhibition and Conference 2016, Abu Dhabi, UAE, 7 November 2016; Society of Petroleum Engineers: Calgary, AB, Canada, 2016; Volume 2016. [CrossRef]
25. Fatemi, S.M.; Sohrabi, M. Experimental Investigation of Near-Miscible Water-Alternating-Gas Injection Performance in Water-Wet and Mixed-Wet Systems. *SPE J.* **2013**, *18*, 114–123. [CrossRef]
26. Lubaś, J.; Stopa, J.; Wojnicki, M. Możliwości zastosowania zaawansowanych metod wspomagania wydobycia ropy naftowej ze złóż dojrzałych. *Naft.-Gaz* **2019**, *75*, 24–28. [CrossRef]
27. Ma, T.D.; Youngren, G.K. Performance of Immiscible Water-Alternating-Gas (IWAG) Injection at Kuparuk River Unit, North Slope, Alaska. In Proceedings of the SPE Annual Technical Conference and Exhibition, New Orleans, LA, USA, 25 September 1994; Society of Petroleum Engineers: Calgary, AB, Canada, 1994. [CrossRef]
28. Ramachandran, K.P.; Gyani, O.N.; Sur, S. Immiscible Hydrocarbon WAG: Laboratory to Field. In Proceedings of the SPE Oil and Gas India Conference and Exhibition, Mumbai, India, 20–22 January 2010; Society of Petroleum Engineers (SPE): Calgary, AB, Canada, 2010. [CrossRef]
29. Agbalaka, C.C.; Dandekar, A.Y.; Patil, S.L.; Khataniar, S.; Hemsath, J. The Effect Of Wettability On Oil Recovery: A Review. In Proceedings of the SPE Asia Pacific Oil and Gas Conference and Exhibition, Perth, Australia, 20–22 October 2008; Society of Petroleum Engineers: Calgary, AB, Canada, 2008. [CrossRef]
30. Huang, E.T.S.; Holm, L.W. Effect of WAG injection and rock wettability on oil recovery during CO_2 flooding. *SPE Reserv. Eng.* **1988**, *3*, 119–129. [CrossRef]
31. Hoare, G.; Coll, C. Effect of small/medium scale reservoir heterogeneity on the effectiveness of water, gas and water alternating gas WAG injection. In Proceedings of the Society of Petroleum Engineers—SPE Europec featured at 80th EAGE Conference and Exhibition 2018, Copenhagen, Denmark, 11–14 June 2018; Society of Petroleum Engineers: Calgary, AB, Canada, 2018. [CrossRef]
32. Speight, J.G. *Introduction to Enhanced Recovery Methods for Heavy Oil and Tar Sands*; Elsevier: Amsterdam, The Netherlands, 2016. [CrossRef]
33. van Lingen, P.P.; Barzanji, O.H.M.; van Kruijsdijk, C.P.J.W. WAG Injection to Reduce Capillary Entrapment in Small-Scale Heterogeneities. In Proceedings of the SPE Annual Technical Conference and Exhibition, Denver, CO, USA, 6 October 1996; Society of Petroleum Engineers: Calgary, AB, Canada, 1996. [CrossRef]
34. Kharrat, R.; Mahdavi, S.; Ghorbani, D. A Comprehensive EOR Study of a Highly Fractured Matured Field-Case Study. In Proceedings of the SPE Europec/EAGE Annual Conference, Copenhagen, Denmark, 4–7 June 2012; Society of Petroleum Engineers: Calgary, AB, Canada, 2012. [CrossRef]

35. Dang, C.; Nghiem, L.; Nguyen, N.; Chen, Z.; Nguyen, Q. Evaluation of CO_2 Low Salinity Water-Alternating-Gas for enhanced oil recovery. *J. Nat. Gas Sci. Eng.* **2016**, *35*, 237–258. [CrossRef]
36. Motealleh, M.; Kharrat, R.; Hashemi, A. An experimental investigation of water-alternating-CO_2 coreflooding in a carbonate oil reservoir in different initial core conditions. *Energy Sources Part A Recovery Util. Environ. Eff.* **2013**, *35*, 1187–1196. [CrossRef]
37. Teklu, T.W.; Alameri, W.; Graves, R.M.; Kazemi, H.; AlSumaiti, A.M. Low-salinity water-alternating-CO_2 EOR. *J. Pet. Sci. Eng.* **2016**, *142*, 101–118. [CrossRef]
38. Ghafoori, A.; Shahbazi, K.; Darabi, A.; Soleymanzadeh, A.; Abedini, A. The experimental investigation of nitrogen and carbon dioxide water-alternating-gas injection in a carbonate reservoir. *Pet. Sci. Technol.* **2012**, *30*, 1071–1081. [CrossRef]
39. Al-Shuraiqi, H.S.; Muggeridge, A.H.; Grattoni, C.A. Laboratory Investigations of First Contact Miscible Wag Displacement: The Effects of Wag Ratio And Flow Rate. In Proceedings of the SPE International Improved Oil Recovery Conference in Asia Pacific, Kuala Lumpur, Malaysia, 20 October 2003; Society of Petroleum Engineers: Calgary, AB, Canada, 2003. [CrossRef]
40. Juanes, R.; Blunt, M.J. Impact of viscous fingering on the prediction of optimum WAG ratio. *SPE J.* **2007**, *12*, 486–495. [CrossRef]
41. Rahimi, V.; Bidarigh, M.; Bahrami, P. Experimental Study and Performance Investigation of Miscible Water-Alternating-CO_2 Flooding for Enhancing Oil Recovery in the Sarvak Formation. *Oil Gas Sci. Technol.–Rev. d'IFP Energies Nouv.* **2017**, *72*, 35. [CrossRef]
42. Wojnicki, M. Experimental investigations of oil displacement using the WAG method with carbon dioxide. *Naft.-Gaz* **2017**, *73*, 864–870. [CrossRef]
43. Jiang, H.; Nuryaningsih, L.; Adidharma, H. The study of timing of cyclic injections in miscible CO_2 WAG. In Proceedings of the Society of Petroleum Engineers Western Regional Meeting 2012, Bakersfield, CA, USA, 21 March 2012; Society of Petroleum Engineers: Calgary, AB, Canada, 2012. [CrossRef]
44. Shahverdi, H.; Sohrabi, M.; Fatemi, M.; Jamiolahmady, M. Three-phase relative permeability and hysteresis effect during WAG process in mixed wet and low IFT systems. *J. Pet. Sci. Eng.* **2011**, *78*, 732–739. [CrossRef]
45. Fatemi, S.M.; Sohrabi, M. Cyclic hysteresis of three-phase relative permeability curves applicable to WAG injection under low gas/oil IFT: Effect of immobile water saturation, injection scenario and rock permeability. In Proceedings of the 75th European Association of Geoscientists and Engineers Conference and Exhibition 2013 Incorporating SPE EUROPEC 2013: Changing Frontiers, European Association of Geoscientists and Engineers, EAGE, London, UK, 10 June 2013. [CrossRef]
46. Sohrabi, M.; Danesh, A.; Jamiolahmady, M. Visualisation of residual oil recovery by near-miscible gas and SWAG injection using high-pressure micromodels. *Transp. Porous Media* **2008**, *74*, 239–257. [CrossRef]
47. Sohrabi, M.; Tehrani, D.H.; Danesh, A.; Henderson, G.D. Visualization of oil recovery by water-alternating-gas injection using high-pressure micromodels. *SPE J.* **2004**, *9*, 290–301. [CrossRef]
48. Dong, M.; Foraie, J.; Huang, S.; Chatzis, I. Analysis of immiscible Water-Alternating-Gas (WAG) injection using micromodel tests. *J. Can. Pet. Technol.* **2005**, *44*, 17–24. [CrossRef]
49. Larsen, J.K.; Bech, N.; Winter, A. Three-Phase Immiscible WAG Injection: Micromodel Experiments and Network Models. In Proceedings of the SPE/DOE Improved Oil Recovery Symposium, Tulsa, OK, USA, 3–5 April 2000; Society of Petroleum Engineers (SPE): Calgary, AB, Canada, 2000. [CrossRef]
50. Wojnicki, M.; Lubás, J.; Warnecki, M.; Kúsnierczyk, J.; Szuflita, S. Experimental studies of immiscible high-nitrogen natural gas WAG injection efficiency in mixed-wet carbonate reservoir. *Energies* **2020**, *13*, 2346. [CrossRef]
51. Luchian, H.; Băutu, A.; Băutu, E. Genetic programming techniques with applications in the oil and gas industry. In *Artificial Intelligent Approaches in Petroleum Geosciences*; Springer International Publishing: Berlin/Heidelberg, Germany, 2015. [CrossRef]
52. Velez-Langs, O. Genetic algorithms in oil industry: An overview. *J. Pet. Sci. Eng.* **2005**, *47*, 15–22. [CrossRef]
53. Jensen, J.P.; Andersen, M.G. *Reservoir Production Optimization Using Genetic Algorithms and Artificial Neural Networks*; Downing, K., Ed.; Institutt for datateknologi og informatikk: Trondheim, Norway, 2009.
54. Soleng, H.H. Oil reservoir production forecasting with uncertainty estimation using genetic algorithms. In Proceedings of the 1999 Congress on Evolutionary Computation, CEC 1999, Washington, DC, USA, 6–9 July 1999; IEEE Computer Society: Washington, DC, USA, 1999; Volume 2. [CrossRef]
55. Vo Thanh, H.; Sugai, Y.; Nguele, R.; Sasaki, K. Robust optimization of CO_2 sequestration through a water alternating gas process under geological uncertainties in Cuu Long Basin, Vietnam. *J. Nat. Gas Sci. Eng.* **2020**, *76*, 103208. [CrossRef]
56. Safarzadeh, M.A.; Motahhari, S.M. Co-optimization of carbon dioxide storage and enhanced oil recovery in oil reservoirs using a multi-objective genetic algorithm (NSGA-II). *Pet. Sci.* **2014**, *11*, 460–468. [CrossRef]
57. Mohagheghian, E.; James, L.A.; Haynes, R.D. Optimization of hydrocarbon water alternating gas in the Norne field: Application of evolutionary algorithms. *Fuel* **2018**, *223*, 86–98. [CrossRef]
58. Nait Amar, M.; Zeraibi, N.; Redouane, K. Optimization of WAG Process Using Dynamic Proxy, Genetic Algorithm and Ant Colony Optimization. *Arab. J. Sci. Eng.* **2018**, *43*, 6399–6412. [CrossRef]
59. Janiga, D.; Czarnota, R.; Stopa, J.; Wojnarowski, P. Huff and puff process optimization in micro scale by coupling laboratory experiment and numerical simulation. *Fuel* **2018**, *224*, 289–301. [CrossRef]
60. Khan, M.R.; Kalam, S.; Khan, R.A.; Tariq, Z.; Abdulraheem, A. Comparative analysis of intelligent algorithms to predict the minimum miscibility pressure for hydrocarbon gas flooding. In Proceedings of the Society of Petroleum Engineers—Abu Dhabi International Petroleum Exhibition and Conference 2019, ADIP 2019, Abu Dhabi, UAE, 11–14 November 2019; Society of Petroleum Engineers: Calgary, AB, Canada, 2019. [CrossRef]

61. Mahdiani, M.R.; Kooti, G. The most accurate heuristic-based algorithms for estimating the oil formation volume factor. *Petroleum* **2016**, *2*, 40–48. [CrossRef]
62. Nasery, S.; Hoseinpour, S.; Phung, L.T.K.; Bahadori, A. Prediction of the viscosity of water-in-oil emulsions. *Pet. Sci. Technol.* **2016**, *34*, 1972–1977. [CrossRef]
63. Tukur, A.D.; Nzerem, P.; Nsan, N.; Okafor, I.S.; Gimba, A.; Ogolo, O.; Oluwaseun, A.; Andrew, O. Well placement optimization using simulated annealing and genetic algorithm. In Proceedings of the Society of Petroleum Engineers—SPE Nigeria Annual International Conference and Exhibition 2019, NAIC 2019, Lagos, Nigeria, 5–7 August 2019; Society of Petroleum Engineers: Calgary, AB, Canada, 2019. [CrossRef]
64. Li, H.; Yu, H.; Cao, N.; Tian, H.; Cheng, S. Applications of Artificial Intelligence in Oil and Gas Development. *Arch. Comput. Methods Eng.* **2020**, *28*, 937–949. [CrossRef]
65. López, S.; Koç, U.; Bakker, E.; Rahmani, J. Optimization of Lift Gas Allocation Using Evolutionary Algorithms. *Int. J. Comput. Appl. Technol. Res.* **2019**, *8*, 353–357. [CrossRef]
66. Andreasen, A. Applied Process Simulation-Driven Oil and Gas Separation Plant Optimization Using Surrogate Modeling and Evolutionary Algorithms. *ChemEngineering* **2020**, *4*, 11. [CrossRef]
67. De Pádua, L.; Sales, A.; Ramalho Pitombeira-Neto, A.; De Athayde Prata, B. A genetic algorithm integrated with Monte Carlo simulation for the field layout design problem. *Oil Gas Sci. Technol.–Rev. d'IFP Energies Nouv.* **2018**, *73*, 24. [CrossRef]
68. American Petroleum Institute (API). *Sampling Petroleum Reservoir Fluids*; API Recommended Practice 2003; American Petroleum Institute (API): Washington, DC, USA, 2003; Volume 44.
69. Langaas, K.; Ekrann, S.; Ebeltoft, E. A criterion for ordering individuals in a composite core. *J. Pet. Sci. Eng.* **1998**, *19*, 21–32. [CrossRef]
70. Piñerez, I.; Puntervold, T.; Strand, S.; Hopkins, P.; Aslanidis, P.; Yang, H.S.; Kinn, M.S. Core wettability reproduction: A new solvent cleaning and core restoration strategy for chalk cores. *J. Pet. Sci. Eng.* **2020**, *195*, 107654. [CrossRef]
71. Holtz, M.H. Immiscible water alternating gas (IWAG) EOR: Current state of the art. In Proceedings of the SPE—DOE Improved Oil Recovery Symposium, Tulsa, OK, USA, 11 April 2016; Society of Petroleum Engineers (SPE): Calgary, AB, Canada, 2016; Volume 2016. [CrossRef]
72. Holm, L.W.; Josendal, V.A. Effect of Oil Composition on Miscible-Type Displacement by Carbon Dioxide. *Soc. Pet. Eng. J.* **1982**, *22*, 87–98. [CrossRef]
73. Holm, L.W. Miscibility and Miscible Displacement. *J. Pet. Technol.* **1986**, *38*, 817–818. [CrossRef]
74. Whitson, C.H.; Brulé, M.R. Phase behavior. *SPE Monogr. Ser.* **2000**, *20*, 233.
75. Koza, J.R. Genetic programming as a means for programming computers by natural selection. *Stat. Comput.* **1994**, *4*, 87–112. [CrossRef]
76. Koza, J.R.; Poli, R. Genetic Programming. In *Search Methodologies: Introductory Tutorials in Optimization and Decision Support Techniques*; Springer: Berlin/Heidelberg, Germany, 2005; pp. 127–164. [CrossRef]
77. Sette, S.; Boullart, L. Genetic programming: Principles and applications. *Eng. Appl. Artif. Intell.* **2001**, *14*, 727–736. [CrossRef]
78. Poli, R.; Langdon, W.B.; Mcphee, N.F.; Koza, J.R. A Field Guide to Genetic Programming. 2008. Available online: http://www.gp-field-guide.org.uk (accessed on 15 October 2021).
79. Madár, J.; Abonyi, J.; Szeifert, F. Genetic Programming for the Identification of Nonlinear Input−Output Models. *Ind. Eng. Chem. Res.* **2005**, *44*, 3178–3186. [CrossRef]
80. Glaso, O. Miscible displacement: Recovery Tests Nitrogen. *SPE Reserv. Eng. Soc. Pet. Eng.* **1990**, *5*, 61–68. [CrossRef]
81. Sebastian, H.M.; Lawrence, D.D. Nitrogen Minimum Miscibility Pressures. In Proceedings of the SPE/DOE Enhanced Oil Recovery Symposium, Tulsa, OK, USA, 22–24 April 1992; Society of Petroleum Engineers: Calgary, AB, Canada, 1992. [CrossRef]
82. Anada, H.R. *State-of-the-Art Review of Nitrogen and Flue Gas Flooding in Enhanced Oil Recovery*; Final Report; Science Applications, Inc.: Pittsburgh, PA, USA; Morgantown, WV, USA, 1980. [CrossRef]
83. Hudgins, D.A.; Llave, F.M.; Chung, F.T.H. Nitrogen Miscible Displacement of Light Crude Oil: A Laboratory Study. *SPE Reserv. Eng.* **1990**, *5*, 100–106. [CrossRef]

energies

MDPI

Article

Effect of Cleaning the Annular Space on the Adhesion of the Cement Sheath to the Rock

Marcin Kremieniewski [1,*], Sławomir Błaż [1], Stanisław Stryczek [2], Rafał Wiśniowski [2] and Andrzej Gonet [2]

1 Oil and Gas Institute—National Research Institute, 31-503 Kraków, Poland; blaz@inig.pl
2 Department of Drilling and Geoengineering, Faculty of Drilling, Oil and Gas, AGH University of Science and Technology, 30 Mickiewicza Ave., 30-059 Kraków, Poland; stryczek@agh.edu.pl (S.S.); wisniows@agh.edu.pl (R.W.); gonet@agh.edu.pl (A.G.)
* Correspondence: kremieniewski@inig.pl

Abstract: Drilling boreholes in gas zones and in zones with the possibility of migration or gas exhalation requires a high index of well tightness. An important parameter determining the effectiveness of sealing the annular space is the adhesion of the cement sheath to the rock formation. Low values of adhesion of the cement sheath to the rock formation and to the casing surface result in the formation of uncontrolled gas flows. The lack of adhesion also reduces the stabilization of the pipe column. To obtain the required adhesion, the annular space should be properly cleaned. Thorough removal of filter cake from the drilling fluid increases adhesion and reduces gas migration from the annular space. Therefore, in this work, the authors focus on determining the effect of cleaning the annular space on the adhesion of the cement sheath to the rock formation. The results of the research work allow for further research on the modification of spacers and cement slurries in order to obtain the required increase in adhesion. The article presents the issues related to the preparation of the borehole for cementing by appropriate cleaning of the rock formation from the residue of the mud cake. During the implementation of the works, tests of cleaning the rock surface are performed. The obtained results are correlated with the results of adhesion on the rock–cement sheath cleaned of the wash mud cake contact. When analyzing the obtained test results, a relationship is found between the cleaning of the rock surface and the adhesion of the cement sheath to it.

Keywords: improving the sealing of the borehole; rational selection of drilling fluids; well cementing; cleaning the borehole; adhesion; cement sheath; spacer fluid

Citation: Kremieniewski, M.; Błaż, S.; Stryczek, S.; Wiśniowski, R.; Gonet, A. Effect of Cleaning the Annular Space on the Adhesion of the Cement Sheath to the Rock. *Energies* **2021**, *14*, 5187. https://doi.org/10.3390/en14165187

Academic Editor: Reza Rezaee

Received: 16 July 2021
Accepted: 17 August 2021
Published: 22 August 2021

1. Introduction

1.1. Purpose of the Work and Literature Background

Cementing of the casing column is performed in order to seal the annular space, i.e., to isolate the gas, oil and aquifers levels. The lack of such tightness makes it possible to create out-of-pipe flows of reservoir media or prevent the exploitation of natural gas [1–4]. In order to properly insulate the annular space, the cement slurry is pressed in. After it is bonded to the rock formation and to the surface of the casing, it guarantees the required tightness. However, this tightness also depends on the adhesion of the cement sheath to the contact surface. According to the literature [5,6], good grip means that there is no gap or no free movement of two bodies in relation to each other. It is interpreted in the contact plane with simultaneous action of tearing, compressive or shear stresses from the outside [5,7,8].

1.2. Factors Influencing the Adhesion of the Cement Sheath

When analyzing the conditions in the borehole, the resulting types and directions of force action in relation to the contact surface of the cement sheath and pipes are distinguished [9]. In the borehole, there are forces with the direction of action perpendicular to the contact surface and the forces acting in the direction parallel to these surfaces, analyzed later in this publication. These forces are caused by the elongation or contraction of the pipes due to temperature changes. They can also be the result of the formation of resistances preventing the mutual displacement of pipes [9–11]. The parameter of adhesion is the slip resistance determined in the tests by the vertical movement of the steel pipe or the rock core in the hardened cement slurry, as shown in Figure 1.

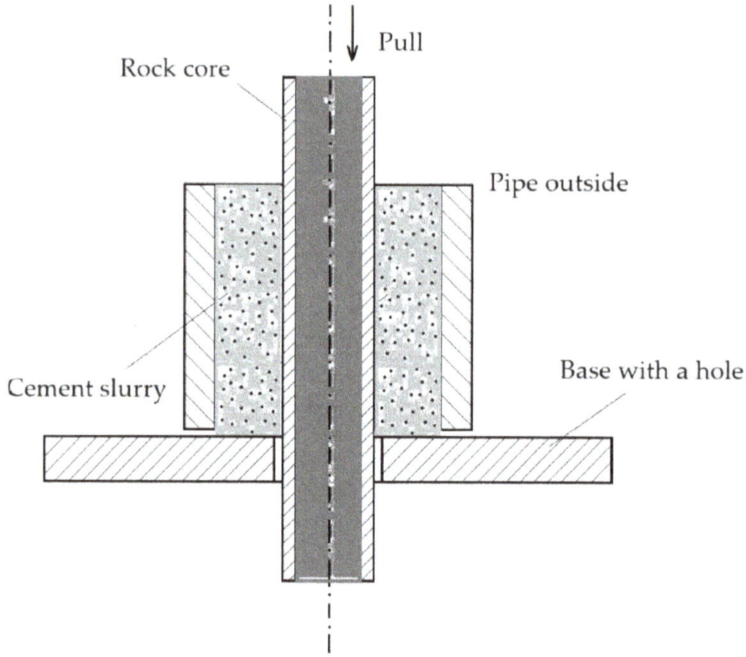

Figure 1. Scheme for determining the adhesion of a cement sheath to steel or rock formation.

The factors that affect the adhesion of the hardened cement slurry to the contact surface are mainly [9,12–14]:

- √ Tight adhesion of the cement sheath to the contact surface;
- √ Positive volume increase of the setting cement slurry;
- √ Gelling of the cement slurry on contact with the hydrated subsurface layer of the rock formation;
- √ The pressure exerted by the casing pipes on the setting cement slurry by cooling the pipes with the pumped liquids and then expanding them due to the temperature increase during the setting of the cement slurry;
- √ Adhesion forces between cement slurry and pipes;
- √ Bonding of the cement slurry to the surface of the pipes and the surface of the rock formation due to the penetration of slurry into the cavities resulting from the roughness of the contact surface.

In turn, the reduction of the adhesion value of the cement sheath to the contact surface is influenced by:

- √ Shrinkage of cement slurry during setting;
- √ Fluctuations in pressure and temperature during the setting of the cement slurry resulting, inter alia, from closing the cement head after the completion of cementing and leaving the pipe column at the final pressure of the treatment;
- √ Injection into the borehole of a cold scrubber when drilling cement plug;
- √ Low surface roughness coefficient of pipes [15–18].

However, from the point of view of preparing the borehole for cementing, the non-displaced mud layer has a significant influence on the adhesion of the cement sheath. It blocks the wetting of the surfaces in contact with the cement slurry due to the formation of adhesion forces at the interface [19]. Moreover, the unremoved mud cake after mixing with the cement slurry in the annular space reduces the mechanical strength of the cement sheath [20–23]. It should be borne in mind that obtaining adequate tightness at the contact of the cement sheath with the rock formation is influenced not only by the close adhesion of individual layers (cement slurry to the wall of the borehole). Adhesion at the contact of these layers also shows a strong influence [24,25]. Only the combination of both coface and adhesion (adhesive and mechanical) allows the total to obtain the required contact tightness effect of the cement sheath and the rock formation. Obtaining proper coface is not a problem because the cement slurry with appropriately designed rheological parameters penetrates the rock mass discontinuities resulting from drilling [8,26–28]. However, obtaining the appropriate values of the second parameter, which is adhesion, requires appropriate preparation of the annular space.

1.3. *Preparation of the Borehole for Cementing*

In the first stage, preparation of the borehole for cementing consists in lowering the rheological parameters of the washer during rinsing [29,30]. Then, the spacer (washing liquid) is pumped into the annular space in order to displace the drilling fluid and the mud residue [31–33]. A very advantageous condition is that the casing column and the drilled rock formation are moistened with the spacer. Such action improves the setting of the cement slurry pumped over these liquids [34–36]. In order to thoroughly remove the drilling fluid residues, washings consisting of a mixture of dispersants and surfactants are used. The parameters of these spacers must be very precisely designed [37–41]. A number of factors influence the efficiency of washing mud cake. These are, among others, contact time of the spacer, spacer pumping rate, chemical composition of the spacer, concentration of the agents used to prepare the spacer. The type of surface from which the mud cake is removed, the type of drilling mud used for drilling and its technological parameters as well as borehole conditions (temperature and pressure) are also important [37,42–46]. It has been observed that obtaining an adequate cleaning of the annular space sometimes contributes to the improvement of the tightness of the borehole.

The aim of the study is to determine the effect of cleaning the annular space on the adhesion of the cement sheath to the rock formation. In order to obtain the required result, it is necessary to thoroughly study the conditions in the cement sheath–rock formation system. The obtained results are helpful in indicating the leading factor (adhesion or adhesion) which determines the tightness. This allows to determine the effect of cleaning the annular space on adhesion and introduce modifications to improve adhesion and obtain adequate tightness of the borehole.

2. Materials and Methods

2.1. *Materials*

Four spacers and water as a control liquid were used to determine the effect of cleaning the annular space on the adhesion of the cement sheath to the rock. The first liquid is a 0.5% solution of SL372-ethoxylated alcohol C12–C15 (anionic surfactant from alkyl ether sulfate group). The second liquid is a solution of 0.5% RL22-ethoxylated alcohol C12–C14 (non-ionic surface-active compound used as emulsifier). The third liquid is CD-fatty alcohol alkyl polyglucoside C8–C10. The fourth liquid called RL80 is ethoxylated alcohol C12–C14 (surface active agent used for wetting and as a non-ionic component of emulsifiers).

CEM I 42.5R Portland cement was used to prepare the cement slurry. This cement is composed of 2.65% SO_3 and 0.064% Cl^-. Materials were also used to regulate the parameters of the cement slurry. PSP 046 plasticizer is a dispersant based on modified lignosulfonates and naphthalene with a bulk density of 440–550 kg/m^3 and a pH value ranging from 6.6 to 8.5. In order to remove air from the cement slurry, a defoaming agent was used—a mixture of esters of unsaturated fatty acids and refined hydrocarbons. The start also included an antifiltrating agent and setting accelerator. Latex, which is a water dispersion of styrene-butadiene copolymer, was used to reduce the porosity of the cement sheath. Additionally, a latex stabilizer was added to the cement slurry. The matrix of the cement sheath was sealed with a 10% addition of microcement, which comes from Halliburton Micro Matrix. This type of microcement has grains smaller than or equal to 10 μm, and its specific surface area is approximately 1380 m^2/kg. The percentage of means to prepare the cement slurry is summarized in Table 1.

Table 1. Recipe and parameters of the cement slurry used in the test of adhesion on the hardened cement slurry–rock contact.

Ingredients	% by Mass of Cement
Water–cement ratio	0.46
Plasticizer	0.2
Latex	8.0
Stabilizer	1.0
Defoaming agent	0.48
Antifiltrating agent	0.22
Setting accelerator	4.2
Microcement	10.0
Cement CEM I 42.5R	100.0

All components in % by mass of cement.

2.2. Methods

2.2.1. Preparation of the Spacer (Washing Liquid)

To prepare the spacer, the following wre used:

- √ The control spacer is water;
- √ Liquid No. 1 to 0.5% solution SL372-ethoxylated alcohol C12-C15 (anionic surfactant from alkyl ether sulfate group);
- √ Liquid No. 2 is a 0.5% solution of RL22-ethoxylated alcohol C12-C14 (non-ionic surface-active compound used as emulsifier);
- √ Liquid No. 3 to 0.5% solution CD-fatty alcohol alkyl polyglucoside C8-C10;
- √ Liquid No. 4 is a 0.5% solution of RL80-ethoxylated alcohol C12-C14 (surface active agent used for wetting and as a non-ionic component of emulsifiers).

To prepare the spacer, a certain amount of water was measured with a measuring cylinder. The water was poured into the mixer. The rotation speed was then set to 500 rpm and the specified amount of rinse aid was added.

2.2.2. Preparation of the Cement Slurry

For the effect of cleaning the annular space on the adhesion of the cement sheath to the rock, the samples of the cores after washing were sealed with cement slurry. The cement slurry was prepared as follows. A certain amount of water was measured out with a measuring cylinder. The water was placed in the mixer. The speed was set to 1600 rpm. The cement slurry agents were added to the mixing water and mixed for 10 min. Later, loose materials (microcement, cement) were dosed into the mixing water with chemicals and mixed for another 20 min. Mixing at low speed reflects slurry at well conditions.

2.2.3. Experimental Procedures

The properties of the cement slurry were tested in accordance with the standard:

- √ PN–EN ISO 10426-2. Oil and gas industry. Cements and materials for cementing holes. Lot 2: Testing of drilling cements. These tests include the following measurements: slurry density, filtration and thickening time.

The adhesion test was carried out according to the standard.

√ PN–EN 196-1: 2006 Cement testing methods. Compressive strength determination

The test plan was adapted to the needs of the tests, the results of which allow to determine the effect of cleaning the annular space on the adhesion of the cement sheath to the rock. The drilling fluid flow simulator constructed at the Oil and Gas Institute-National Research Institute (Patent P.423842) was used for the tests. The device (Figure 2) allows to simulate the mud flow in the annular space. In this way, a washing mud cake was formed on the surface of the rock cores. The mud cake removal tests were carried out on four selected spacers, which were pumped at a rate of 11.2 L/min. This gives a Reynolds number of 3100. Water was used as the base spacer to determine the deposit removal checkpoint. The washing time was 4 min. The samples on which the mud cake was formed were three sandstone cores with an outer diameter of 25 mm and a length of 60 mm. They were fixed in a tripod inside a PVC pipe (Figure 2-left part of the photo) in a drilling fluid flow simulator.

Figure 2. Drilling fluid flow simulator-view of the spacer pumping device.

In the first stage, the degree of surface cleaning of the cores was determined by measuring the mass of the mud cake remaining on the cores after washing. The cores reflect the annular space of the borehole. The mass of the core before mud cake formation (m_R) was determined, then the core with the mud cake (m_{Ros}) was weighed, and after pumping the spacer, the weight of the rock core with the residue of the spacer was determined again after pumping the spacer (m_{Rpo}). On the basis of the obtained results, the percentage effectiveness of the mud cake removal from the core surface was calculated according to formula:

$$\% = 100 \cdot \frac{m_{Ros} - m_{Rpo}}{m_{Ros} - m_R} \quad (1)$$

where:

%—percent of sediment washout;
m_R—core mass before the test (without mud cake);
m_{Ros}—core mass with mud cake;
m_{Rpo}—mass of the core with the residue of the mud cake (after washing);

In the tests, a mud cake from the PSK borehole was used to create a mud, which contained contaminants from the interval drilled into a 7-inch pipe section. Four spacers and a standard liquid–water spacer were used to determine the degree of deposit removal. Table 2 shows the results of the mud cake removal tests.

Table 2. The results of the research on the efficiency of the mud cake removal from the core surface and the adhesion of the cement sheath to the rock formation.

Spacer	Measurement No.	Water	Spacer 1	Spacer 2	Spacer 3	Spacer 4
Core weight before test m_R, g	1	68.65	68.64	68.65	68.66	68.66
	2	68.68	68.65	68.64	68.64	68.66
	3	68.65	68.65	68.66	68.65	68.65
Core weight after mud m_{Ros}, g	1	72.58	72.52	72.5	72.47	72.44
	2	72.55	72.5	72.52	72.4	72.89
	3	72.54	72.44	72.47	72.43	72.75
Core weight after spacer m_{Rpo}, g	1	70.44	69.52	69.49	69.65	69.44
	2	70.38	69.6	69.49	69.72	69.43
	3	70.37	69.58	69.47	69.72	69.4
Washing percentage, %	1	54.45	77.32	78.18	74.02	79.37
	2	56.07	75.32	78.09	71.28	81.8
	3	55.78	75.46	78.74	71.69	81.71
Washing percentage (removal of mud cake) average value, %	$\overline{x}$	55.44	76.04	78.34	72.33	80.96
Percentage increase in washing efficiency compared to the base value (washing with water), %	$\overline{x}$	0.00	37.16	41.31	30.47	46.03
The average value of the adhesion of the cement sheath to the rock, MPa	$\overline{x}$	0.696	0.928	0.957	0.870	0.986
Percentage of adhesion increase in relation to the base value (adhesion after washing with water), %	$\overline{x}$	0.00	33.33	37.50	25.00	41.67
The difference between the percentage increase in washing efficiency and the percentage increase in adhesion, %	$\overline{x}$	0.00	3.83	3.81	5.47	4.36

In the second stage of the research, the values of adhesion at the contact between the cement sheath and the rock core were determined. The tests were carried out for previously washed core samples and for the control sample, i.e., the core washed with water as the spacer. In order to determine the adhesion of the cement sheath to the sandstone core, after the mud cake was formed on it, and then, after washing with selected liquids, it was placed in a mold (Figure 3) and poured with cement slurry (Figure 4).

Figure 3. Core sample ready to be filled with cement slurry.

Figure 4. Sample prepared for adhesion tests.

After 48 h of hydration, adhesion tests were performed on the contact between the hardened cement slurry and the rock core (cleaned of the previously produced mud cake). The adhesion test consists of placing a core sample sealed with cement slurry between the plates of a testing machine (Figure 5) and determining the breaking strength of adhesion on the hardened cement slurry–rock contact under the influence of a load applied to the sample.

The adhesion (MPa) on the contact between the hardened cement slurry and the rock core was calculated according to the Formula (2):

$$\sigma_p = \frac{P}{s} \cdot 10^{-3}\ [\mathrm{MPa}] \tag{2}$$

where:

σ_p—contact adhesion hardened cement slurry–rock core (MPa),
P—pressure force causing the connection to be broken at the contact of the hardened cement slurry with the rock (kN)—original record from a testing machine,
s—contact surface of the rock sample with the cement slurry (m^2).

Figure 5. Machine for testing strength and adhesion.

The contact force (P) was read from the testing machine. The contact surface of the rock with the hardened cement slurry was equal to the value of the outer surface area of the core and the height of the cement slurry in the mold. Figure 6 shows the dimensions necessary to determine the contact surface of the rock with the hardened cement slurry. In order to obtain reliable results, all tests were carried out using a cement slurry of the same composition. The results of the adhesion tests obtained on the contact between the hardened cement slurry and the rock core are presented in Table 2.

$$s = \pi \cdot d \cdot h \left[\mathrm{m}^2\right] \tag{3}$$

where:

d—core diameter (m) for the conducted research is equal to 25 mm = 0.025 m;
h—the height of the cemented part of the core for the conducted research is 44 mm = 0.044 m.

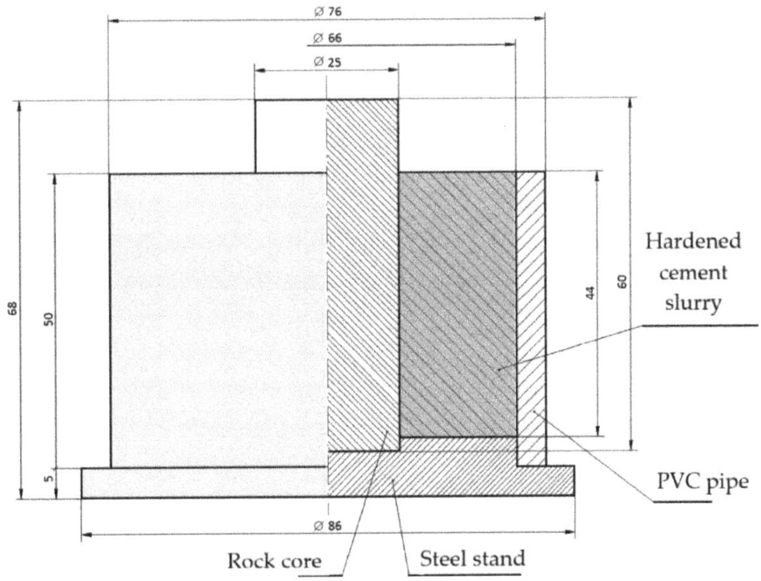

Figure 6. Diagram showing the core sample placed in the adhesion test mold.

3. Results

The study of the mud cake removal efficiency begins with the determination of the value of the mud cake removal from the core surface with water (control test). This sample achieved a 55.44% removal of mud cake from the core surface. Then, tests of mud cake removal by various spacers were carried out. The obtained values of mud cake removal range from 72.33% for washing the cores with the spacer No. 3 to the maximum value, which is 80.96% after using the washing spacer No. 4 (Table 1). On the basis of the obtained results, the percentage increase in the washing efficiency in relation to the control sample is estimated at 55.44% (washing with water). The range of percentage increase in the mud cake leaching efficiency in relation to the base value, presented in Table 2, ranges from 30.47% to 46.03%. The results of the percent mud cake removal and the increase compared to the baseline value are summarized in Table 2 and Figure 7.

Another test is to determine the adhesion on the contact between the hardened cement slurry and the rock core cleaned of the produced washing mud cake. As in the previous test, first the adhesion of the control sample was determined. This is the adhesion after washing the core set with water as the spacer. The average of the adhesion values is 0.696 MPa (Table 2). Then, adhesion tests were carried out for samples washed by selected spacers. The average values of adhesion range from 0.87 MPa for the sample after the application of spacer No. 3 to the maximum value of adhesion, which is 0.986 MPa after washing the samples with the spacer No. 4 (Table 2 and Figure 7). The percentage increase in adhesion compared to the control sample ranges from 25% (spacer No. 3) to 41.67% (spacer No. 4). The results are presented in Table 2 and in Figure 7.

Additionally, an important interpretation is the difference between the percentage increase in the washing efficiency and the percentage increase in adhesion presented in the last row of Table 2. Values ranging from 3.81% to 5.47% were obtained. For such an interpretation, the percentage increase of the analyzed parameters (washing efficiency and adhesion) in relation to the base value was selected. This made it possible to determine the convergence of the analyzed values.

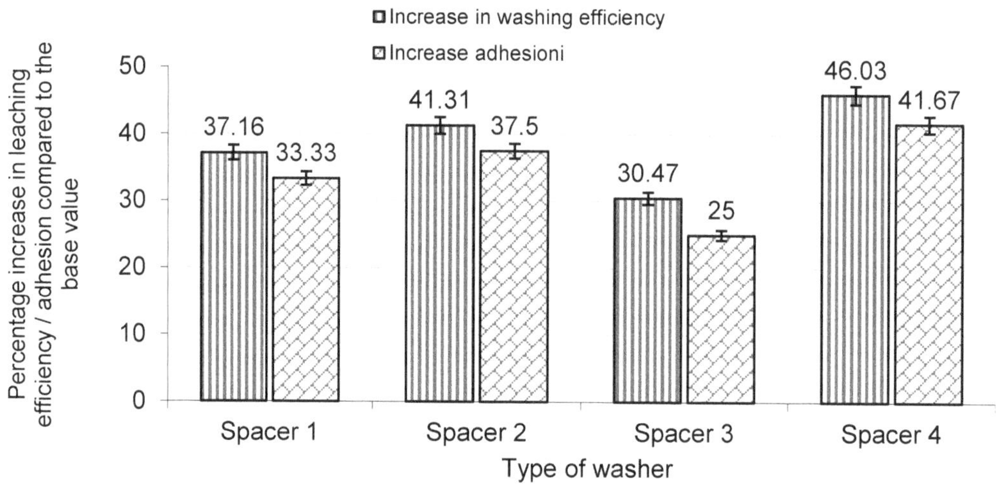

Figure 7. List of the percentage increase of the analyzed parameters (the effectiveness of the removal of mud cake and adhesion on the contact between the hardened cement slurry–the rock core) in relation to the base value.

When analyzing the test results, it is stated that the values of the mud cake removal efficiency, i.e., the efficiency of cleaning the annular space, is proportional to the adhesion of the cement sheath to the rock formation. Therefore, a correlation analysis was performed to determine the degree of convergence between the compared features on the basis of the Pearson's correlation coefficient. Pearson's correlation coefficient is calculated on the basis of Formula (4), then according to Formula (5), the covariance (cov) was calculated, which determines the linear relationship between the analyzed variables x and y, and the maximum likelihood estimator Sdx, Sdy giving the smallest deviation values (Formulas (6), (7), and Table 3). In the final step, the linear determination index (8) was calculated, which informs about the percentage of linearly expressed variability of the dependent variable by the independent variable. The interpretation of the strength of the correlation relationships is as follows:

Pearson's correlation coefficient:

- below 0.2—weak correlation (practically no relationship);
- 0.2–0.4—low correlation (clear relationship);
- 0.4–0.6—moderate correlation (significant relationship);
- 0.6–0.8—high correlation (significant dependence);
- 0.8–0.9—very high correlation (very high correlation);
- 0.9–1.0—total correlation (practically full relationship).

$$r_{x,y} = \frac{\mathrm{cov}(x,y)}{Sd_x \cdot Sd_y} \tag{4}$$

where:

x—percentage values of elution (removal of mud cake) mean values (%);
y—value of the average adhesion of the cement sheath to the rock (MPa);
Sd_x, Sd_y—maximum likelihood estimators;
n—number of attempts.

$$\text{cov}(x,y) = \frac{\sum(x_i - \overline{x})(y_i - \overline{y})}{n} \quad (5)$$

$$Sd_x = \sqrt{\frac{\sum(x_i - \overline{x})^2}{n}} \quad (6)$$

$$Sd_y = \sqrt{\frac{\sum(y_i - \overline{y})^2}{n}} \quad (7)$$

$$WD = {r_{xy}}^2 \cdot 100\% \quad (8)$$

After substituting to formulas

$$\text{cov}(x,y) = \tfrac{4.6623}{5} = 0.930517$$

$$Sd_x = \sqrt{\tfrac{409.207}{5}} = 9.05$$

$$Sd_y = \sqrt{\tfrac{0.053}{5}} = 0.10$$

$$r_{x,y} = \tfrac{0.930517}{9.05 \cdot 0.10} = 0.997623$$

$$r = 0.99 \in \langle 0.9; 1.0 \rangle$$

$$WD = (0.997623)^2 \cdot 100\%$$

$$WD = 99.53\%$$

Table 3. Calculation data from the correlation of the mud cake removal efficiency to the adhesion of the cement sheath in contact with the rock formation.

n	x_i	y_i	$(x_i - \overline{x})$	$(y_i - \overline{y})$	$(x_i - \overline{x})(y_i - \overline{y})$	$(x_i - \overline{x})^2$	$(y_i - \overline{y})^2$
1	55.44	0.696	−17.2	−0.2	3.29	295.22	0.04
2	76.04	0.928	3.4	0.0	0.14	11.68	0.00
3	78.34	0.957	5.7	0.1	0.40	32.70	0.00
4	72.33	0.87	−0.3	0.0	0.01	0.09	0.00
5	80.96	0.986	8.3	0.1	0.82	69.52	0.01
	Σ = 363.11	Σ = 4.437			Σ = 4.653	Σ = 409.207	Σ = 0.053

x_i—readings of the percentage of elution (removal of mud cake) average values (%). y_i—value of the average adhesion of the cement sheath to the rock (MPa). Average values, $\overline{x} = 72.6$, $\overline{y} = 0.90$.

Based on the correlation analysis of the obtained results of the research, the percentage increase in the effectiveness of mud cake removal to the percentage increase in adhesion, a strong match of the analyzed features is found. The value of linear regression R2 is in the range of the total correlation (Figure 8). Correlation analysis confirms the strong correlation between the increase in adhesion and the effectiveness of the removal of mud cake from the surface of the rock formation. The confirmation of the dependence of the analyzed features obtained on the basis of the test results allows us to state how strong the adhesion growth will be depending on the increase in the efficiency of cleaning the annular space. Additionally, the reverse relationship, i.e., how the efficiency of cleaning the space will increase on the basis of the obtained results of adhesion of the hardened cement slurry to the rock formation.

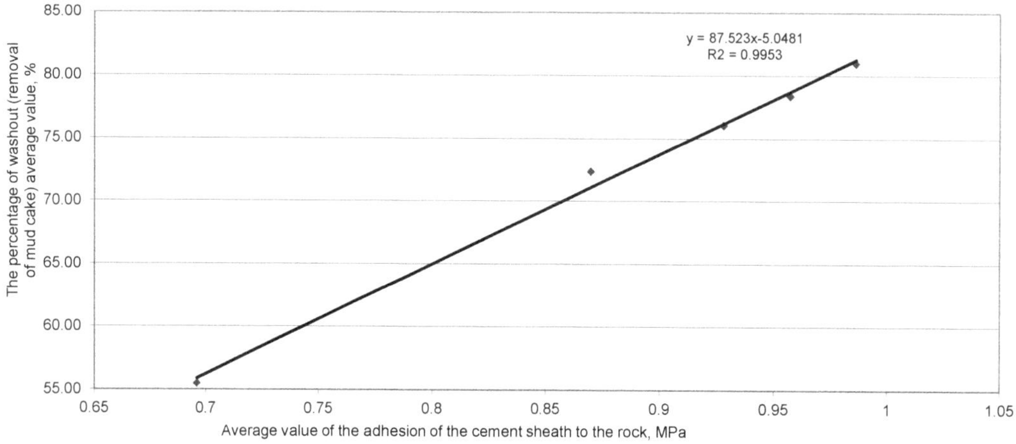

Figure 8. The relationship between the efficiency of mud cake removal and the adhesion of the cement sheath on contact with the rock formation.

4. Discussion

The performed tests confirmed the necessity to properly prepare the annular space for the cementing procedure by effectively removing the residues of the washing mud cake. It was found that one of the most important factors determining the tightness and durability of the cement sheath is the proper cleaning of the annular space from the mud cake. A number of other factors that determine the effectiveness of sealing should be taken into account. These include the parameters of the drilling mud used for drilling, the composition, the type and parameters of the cement slurry, and the geological and technical conditions. Only a comprehensive analysis of all factors contributes to the expected results of work on improving the effectiveness of cementing.

5. Conclusions

Based on the research on the effect of cleaning the annular space on the adhesion of the cement sheath to the formation, it is concluded that:

1. The use of water for washing the geological structure from the washing mud cake formed on its surface results in the removal of approx. 55% of the washing mud cake.
2. Spacers selected for testing remove the washing mud cake formed on the surface of the cores in the range from 72.3% to 81%.
3. The obtained values of cleaning the surface of the core represent an increase in removal from 30.5% to 46% compared to cleaning the surface of the cores with water, which is used as the standard spacer.
4. The adhesion of the hardened cement slurry formulated for the tests to the rock core washed with water is 0.7 MPa.
5. The use of spacer selected for the purpose of testing allows to obtain adhesion on the contact between the hardened cement slurry–rock formation in the range from 0.87 to 0.99 MPa.
6. The percentage increase in adhesion compared to the base value ranges from 25% to 42%.
7. The difference between the percentage increase in washing efficiency and the percentage increase in adhesion shall not exceed the value of 5.5%.
8. On the basis of the obtained test results and the correlation analysis carried out, a very strong relationship is found between the cleaning of the annular space from the slurry mud cake and the adhesion of the hardened slurry to the rock formation.

It should be borne in mind that despite the strong correlation between the cleaning of the annular space from the drilling mud cake and the adhesion of the cement sheath to the rock formation, a number of additional parameters determining the effectiveness of sealing the borehole should be taken into account. Therefore, each time before the cementing procedure is performed, specialized laboratory tests are carried out for specific geological and technological conditions.

Author Contributions: Data curation, M.K.; Formal analysis, M.K., S.B., S.S. and R.W.; Funding acquisition, M.K.; Investigation, M.K., S.S., R.W. and A.G.; Methodology, M.K. and S.S.; Project administration, S.B. and R.W.; Software, S.B., R.W. and A.G.; Validation, S.S., R.W. and A.G.; Visualization, S.B., S.S. and A.G.; Writing—original draft, M.K.; Writing—review & editing, S.S., R.W. and A.G. All authors have read and agreed to the published version of the manuscript.

Funding: The work was financially supported by Ministry of Science and Higher Education Warsaw (Internal order Oil and Gas Institute—National Research Institute Project No. 0015/KW/21).

Institutional Review Board Statement: Not applicable.

Informed Consent Statement: Not applicable.

Data Availability Statement: Not applicable.

Acknowledgments: The author thanks the anonymous reviewers for their constructive comments and the editor for handling the paper.

Conflicts of Interest: The author declares no conflict of interest.

Nomenclature

Symbol	Explanation
%	percent of sediment washout;
m_R	core mass before the test (without mud cake);
m_{Ros}	core mass with mud cake;
m_{Rpo}	mass of the core with the residue of the mud cake (after washing);
d	core diameter (m) for the conducted research is equal to 25 mm = 0.025 m;
h	the height of the cemented part of the core for the conducted research is 44 mm = 0.044 m;
x	percentage values of elution (removal of mud cake) mean values (%);
y	value of the average adhesion of the cement sheath to the rock (MPa);
Sd_x, Sd_y	maximum likelihood estimators;
n	number of attempts;
x_i	readings of the percentage of elution (removal of mud cake) average values (%);
y_i	value of the average adhesion of the cement sheath to the rock (MPa);
MPa	megapascal;
KN	kilonewton.

References

1. Kremieniewski, M. Cleaning of the casing string before cementation, based on research using a rotational viscometer. *Nafta-Gaz* **2018**, *74*, 676–683. [CrossRef]
2. Mitchell, R.F.; Miska, S. *Fundamentals of Drilling Engineering*; Spe Textbook Series; Society of Petroleum Engineers: DallaS, TX, USA, 2011; Volume 12, ISBN 978-1-55563-207-6. Available online: https://store.spe.org/Fundamentals-of-Drilling-Engineering-P468.aspx (accessed on 5 May 2021).
3. Kremieniewski, M.; Rzepka, M.; Stryczek, S.; Wiśniowski, R. Comparison of Gas Migration Test and Building Structural Strength of Slurry in the Aspect of Limiting Gas Invasion. *AGH Drill. Oil Gas* **2016**, *33*, 595–608. [CrossRef]
4. Kremieniewski, M.; Wiśniowski, R.; Stryczek, S.; Orłowicz, G. 2021 Possibilities of Limiting Migration of Natural Gas in Boreholes in the Context of Laboratory Studies. *Energies* **2021**, *14*, 4251. [CrossRef]
5. Kremieniewski, M.; Stryczek, S. Zastosowanie cementu wysokoglinowego do sporządzania zaczynów uszczelniających w technologiach wiertniczych. *Cem. Wapno Beton* **2019**, *3*, 215–226.
6. Sauer, C.W. *Mud Displacement During Cementing: A State of the Art*. JPT: 1987. Available online: https://onepetro.org/JPT/article-abstract/39/09/1091/73826/Mud-Displacement-During-Cementing-State-of-the-Art (accessed on 5 May 2021).

7. Nelson, E.B. *Well Cementing*; Schlumberger Educational Service: Houston, TX, USA, 1990. Available online: https://books.google.pl/books?hl=pl&lr=&id=1TRGZEnXOKMC&oi=fnd&pg=PP1&dq=7.%09Nelson,+E.B.+%5Bet+al.%5D,+Well+Cementing%3B+Schlumberger+Educational+Service:+Houston,+TX,+USA,+1990&ots=gnpCi0ZE-l&sig=ByhWpqGDzUc8vVRGGAuhQ7OvRWw&redir_esc=y#v=onepage&q&f=false (accessed on 9 August 2021).
8. Lavrov, A.; Torsæter, M. *Physics and Mechanics of Primary Well Cementing*; Springer International Publishing: Houston, TX, USA, 2016; ISBN 978-3-319-43165-9. Available online: https://www.springer.com/gp/book/9783319431642 (accessed on 9 August 2021).
9. Habrat, S.; Raczkowski, J.; Zawada, S. *Technika i Technologia Cementowań w Wiertnictwie*; Wydawnictwa Geologiczne: Warsaw, Poland, 1980.
10. McLean, R.H.; Manry, C.W.; Whitaker, W.W. Displacement Mechanics in Primary Cementing. *J. Pet. Technol.* **1967**, *19*, 251–260. [CrossRef]
11. Zhang, F.; Miska, S.; Yu, M.; Ozbayoglu, E.; Takach, N.; Osgouei, R.E. Is Well Clean Enough? A Fast Approach to Estimate Hole Cleaning for Directional Drilling. In Proceedings of the SPE/ICoTA Coiled Tubing & Well Intervention Conference & Exhibition, The Woodlands, TX, USA, 24 March 2015. [CrossRef]
12. Knez, D.; Gonet, A.; Macuda, J.; Stryczek, S. *Selected Issues of Wellbore Hydraulics and Cementing*; AGH University of Science and Technology Press: Cracow, Poland, 2017; ISBN 978-83-7464-885-1. Available online: https://www.wydawnictwoagh.pl/pliki/265960783.pdf (accessed on 9 August 2021).
13. Kremieniewski, M. Recipe of Lightweight Slurry with High Early Strength of the Resultant Cement Sheath. *Energies* **2020**, *13*, 1583. [CrossRef]
14. Bizhani, M.; Corredor, F.E.R.; Kuru, E. Quantitative Evaluation of Critical Conditions Required for Effective Hole Cleaning in Coiled-Tubing Drilling of Horizontal Wells. *SPE Drill. Compl.* **2016**, *31*, 188–199. [CrossRef]
15. Stryczek, S.; Wiśniowski, R.; Uliasz-Misiak, B.; Złotkowski, A.; Kotwica, Ł.; Rzepka, M.; Kremieniewski, M. Studia nad doborem zaczynów uszczelniających w warunkach wierceń w basenie pomorskim. *Wydaw. AHG Kraków* **2016**, 1–232. Available online: https://www.wydawnictwoagh.pl/STUDIA-NAD-DOBOREMZACZYNOW-USZCZELNIAJACYCHW-WARUNKACH-WIERCENW-BASENIE-POMORSKIM;s,karta,id,1172 (accessed on 9 August 2021).
16. Stryczek, S.; Wiśniowski, R.; Gonet, A.; Rzyczniak, M.; Sapińska-Śliwa, A. Wpływ wybranych superplastyfikatorów na właściwości reologiczne zaczynów cementowych stosowanych podczas cementowania kolumn rur okładzinowych w otworach wiertniczych. *Przemysł Chem.* **2018**, *97*, 903–905. [CrossRef]
17. Kremieniewski, M.; Rzepka, M. Przyczyny i skutki przepływu gazu w zacementowanej przestrzeni pierścieniowej otworu wiertniczego oraz metody zapobiegania temu zjawisku. *Nafta-Gaz* **2016**, *9*, 722–728. [CrossRef]
18. Kremieniewski, M. Ultra-Lightweight Cement Slurry to Seal Wellbore of Poor Wellbore Stability. *Energies* **2020**, *13*, 3124. [CrossRef]
19. Zima, G. Wpływ właściwości płuczek wiertniczych na jakość cementowania w gazonośnych poziomach miocenu. *Nafta-Gaz* **2014**, *12*, 899–907.
20. Kremieniewski, M. Korelacja skuteczności usuwania osadu za pomocą cieczy na osnowie jonowych (anionowych) i niejonowych SPCz. *Nafta-Gaz* **2019**, *3*, 38–48. [CrossRef]
21. Kremieniewski, M. Porównanie wyników metod pomiaru skuteczności usuwania osadu płuczkowego. *Nafta-Gaz* **2021**, *1*, 34–46. [CrossRef]
22. Jasiński, B. Ocena wpływu cieczy przemywającej na jakość zacementowania rur w otworze wiertniczym po użyciu płuczki glikolowo-potasowej. *Nafta-Gaz* **2016**, *6*, 413–421. [CrossRef]
23. Hirpa, M.M.; Arnipally, S.K.; Kuru, E. Effect of the Particle Size on the Near-Wall Turbulence Characteristics of the Polymer Fluid Flow and the Critical Velocity Required for Particle Removal from the Sand Bed Deposited in Horizontal Wells. *Energies* **2020**, *13*, 3172. [CrossRef]
24. Jamrozik, A.; Gonet, A.; Czekaj, L.; Fijał, J.; Ziaja, J.; Wiśniowski, R.; Stryczek, S. Salinity of drilling wastes as a factor determining their management. *Przemysł Chem.* **2017**, *96*, 972–976. [CrossRef]
25. Martin, M.; Latil, M. *Mud Displacement by Slurry During Primary Cementing Jobs—Predicting Optimum Conditions*; SPE: Houston, TX, USA, 1992.
26. Śliwa, T.; Stryczek, S.; Wysogląd, T.; Skakuj, A.; Wiśniowski RSapińska-Śliwa, A.; Bieda, A.; Kowalski, T. Wpływ grafitu i diatomitu na parametry wytrzymałościowe stwardniałych zaczynów cementowych. *Przemysł Chem.* **2017**, *96*, 960–963. [CrossRef]
27. Kremieniewski, M. Influence of Graphene Oxide on Rheological Parameters of Cement Slurries. *Energies* **2020**, *13*, 5441. [CrossRef]
28. Wiśniowski, R.; Skrzypaszek, K.; Małachowski, T. Selection of a Suitable Rheological Model for Drilling Fluid Using Applied Numerical Methods. *Energies* **2020**, *13*, 3192. [CrossRef]
29. Kremieniewski, M. Korelacja skuteczności działania środków dyspergujących o różnym mechanizmie upłynniania. *Nafta-Gaz* **2020**, *11*, 816–826. [CrossRef]
30. Adari, R.B.; Miska, S.; Kuru, E.; Bern, P.; Saasen, A. Selecting Drilling Fluid Properties and Flow Rates For Effective Hole Cleaning in High-Angle and Horizontal Wells. In Proceedings of the SPE Annual Technical Conference and Exhibition, Dallas, TX, USA, 1 October 2000. [CrossRef]
31. Kremieniewski, M. Hybrid Washer Fluid for Primary Cementing. *Energies* **2021**, *14*, 1295. [CrossRef]
32. Błaż, S. Nowe rodzaje cieczy przemywających osady z płuczki inwersyjnej przed zabiegiem cementowania otworów wiertniczych. *Nafta-Gaz* **2017**, *5*, 302–311. [CrossRef]

33. Peterson, G. Die Haftung von Tiefbohrzementen bei Ringraum-zementatnion. *Erdol Z.* **1962**, *1*, 339–367.
34. Stryczek, S.; Gonet, A. Kierunki ograniczania migracji gazu z przestrzeni pierścieniowej otworu wiertniczego. *WUG* **2005**, *3*, 10–13.
35. Kremieniewski, M.; Rzepka, M. Wpływ superplastyfikatora na hydrofilowość zaczynów cementowych. *Nafta-Gaz* **2018**, *10*, 745–751. [CrossRef]
36. Kremieniewski, M.; Wiśniowski, R.; Stryczek, S.; Łopata, P. Comparison of Efficient Ways of Mud Cake Removal from Casing Surface with Traditional and New Agents. *Energies* **2021**, *14*, 3653. [CrossRef]
37. Kremieniewski, M.; Rzepka, M.; Kędzierski, M. Influence of the Contact Time of the Preflush Fluid with the Filter Cake on the Effectiveness of the Purification of the Annular Space. *Nafta-Gaz* **2018**, *74*, 29–36. [CrossRef]
38. Li, L.; Alegria, A.; Doan, A.A.; Kellum, M.G.; Hughes, B. A Novel Spacer System to Prevent Lost Circulation in Cementing Applications. In Proceedings of the 2016 AADE Fluids Technical Conference and Exhibition, Houston, TX, USA, 12–13 April 2016. Available online: https://www.aade.org/application/files/2715/7131/8487/AADE-16-FTCE-13_-_Li.pdf (accessed on 11 May 2021).
39. Li, L.; Padilla, F.; Doan, A.A.; Kellum, M.G.; Hughes, B. Evaluation of a New Spacer System Mixed On-The-Fly. In Proceedings of the 2016 AADE Fluids Technical Conference and Exhibition, Houston, TX, USA, 2016.
40. Ryan, D.F.; Browne, S.V.; Burnham, M.P. Mud Clean-Up in Horizontal Wells: A Major Joint Industry Study. In Proceedings of the SPE Annual Technical Conference and Exhibition, Dallas, TX, USA, 1 October 1995. [CrossRef]
41. Zamora Mario Jefferson, D.T.; Powell, J.W. Hole-Cleaning Study of Polymer-Based Drilling Fluids. In Proceedings of the SPE Annual Technical Conference and Exhibition, Houston, TX, USA, 1 October 1993. [CrossRef]
42. Kremieniewski, M. Korelacja wyników badań wytrzymałości na ściskanie i przyczepności do rur stalowych płaszcza cementowego z zaczynu o obniżonej gęstości. *Nafta-Gaz* **2019**, *10*, 613–624. [CrossRef]
43. Sanchez, R.A.; Azar, J.J.; Bassal, A.A.; Martins, A.L. The Effect of Drillpipe Rotation on Hole Cleaning During Directional Well Drilling. In Proceedings of the SPE/IADC Drilling Conference, Amsterdam, The Netherlands, 1 March 1997. [CrossRef]
44. Gbadamosi, A.O.; Junin, R.; Oseh, J.O.; Agi, A.; Yekeen, N.; Abdalla, Y.; Ogiriki, S.O.; Yusuff, A.S. Improving Hole Cleaning Efficiency using Nanosilica in Water-Based Drilling Mud. In Proceedings of the SPE Nigeria Annual International Conference and Exhibition, Lagos, Nigeria, 1 August 2018. [CrossRef]
45. Kremieniewski, M.; Kędzierski, M. Research into the development of a hybrid spacer. *Nafta-Gaz* **2020**, *76*, 517–526. [CrossRef]
46. Xu, D.; Guo, J.; Yuan, B.; Wen, D.; Fang, X.; Li, H.; Ling, B. A Minimum Volume Prediction of Spacer Based on Turbulent Dispersion Theory: Model and Example. *Petroleum* **2019**, *5*, 397–401. [CrossRef]

energies

MDPI

Article

Use of Cluster Analysis to Group Organic Shale Gas Rocks by Hydrocarbon Generation Zones

Tadeusz Kwilosz *, Bogdan Filar and Mariusz Miziołek

Oil and Gas Institute-National Research Institute, 25A Lubicz Str., 31-503 Krakow, Poland; filar@inig.pl (B.F.); miziolek@inig.pl (M.M.)
* Correspondence: kwilosz@inig.pl

Abstract: In the last decade, exploration for unconventional hydrocarbon (shale gas) reservoirs has been carried out in Poland. The drilling of wells in prospective shale gas areas supplies numerous physicochemical measurements from rock and reservoir fluid samples. The objective of this paper is to present the method that has been developed for finding similarities between individual geological structures in terms of their hydrocarbon generation properties and hydrocarbon resources. The measurements and geochemical investigations of six wells located in the Ordovician, Silurian, and Cambrian formations of the Polish part of the East European Platform are used. Cluster analysis is used to compare and classify objects described by multiple attributes. The focus is on the issue of generating clusters that group samples within the gas, condensate, and oil windows. The vitrinite reflectance value (R_o) is adopted as the criterion for classifying individual samples into the respective windows. An additional issue was determining other characteristic geochemical properties of the samples classified into the selected clusters. Two variants of cluster analysis are applied—the furthest neighbor method and Ward's method—which resulted in 10 and 11 clusters, respectively. Particular attention was paid to the mean Ro values (within each cluster), allowing the classification of samples from a given cluster into one of the windows (gas, condensate, or oil). Using these methods, the samples were effectively classified into individual windows, and their percentage share within the Silurian, Ordovician, and Cambrian units is determined.

Keywords: unconventional resources; shale gas; oil gas; total organic carbon (TOC); cluster analysis; genetic type of kerogen

Citation: Kwilosz, T.; Filar, B.; Miziołek, M. Use of Cluster Analysis to Group Organic Shale Gas Rocks by Hydrocarbon Generation Zones. *Energies* **2022**, *15*, 1464. https://doi.org/10.3390/en15041464

Academic Editor: Nikolaos Koukouzas

Received: 3 January 2022
Accepted: 11 February 2022
Published: 17 February 2022

1. Introduction

In nature, there are no two identical natural gas and crude oil reservoirs. In spite of this, all natural gas reservoirs can be divided into two groups:

- conventional (traditional) reservoirs,
- unconventional reservoirs [1].

The classification of a reservoir as belonging to a specific group requires adopting selection criteria. Because the profitability of an investment depends largely on the income from gas sales, the primary selection criterion is the permeability of reservoir rock, since it directly affects the well output. In the petroleum industry, it has been assumed that reservoirs with permeability exceeding 0.1–0.5 mD are categorized as conventional reservoirs. The remaining reservoirs are in turn included in the group of unconventional reservoirs. The group of unconventional reservoirs can be divided into the following types of reservoirs:

- tight gas,
- shale gas,
- gas reservoirs in coal seams,
- reservoirs of gas trapped in hydrates [2].

In order to perform the challenges that are faced by researchers, it should be pointed out that the term: 'shale gas' is used very broadly in the world. Lithological diversity in reservoirs of the shale gas-type indicates that natural gas is present not just in shales, but also in a broad range of rocks with diverse lithology and texture, from siltstones to very fine-grained sandstones; each of them can have silicates or carbonates in its composition [3]. What is being generally referred to as shale is very often siltstone or fine-grained sandstone, or a heterogeneous type of rock, such assiltstone laminae interlayered with shale laminae or located in a shale "matrix" [4]. The presence of different types of rocks rich in organic matter indicates that there are various mechanisms of gas storage in these rocks [5]. Gas can be adsorbed in organic matter, and it can be present as free gas in micropores.

Gasiferous shales are characterized by:

- variable lithology, from 'pure' shales, to shales with siltstone insets,
- variable porosity, from relatively high, to low,
- variable TOC, from high to low,
- variable ratio of adsorbed to free gas, from high, to low values,
- variable quartz content and type,
- the rock can be solid or/and naturally fractured [6].

Conventional reservoirs, which are virtually the only ones currently undergoing extraction in Poland, are characterized by the following parameters:

- medium or small surface area,
- medium or small thickness,
- very good and good porosity of the reservoir rock,
- very good and good permeability of the reservoir rock,
- high and medium well outputs [7].

Very good physical and petrophysical parameters of conventional reservoirs allow for their relatively easy development and extraction. As a consequence, economic extraction of conventional reservoirs requires the use of traditional extraction technology, based on relatively cheap vertical boreholes. Unfortunately, natural gas resources from conventional reservoirs are close to depletion, since these reservoirs have been undergoing extraction for several decades. On the other hand, there are unconventional reservoirs, the parameters of which are radically different from that of conventional reservoirs. Unconventional reservoirs are characterized by the following:

- very large or large surface area,
- large and medium thickness,
- low porosity of the reservoir rock,
- very low permeability of the reservoir rock,
- low well outputs [8–11].

The weak parameters of the reservoir rock result in the infeasibility of extraction of this type of reservoirs in a traditional manner, i.e., using vertical or directional boreholes. This situation is caused primarily by the insufficiently high output of vertical wells. Because of this, the only method allowing for economic extraction of reservoirs of the shale and tight gas-type involves the use of the latest approaches, which are based on:

- horizontal well technology,
- slim hole technology,
- multi-section fracturing technology [12].

Considering that unconventional natural gas reservoirs of the shale and tight gas-type stand out due to their large variability of all reservoir parameters, it is very difficult to make the investment decisions that are necessary to access, and, therefore, extract unconventional reservoirs located in Poland. The making of investment decisions based on limited amounts of data is burdened with a considerable risk of failure [13]. Currently, only several dozen boreholes have been drilled in Poland in order to search for gasiferous shale formations. Because of this, there is little information allowing for proper estimation of both the resources and the investment possibilities. It should be pointed out that the existence of a very large number of wells is necessary for the performed estimations to have acceptable precision. On the other hand, there is an extensive amount of information that was acquired during the prospecting, drilling exploration and extraction of unconventional reservoirs of the shale and tight gas-type located in the USA and in Canada [14]. The ability to use the experience gained in the USA and in Canada would be very helpful as regards the implementation of investments related to the exploration, development, and extraction of gas from shale reservoirs in Poland, especially in the initial phase of searching, when there is no sufficient data. Unfortunately, it is difficult to directly benefit from this experience, since there are no two identical reservoirs of the shale gas-type in Poland, the USA, or Canada. Basically, one can only notice certain similarities, which can be helpful in the process of exploring, accessing, and extracting gas from shale formations.

An important emerging research challenge involves the development of methods and tools that would be helpful when analysing the research results originating from the explored geological structures in terms of seeking and assessing similarities to data originating from identified and documented geological structures. These tools, supplemented by the developed investment risk assessment methods, will be helpful in decision-making, whether in the stage of exploration, accessing or developing unconventional reservoirs. Vast possibilities related to the characterization of unconventional reservoirs are offered by artificial intelligence methods. Simple summaries and the comparison of data in tables, based on an intuitive quality assessment of the analyzed data, can be unreliable, and often impossible. This is caused by the necessity to perform an analysis involving a very large data set. The use of a more intuitive approach in a comparative analysis, without the application of quantitative measure and similarity criteria, can result in a failure to identify numerous significant similarities in the analyzed datasets.

Artificial intelligence (AI) can be considered as being a set of man-made analytical tools that imitate natural intelligent behavior [15]. AI techniques exhibit the ability to learn and cope in new situations. Artificial neural networks (ANN), programming based on evolution algorithms and fuzzy set logic function among models that have been categorized as AI. These techniques have one or more 'sentient' features, such as the possibility of learning, discovering, assembling, and abstraction. Over the last decade, as part of the development of artificial intelligence, sets of analytical tools were developed to facilitate the solving of problems that had previously been difficult or impossible to solve [1].

AI is already being used in multiple branches of economy. Currently, there is a general trend to integrate individual AI tools into more complex systems. In the petroleum industry, individual AI methods are used in various ways, among others, as:

- neural networks, which are applied to analyzea large amount of data on geology, geophysics, and extraction,
- genetic algorithms, which are used to analyze geological and petrophysical data, for reservoir simulations and to plan the fracturing procedures,
- fuzzy set logic, which is employed for petrophysical analyses, characterization of reservoir parameters, drilling exploration of reservoirs, planning the stimulation procedures, increasing the depletion ratio of the reservoirs, and for analyses supporting the making of investment decisions.

Computer systems utilizing AI support exploration, drilling, planning of production wells, the performance of stimulation procedures and the making of investment decisions. They are being applied in the petroleum industry with an increasing frequency, since they allow for reducing the operating risk of companies.

The area of research is located in the N part of Poland and includes Lower Paleozoic formations occurring in the area of Peribaltic syneclysis. Peribaltic syneclysis is a subpermian unit bordered on the south by the Masurian-Belarusian elevation. In the substrate of Peribaltic syneclysis, there are crystalline Precambrian rocks, covered by sedimentary and meta-sedimentary Precambrian-Paleozoic rocks. These rocks consist of formations of the Vendian, Cambrian, Ordovician, locally Ordovician, and Permian. The subject of research included: Cambrian sediments developed as detrital, sandstone and siltstone formations, Ordovician sediments developed as carbonate and loamy formations and Silurian sediments developed as graptolites. Above the Paleozoic rocks, there are formations belonging to other structural and tectonic units: Triassic, Jurassic, Cretaceous, Tertiary, and Quaternary [16].

The main task is to assess the potential of unconventional hydrocarbon resources in selected geological structures in Poland. The assessment of the potential of hydrocarbon resources is based on total organic carbon (TOC) data, the thermal maturity of the organic matter and the genetic type of kerogen [17,18]. The carried-out studies show that shale type of geological structures are characterized by high sedimentological variability and high variability of the physicochemical properties of the organic matter [19–22]. Geochemical and petrophysical studies analyzed by other authors have shown that the upper parts of the Silurian are characterized by a low level of the thermal maturity [23–25]. In turn, high maturity for the generation of hydrocarbons level is achieved by the lower parts of the Silurian, Ordovician, and Cambrian deposits [26–28]. The results presented in this article, obtained through methods using artificial intelligence, may be a valuable supplement to these results [29].

The research method, cluster analysis, can be used to search for analogies between the reservoirs under study and other geological structures with unconventional natural gas deposits which are known and well documented in terms of measurements and research results. This article not only attempts to isolate clusters that group samples with similar geochemical parameters, but it also focuses on identifying clusters that are characteristic of particular windows featuring the thermal maturity of kerogen, namely gas, condensate, and oil windows. The method of generating clusters, in the multidimensional space of geochemical parameters to describe rocks, has never been combined with the method of assigning them to windows (condensate, oil, and gas) in studies on this topic. The use of cluster analysis to describe reservoirs has a long history [30–32]. This method has been used to determine the permeability of rocks based on analogies with other petrophysical parameters (e.g., porosity) of rocks with well-recognized geological structures [33] or the separation of electrofacies based on clusters grouped by similar petrophysical parameters [34,35]. Interesting results related to the identification of gas, condensate, and oil windows, corresponding to the methods presented in this article, have already been presented by other authors [36–38]. Cluster analysis methods have also been applied to investigate the similarity of samples described with geochemical parameters. One of the key issues presented in these works was the identification of rocks with similar total organic carbon (TOC) values [39–41]. The results demonstrated that it is possible to group the samples assigned to gas, condensate, and oil windows through cluster analysis. Additionally, information was obtained on the remaining geomechanical properties of the samples assigned to the selected windows.

2. Materials and Methods

2.1. Materials

The generation potential of rocks can be determined using an appropriately selected set of physicochemical parameters. Geochemical data were selected at the laboratory stage. The results of measurements taken on cores from six wells of Silurian, Ordovician, and Cambrian units were selected: Gd-1 (36 samples), Go-1 (13 samples), Ke1 (7 samples), Ma-1 (10 samples), Ol-1 (8 samples), and Pr-1 (12 samples).

The test samples are described by the set of seven parameters: Total organic carbon TOC (% by weight); the temperature at which the maximum quantity of hydrocarbons is produced during kerogen cracking T_{max}(°C); free hydrocarbon content S_1(mg HC/g of rock); the amount of hydrocarbons released during kerogen cracking S_2(mg HC/g of rock); hydrogen index HI (mg HC/g TOC); oxygen index OI (mg CO_2/g TOC); and vitrinite reflectance R_o(%).The parameters listed are from the Rock-Eval pyrolysis analysis with the exception of R_o.

2.2. Methods

A wide group of data segmentation methods is executed by dividing a set of data (observations, test results) into subsets (classes) containing 'similar' elements (according to a predetermined similarity principle). This task is being implemented via identification of natural groups that contain elements similar in terms of a predetermined measure of similarity. Mutually similar objects are placed in one group, while objects that vary significantly are in different groups. The number of groups created in this manner is not known *a priori*, which distinguishes this method from standard classification, in which objects are assigned to groups with predetermined properties.

The choice of an efficient data segmentation method has been made in order to answer the question whether the examined geological structures are similar in terms of generation properties and hydrocarbon resources. An assumption has been made that the studied geological structures can be considered as analogues, if the samples representing them (described by combining the results of various types of measurements, but affecting the generation properties and hydrocarbon resources) are put in the same groups (subsets) generated using the data segmentation method.

The group of methods referred to by their collective name 'cluster analysis' is one of the most efficient data segmentation methods. These methods are characterized by the ability to compare and categorize objects described by means of multiple attributes. Depending on detailed solutions, these procedures allow for the creation of groups (clusters) of objects which are 'least distant from each other' or 'most similar to each other'. These objects are considered to be points in a multidimensional space, wherein the dimension of the space is determined by the number of variables describing the given objects.

The following types of cluster analysis can be distinguished based on the applied data processing methods:

- optimizing-iterative, involving the division of a set of objects into a specified number of k subsets, following one of the optimising criteria:
  - K-means—the groups are represented by a 'center of gravity'.
  - K-medoids—the groups are represented by one of the objects.
- hierarchical, under which clusters of a higher level contain clusters of a lower level. Hierarchical methods include agglomerative and divisive techniques.

In agglomerative techniques, the starting point consists of individual objects, each of which constitutes a separate group—a single-element cluster. The objects are combined into more numerous groups, until a single group is developed in the end that includes all objects. The divisive techniques initially assume that the entire set of objects constitutes a single group. This group is subsequently divided into an increasing number of groups, until single-element groups are generated.

Hierarchical agglomerate grouping is considered to be one of the more efficient object-grouping methods. In this method, new clusters are formed by merging existing clusters. The condition of merging the clusters is their adequate distance.

The grouping algorithm involves:

1. selecting the initial set of clusters,
2. finding the closest pair of clusters and merging them into one,
3. repeating step 2 until fulfilling the rule of completion.

The rule of completion is (usually):

- the lack of cluster pairs located less than a given threshold distance apart (d_{max}),
- merging of all clusters into a single set.

A properly defined measure of length is applied in order to determine the mutual distance between individual objects in a multidimensional space. The distance may be defined in multiple ways, depending on the type of attributes describing the individual data (quantitative, qualitative data, ranks). Among the most frequently employed are:

- Euclidean distance described by Formula (1):

$$d_{ik} = \sqrt{\sum_{j=1}^{m} (x'_{ij} - x'_{kj})^2} \tag{1}$$

- City block (Manhattan) distance described by Formula (2):

$$d_{ik} = \sum_{j=1}^{m} \left| x'_{ij} - x'_{kj} \right| \tag{2}$$

- Chebyshev distance described by Formula (3) [42]:

$$d_{ik} = max_j \left| x'_{ij} - x'_{kj} \right| \tag{3}$$

The individual variants of cluster analysis differ in the manner of determining the distance between clusters.

1. The nearest neighbor method (single linkage)—the distance between clusters is the distance between the two closest objects.
2. The farthest neighbor method (complete linkage)—the distance between clusters is the distance between the two most distant objects.
3. The median method—the distance between two clusters is the median of the distance between the units of the first and the second cluster.
4. The group average method—the distance between two clusters is the average distance between the units of the first and the second cluster.
5. The center of gravity method—the distance between two clusters is the distance between the centres of gravity of the first and the second cluster.
6. The Ward method—sampling the merging of all cluster pairs and selecting such merging in which the variance of distance inside a formed cluster is the smallest.

Cluster analysis was chosen to study the similarities of rocks in terms of their geochemical properties. The choice of this method was dictated by its high efficiency in solving problems related to the study of the similarities of objects described in multidimensional state spaces and the simplicity of applying and controlling the results [43,44]. The nominal values of the various parameters describing the objects may vary by orders of magnitude. To avoid negative numerical effects during the calculations, the data were standardized using Formula (4):

$$x'_i = \frac{x_i - \overline{x}}{s_x} \tag{4}$$

where x_i is the attribute value, is the mean, and s_x is the standard deviation from the mean for the data series describing the attribute. The clustering method involves grouping objects into clusters. The criterion for combining objects into clusters is that the distance between the objects does not exceed the predefined d_{max} value. From the various distance measures, the Euclidean distance described by Formula (1) was selected,

$$d_{ik} = \sqrt{\frac{1}{m}\sum_{j=1}^{m} (x'_{ij} - x'_{kj})^2} \tag{5}$$

where m is the size of the space (number of attributes) and x'_{ij} and x'_{kj} are the corresponding attribute values (j) for the objects (i and k). This method has many variants that differ in the way objects are selected for clustering. For this study, the authors selected the furthest neighbor method, which defines the distance between two clusters as the distance between the two furthest members in the clusters, and Ward's method, which groups objects so that the within-cluster distance variance is the lowest while meeting the condition of the maximum distance of objects in one cluster. During numerical experiments on clustering, the maximum distances between clusters were selected using multiples of the standard deviation calculated from a series of all the distances of the test samples. The algorithm was implemented for distances within one, two, or three standard deviations.

3. Results and Discussion

Datasets describing the analyzed samples were prepared and the data were standardized. The following results were obtained from numerical experiments based on the application of cluster analysis. The best results, in the case of the furthest neighbor method, were obtained for a maximum distance d_{max} of 3.2 (two standard deviations from the standardized distance series) and with Ward's method for a d_{max} of 4.8 (on the order of three standard deviations from the standardized distance series).

After many tests, it was found that the best results, due to the optimal number of clusters (10–20), were obtained for a maximum distance d_{max} of 3.2 (two standard deviations from the set of standardized distances) for the furthest neighbor method and a d_{max} of 4.8 (on the order of three standard deviations for the set of standardized distances) in the case of Ward's method. In this analysis variant, particular attention was paid to the mean R_o values (within each cluster), allowing the classification of samples from a given cluster into one of the windows (gas, condensate, or oil). The following assumption was made: $0.6 \leq R_o \leq 0.8\%$ is classified as an oil window, $0.8\% < R_o \leq 1.25\%$ as a condensate window, and $1.25 < R_o \leq 2.4\%$ as a gas window. Intervals < 0.6% = immature window and > 2.4% = overmature deposits [45]. To estimate the uncertainty of classifying samples from a given cluster into the correct window, standard deviations (S) were calculated for the mean values of R_o. It was assumed that if the values of $R_o - S$ and $R_o + S$ are within the limits of a given window, it can be assumed that the samples belonging to this cluster belong to this window with a probability of 68.2%. In addition, the percentage of samples in a given cluster within each unit (Silurian, Ordovician, and Cambrian) was determined. The analysis of the results indicated that the samples in particular clusters (based on similarities in all analyzed parameters), have a high probability of belonging to particular hydrocarbon generation windows. The results obtained are presented in Tables 1 and 2, in which the mean values of the parameters under study are given.

Table 1. Results of furthest neighbor cluster analysis.

Cluster No.	Number of Elements	TOC	T_{max}	S_1	S_2	HI	OI	R_o	Cluster Characteristics
1	14	0.25	432	0.16	2.01	169.85	101.25	0.77	R_o = 0.77% − oil window; R_o + S = 0.91% − part in the condensate window. Silurian = 57%; Ordovician = 7%; Cambrian = 36%
2	12	1.31	449	1.22	1.43	22.02	13.04	1.11	R_o = 1.11%; R_o − S = 0.95% − condensate window; R_o + S = 1.27% − part in the gas window. Silurian = 83%; Ordovician = 17%; Cambrian = 0%
3	23	0.81	446	0.40	0.58	50.08	41.80	1.25	R_o = 1.25%; R_o − S = 1.13% − condensate window; Ro + S = 1.36% − part in the gas window. Silurian = 83%; Ordovician = 13%; Cambrian = 4%
4	5	6.29	456	2.27	5.20	2.96	2.96	1.30	R_o = 1.30%; R_o − S = 1.28% − gas window. Silurian = 80%; Ordovician = 20%; Cambrian = 0%
5	4	0.15	493	0.05	0.50	100.05	133.97	1.44	R_o = 1.44%; R_o − S = 1.33% − gas window. Silurian = 0%; Ordovician = 0%; Cambrian = 100%

Table 1. *Cont.*

Cluster No.	Number of Elements	TOC	T_{max}	S_1	S_2	HI	OI	R_o	Cluster Characteristics
6	7	0.46	425	1.59	28.05	351.17	71.98	0.71	R_o = 0.71%; R_o + S = 0.78% – oil window. Silurian = 0%; Ordovician = 57%; Cambrian = 43%
8	4	0.13	437	0.30	4.50	339.66	275.61	0.77	R_o = 0.77% – oil window; R_o + S = 0.85% – part in the condensate window. Silurian = 25%; Ordovician = 0%; Cambrian = 75%
	Mean	1.34	448	0.86	6.04	147.97	91.52	1.05	

Table 2. Results of Ward's method cluster analysis.

Cluster No.	Number of Elements	TOC	T_{max}	S_1	S_2	HI	OI	R_o	Cluster Characteristics
1	10	0.24	433	0.07	0.47	171.04	105.88	0.72	R_o = 0.72%; R_o + S = 0.79% – oil window. Silurian = 60%; Ordovician = 10%; Cambrian = 30%
2	14	1.28	448	1.14	1.35	32.17	13.24	1.11	R_o = 1.11%; R_o – S = 0.96%; R_o + S = 1.25% – condensate window. Silurian = 86%; Ordovician = 14%; Cambrian = 0%
3	11	0.59	435	0.34	0.46	77.96	81.40	1.20	R_o = 1.20%; R_o – S = 1.07% – condensate window; R_o + S = 1.32% – part in the gas window. Silurian = 73%; Ordovician = 18%; Cambrian = 9%
4	12	0.90	454	0.36	0.62	26.99	28.16	1.29	R_o = 1.29% – gas window; R_o – S = 1.20% – part in the condensate window. Silurian = 92%; Ordovician = 8%; Cambrian = 8%
5	5	6.29	456	2.27	5.20	2.96	2.96	1.30	R_o = 1.30%; R_o – S = 1.28% – gas window. Silurian = 80%; Ordovician = 20%; Cambrian = 0%
6	4	0.15	493	0.05	0.50	100.05	133.97	1.44	R_o = 1.44%; R_o – S = 1.33% – gas window. Silurian = 0%; Ordovician = 0%; Cambrian = 100%
7	8	0.13	422	1.08	19.90	278.67	28.25	0.77	R_o = 0.77%; R_o + S = 0.78% – oil window. Silurian = 0%; Ordovician = 0%; Cambrian = 100%
8	4	0.69	427	1.76	29.84	381.63	105.16	0.67	R_o = 0.67%; R_o + S = 0.73% – oil window. Silurian = 0%; Ordovician = 100%; Cambrian = 0%
10	4	0.13	437	0.30	4.50	339.66	275.61	0.77	R_o = 0.77% – oil window; R_o + S = 0.85% – part in the condensate window. Silurian = 25%; Ordovician = 0%; Cambrian = 75%
11	4	0.21	417	0.03	0.06	137. 06	287.87	1.33	R_o = 1.33% – gas window; R_o – S = 1.02% – part in the condensate window. Silurian = 50%; Ordovician = 0%; Cambrian = 50%
	Mean	1.06	442	0.74	6.29	154.82	106.25	1.06	

The results illustrating the proportion of samples in each of the generated clusters are presented in Figures 1–6.

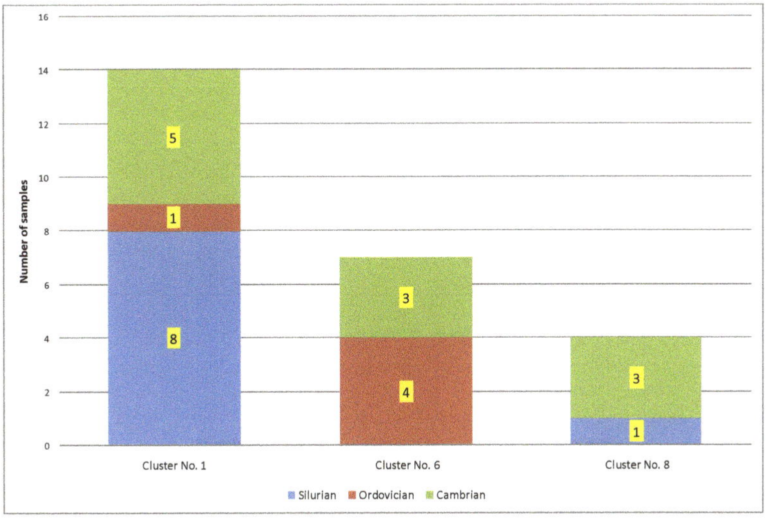

Figure 1. Results of furthest neighbor cluster analysis. Oil window.

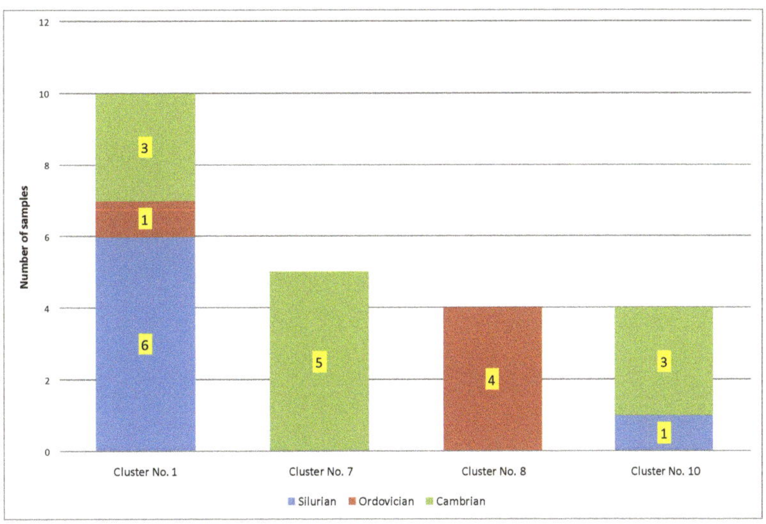

Figure 2. Results of Ward's method cluster analysis. Oil window.

Samples qualified for the oil window were grouped in 3 clusters in the farthest neighbor method (1, 6, 8) and in 4 clusters in Ward's cluster method (1, 7, 8, 10). Some of these clusters contain samples classified for both oil and condensate windows. Only cluster 1 subjected to the farthest neighbor method and clusters 1 and 8 to Ward's method contain samples solely from the oil window. The samples presented in these clusters are characterized by low values for TOC and T_{max} and high values for S_2, HI, and OI. Compared to the other windows (condensate and gas), the clusters generated using both methods contain the largest number of samples from Cambrian formations. Silurian formations are the least represented.

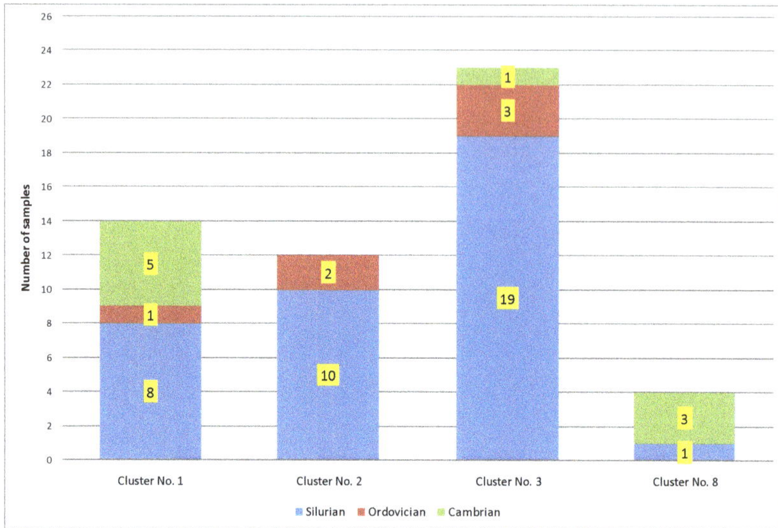

Figure 3. Results of furthest neighbor cluster analysis. Condensate window.

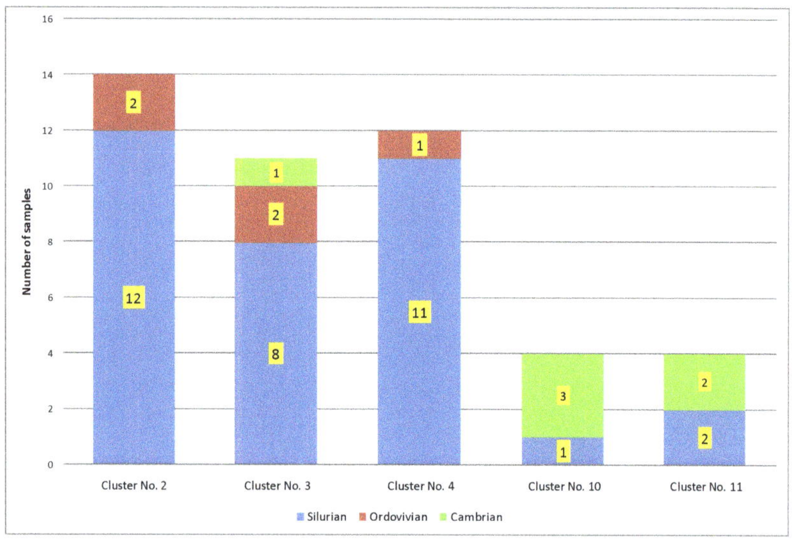

Figure 4. Results of Ward's method cluster analysis. Condensate window.

Samples qualified for the condensate window were grouped in four clusters in the farthest neighbor method (1, 2, 3, 8) and in five clusters in Ward's cluster method (1, 2, 3, 4, 10, 11). Some of these clusters contain samples classified for both oil, condensate, and gas windows. Only cluster 2 subjected to the Ward's method contain samples solely from the condensate window. The samples presented in these clusters are characterized by low values for S_2, OI, and HI and high values for S_1. Compared to the other windows (oil and gas), the clusters generated using both methods contain the largest number of samples from Silurian formations.

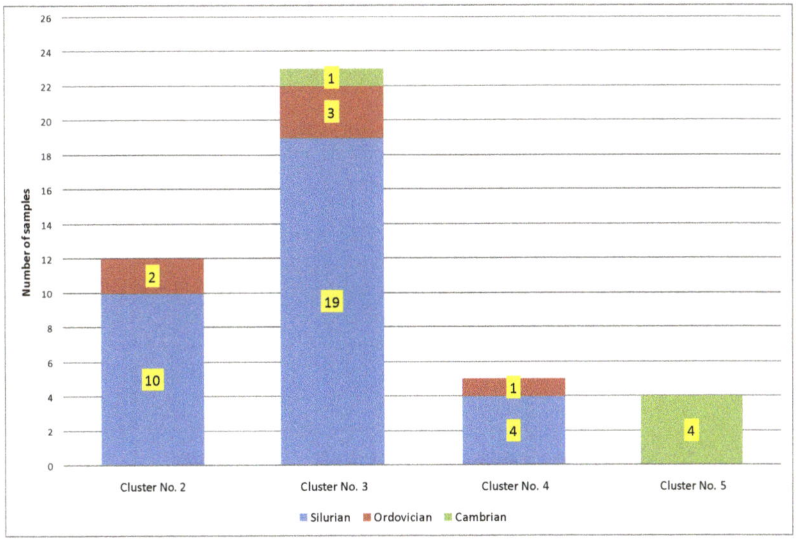

Figure 5. Results of furthest neighbor cluster analysis. Gas window.

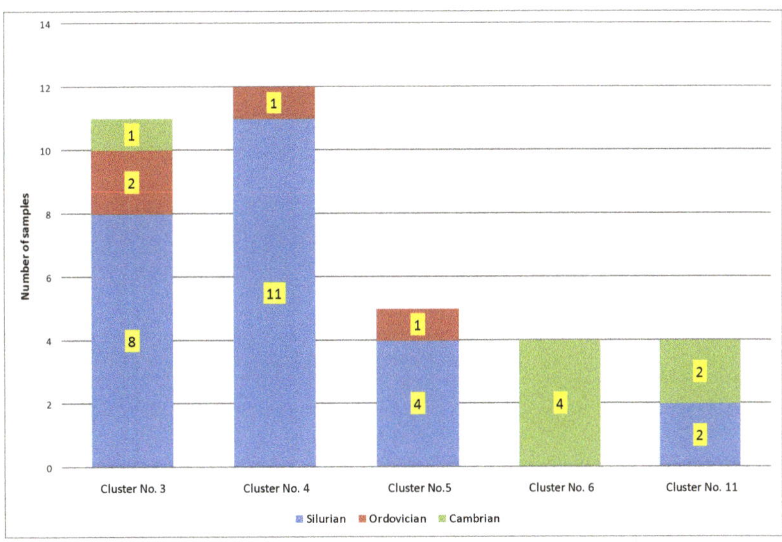

Figure 6. Results of Ward's method cluster analysis. Gas window.

Samples qualified for the gas window were grouped in 4 clusters in the farthest neighbor method (2, 3, 4, 5) and in 5 clusters in Ward's cluster method (3, 4, 5, 6, 11). Some of these clusters contain samples classified for both condensate and gas windows. Only clusters 4 and 5 subjected to the farthest neighbor method and clusters 5 and 6 to Ward's method contain samples solely from the gas window. The samples presented in these clusters are characterised by low values for HI and high values for T_{max}. The clusters generated using both methods contain the largest number of samples from Silurian formations.

4. Conclusions

The cluster analysis method can be an effective tool for grouping samples characterized by geochemical parameters obtained using the Rock-Eval method supplemented by the results of microscopic studies of vitrinite reflectivity. The use of cluster analysis not only allowed the samples to be classified into clusters specific to each window (oil, condensate, and gas) but also allowed the authors of the research to assign characteristic geochemical parameter values to them. Not in all cases was it possible to obtain unambiguous results. This applies particularly to the average TOC and HI values for the gas window where in two different clusters containing samples from this window extremely different average values of the mentioned parameters were obtained. This can be explained by too small a random sample used in the study. An important factor in achieving acceptable results is the adequate calibration of the parameters for the methods used carried out through multiple numerical experiments. One such parameter is to determine optimal (in terms of the effective grouping process)values of maximum distances between objects (in this case samples) that must be maintained in order for the samples to be qualified to the same window. The method used is complementary to the work of other scientists involved in research on the classification of rock samples based on kerogen maturity [16,45–48].

Author Contributions: Conceptualization, T.K., B.F. and M.M.; Formal analysis, T.K. and B.F.; Investigation, M.M.; Methodology, T.K. and B.F.; Project administration, B.F.; Resources, B.F.; Software, T.K.; Supervision, T.K. and B.F.; Validation, T.K. and B.F.; Visualization, M.M.; Writing—original draft, T.K., B.F. and M.M.; Writing—review & editing, B.F. and M.M. All authors have read and agreed to the published version of the manuscript.

Funding: This article is the result of research conducted in connection with the project: Selection of optimal methods for estimation of resources and (geological and commercial) risk at prospecting stage in relation to unconventional 'shale gas', 'shale oil', and 'tight gas' deposits in Poland, and development of methods for documentation of unconventional deposits, co-funded by the National Center for Research and Development as part of the programme BLUE GAS—POLISH SHALE GAS. Contract No. BG1/ŁUPZAS/13.

Conflicts of Interest: The authors declare no conflict of interest.

Nomenclature

TOC	total organic carbon, % by weight
T_{max}	temperature at which the maximum quantity of hydrocarbons is produced during kerogen cracking, °C
S_1	free hydrocarbon content, mg HC/g of rock
S_2	amount of hydrocarbons released during kerogen cracking, mg HC/g of rock
HI	hydrogen index, mg HC/g TOC
OI	oxygen index, mg CO_2/g TOC
R_o	vitrinite reflectance, %
S	standard deviations were calculated for the mean values of R_o, %
d_{max}	maximum distance between the objects described by the standardized data

References

1. Mandal, P.P.; Rezaee, R.; Emelyanova, I. Ensemble Learning for Predicting TOC from Well-Logs of the Unconventional Goldwyer Shale. *Energies* **2021**, *15*, 216. [CrossRef]
2. Holditch, S.A. Tight Gas Sands. *J. Pet. Technol.* **2006**, *58*, 86–93. [CrossRef]
3. Nie, H.; Jin, Z.; Zhang, J. Characteristics of three organic matter pore types in the Wufeng-Longmaxi Shale of the Sichuan Basin, Southwest China. *Sci. Rep.* **2018**, *8*, 7014. [CrossRef] [PubMed]
4. Josh, M.; Esteban, L.; Piane, C.D.; Sarout, J.; Dewhurst, D.; Clennell, M. Laboratory characterisation of shale properties. *J. Pet. Sci. Eng.* **2012**, *88–89*, 107–124. [CrossRef]
5. Ma, L.; Slater, T.; Dowey, P.J.; Yue, S.; Rutter, E.; Taylor, K.G.; Lee, P.D. Hierarchical integration of porosity in shales. *Sci. Rep.* **2018**, *8*, 11683. [CrossRef]
6. Boswell, R.; Collett, T.S. Current perspectives on gas hydrate resources. *Energy Environ. Sci.* **2011**, *4*, 1206–1215. [CrossRef]

7. Piesik-Buś, B.; Filar, B. Analysis of the current state of natural gas resources in domestic deposits and a forecast of domestic gas production until 2030. *Nafta-Gaz* **2016**, *6*, 376–382. [CrossRef]
8. Song, Y.; Li, Z.; Jiang, L.; Hong, F. The concept and the accumulation characteristics of unconventional hydrocarbon resources. *Pet. Sci.* **2015**, *12*, 563–572. [CrossRef]
9. Das, B.; Chatterjee, R. Mapping of porepressure, in-situ stress and brittleness in unconventional shale reservoir of Krishna-Godavari basin. *J. Nat. Gas Sci. Eng.* **2018**, *50*, 74–89. [CrossRef]
10. Piane, C.D.; Almqvist, B.S.; MacRae, C.; Torpy, A.; Mory, A.J.; Dewhurst, D. Texture and diagenesis of Ordovician shale from the Canning Basin, Western Australia: Implications for elastic anisotropy and geomechanical properties. *Mar. Pet. Geol.* **2015**, *59*, 56–71. [CrossRef]
11. Rezaee, R. (Ed.) . *Fundamentals of Gas Shale Reservoirs*; Wiley: Hoboken, NJ, USA, 2015.
12. Xie, J. Rapid shale gas development accelerated by the progress in key technologies: A case study of the Changning–Weiyuan National Shale Gas Demonstration Zone. *Nat. GasInd. B* **2018**, *5*, 283–292. [CrossRef]
13. Adamus, W.; Florkowski, W.J. The evolution of shale gas development and energy security in Poland: Presenting a hierarchical choice of priorities. *Energy Res. Soc. Sci.* **2016**, *20*, 168–178. [CrossRef]
14. Lozano-Maya, J.R. Looking through the prism of shale gas development: Towards a holistic framework for analysis. *Energy Res. Soc. Sci.* **2016**, *20*, 63–72. [CrossRef]
15. Syed, F.I.; Muther, T.; Dahaghi, A.K.; Negahban, S. AI/ML assisted shale gas production performance evaluation. *J. Pet. Explor. Prod. Technol.* **2021**, *11*, 3509–3519. [CrossRef]
16. Sowiżdżał, K.; Słoczyński, T.; Stadtműller, M.; Kaczmarczyk, W. Lower Palaeozoic petroleum systems of the Baltic Basin in northern Poland: A 3D basin modeling study of selected areas (onshore and offshore). *Interpretation* **2018**, *6*, SH117–SH132. [CrossRef]
17. Poprawa, P. Lower Paleozoic oil and gas shale in the Baltic-Podlasie-Lublin Basin (central and eastern Europe)—Areview. *Geol. Q.* **2020**, *64*, 515. [CrossRef]
18. Mroczkowska-Szerszeń, M.; Ziemianin, K.; Brzuszek, P.; Matyasik, I.; Jankowski, L. The organic matter type in the shale rock samples assessed by FTIR-ART analyses. *Nafta-Gaz* **2015**, *6*, 361–369.
19. Botor, D. Hydrocarbon generation in the Upper Cambrian—Lower Silurian source rocks of the Baltic Basin (Poland), implications for shale gas exploration. In Proceedings of the 16-th International Scientific GeoConference SGEM, Vienna, Austria, 2–5 November 2016; Book1, Oil and Gas Section. pp. 127–134. [CrossRef]
20. Botor, D.; Kotarba, M.; Kosakowski, P. Petroleum generation in the Carboniferous strata of the Lublin Trough (Poland): An integrated geochemical and numerical modelling approach. *Org. Geochem.* **2002**, *33*, 461–476. [CrossRef]
21. Kosakowski, P.; Wróbel, M.; Poprawa, P. Hydrocarbon generation and expulsion modelling of the lower Paleozoic source rocks in the Polish part of the Baltic region. *Geol. Q.* **2010**, *54*, 241–256.
22. Jarvie, D.M.; Hill, R.J.; Ruble, T.E.; Pollastro, R.M. Unconventional shale-gas systems: The Mississippian Barnett Shale of north-central Texas as one model for thermogenic shale-gas assessment. *AAPG Bull.* **2007**, *91*, 475–499. [CrossRef]
23. Nehring-Lefeld, M.; Modliński, Z.; Swadowska, E. Thermal evolution of the Ordovician in the western margin of the East-European Platform: CAI and RO data. *Geol. Q.* **1997**, *41*, 129–137.
24. Grotek, I. Origin and thermal maturity of the organic matter in the Lower Paleozoic rocks of the Pomerania Caledonides and their foreland (N Poland). *Geol. Q.* **1999**, *43*, 297–312. [CrossRef]
25. Swadowska, E.; Sikorska, M. Burial history of Cambrian constrained by vitrinite-like macerals in Polish part of the East European Platform. *Przegląd Geol.* **1998**, *46*, 699–706.
26. Zdanaviciute, O. Perspectives of oil field exploration in Middle Cambrian sandstones of Western Lithuania. *Geologija* **2005**, *51*, 10–18.
27. Więcław, D.; Kotarba, M.J.; Kosakowski, P.; Kowalski, A.; Grotek, I. Habitat and hydrocarbon potential of the Lower Palaeozoic source rocks of the Polish part of the Baltic region. *Geol. Q.* **2010**, *54*, 159–182.
28. Więcław, D.; Kosakowski, P.; Kotarba, M.J.; Koltun, Y.V.; Kowalski, A. Assessment of hydrocarbon potential of the Lower Palaeozoic strata in the Tarnogród–Stryi area (SE Poland and western Ukraine). *Ann. Soc. Geol. Pol.* **2012**, *82*, 65–80.
29. Wang, P.; Peng, S. A New Scheme to Improve the Performance of Artificial Intelligence Techniques for Estimating Total Organic Carbon from Well Logs. *Energies* **2018**, *11*, 747. [CrossRef]
30. Klaja, J.; Łykowska, G. Wyznaczenie typów petrofizycznych skał czerwonego spągowca z rejonu południowo-zachodniej części niecki poznańskiej na podstawie analizy statycznej wyników pomiarów laboratoryjnych. *Nafta-Gaz* **2014**, *11*, 757–764.
31. Puskarczyk, E. Application of Multivariate Statistical Methods and Artificial Neural Network for Facies Analysis from Well Logs Data: An Example of Miocene Deposits. *Energies* **2020**, *13*, 1548. [CrossRef]
32. Radzikowski, K.; Nowak, R.; Arabas, J.; Budak, P.; Łętkowski, P. Classification of Polish shale gas boreholes using measurement data. In Proceedings of the XXXVIII-th IEEE-SPIE Joint Symposium on Photonics, Web Engineering, Electronics for Astronomy and High Energy Physics Experiments, Wilga, Poland, 30 May–6 June 2016; Volume 10031. [CrossRef]
33. Khoshbakht, F.; Mohammadnia, M. Assessment of Clustering Methods for Predicting Permeability in a Heterogeneous Carbonate Reservoir. *J. Pet. Sci. Technol.* **2012**, *2*, 50–57.
34. Abdideh, M.; Ameri, A. Cluster An alysis of Petrophysical and Geological Parameters for Separating the Electrofacies of a Gas Carbonate Reservoir Sequence. *Nonrenewable Resour.* **2020**, *29*, 1843–1856. [CrossRef]

35. Mahmoud, A.A.; Elkatatny, S.; Ali, A.Z.; Abouelresh, M.; Abdulraheem, A. Evaluation of the Total Organic Carbon (TOC) Using Different Artificial Intelligence Techniques. *Sustainability* **2019**, *11*, 5643. [CrossRef]
36. Torghabeh, A.K.; Rezaee, R.; Harami, R.M.; Pimentel, N. Using electrofacies cluster analysis to evaluate shale-gas potential: Carynginia Formation, Perth Basin, Western Australia. *Int. J. Oil Gas Coal Technol.* **2015**, *10*, 250. [CrossRef]
37. Farzi, R.; Bolandi, V. Estimation of organic facies using ensemble methods in comparison with conventional intelligent approaches: A case study of the South Pars Gas Field, Persian Gulf, Iran. *Model. Earth Syst. Environ.* **2016**, *2*, 1–13. [CrossRef]
38. Baiyegunhi, T.L.; Liu, K.; Gwavava, O.; Wagner, N.; Baiyegunhi, C. Geochemical Evaluation of the Cretaceous Mudrocks and Sandstones (Wackes) in the Southern Bredasdorp Basin, Offshore South Africa: Implications for Hydrocarbon Potential. *Minerals* **2020**, *10*, 595. [CrossRef]
39. Kadkhodaie, A.; Sfidari, E.; Najjari, S. Comparison of intelligent and statistical clustering approaches to predicting total organic carbon using intelligent systems. *J. Pet. Sci. Eng.* **2012**, *86–87*, 190–205. [CrossRef]
40. Alizadeh, B.; Najjari, S.; Kadkhodaie, A. Artificial neural network modeling and cluster analysis for organic facies and burial history estimation using well log data: A case study of the South Pars Gas Field, Persian Gulf, Iran. *Comput. Geosci.* **2012**, *45*, 261–269. [CrossRef]
41. Łętkowski, P.; Gołąbek, A.; Budak, P.; Szpunar, T.; Nowak, R.; Arabas, J. Determination of the statistical similarity of the physicochemical measurement data of shale formations based on the methods of cluster analysis. *Nafta-Gaz* **2016**, *72*, 910–918. [CrossRef]
42. Prasath, V.S.; Alfeilat, H.A.A.; Hassanat, A.B.; Lasassmeh, O.; Tarawneh, A.S.; Alhasanat, M.B.; Salman, H.S.E. Effects of Distance Measure Choice on KNN Classifier Performance-A Review. *arXiv* **2017**, arXiv:1708.04321.
43. Jain, A.K.; Murty, M.N.; Flynn, P.J. Data clustering: A review. *ACM Comput. Surv.* **1999**, *31*, 264–323. [CrossRef]
44. Jain, A.K.; Dubes, R. *Algorithms for Clustering Data*; Prentice Hall: Hoboken, NJ, USA, 1988. [CrossRef]
45. Papiernik, B.; Botor, D.; Golonka, J.; Porębski, S.J. Insight from three-dimensional modelling of total organic carbon and thermal maturity. *Ann. Soc. Geol. Pol.* **2019**, *89*, 511. [CrossRef]
46. Zhang, Z.; Huang, Y.; Ran, B.; Liu, W.; Li, X.; Wang, C. Chemostratigraphic Analysis of Wufeng and Longmaxi Formation in Changning, Sichuan, China: Achieved by Principal Componentand Constrained Clustering Analysis. *Energies* **2021**, *14*, 7048. [CrossRef]
47. Waszkiewicz, S.; Krakowska-Madejska, P.I. Vitrinite Equivalent Reflectance Estimation from Improved Maturity Indicator and Well Logs Based on Statistical Methods. *Energies* **2021**, *14*, 6182. [CrossRef]
48. Huang, X.; Gu, L.; Li, S.; Du, Y.; Liu, Y. Absolute adsorption of light hydrocarbons on organic-rich shale: An efficient determination method. *Fuel* **2021**, *308*, 121998. [CrossRef]

energies

MDPI

Article

Effect of Viscosity Action and Capillarity on Pore-Scale Oil–Water Flowing Behaviors in a Low-Permeability Sandstone Waterflood

Tao Ning [1], Meng Xi [1], Bingtao Hu [2], Le Wang [3], Chuanqing Huang [4] and Junwei Su [2,*]

[1] The Management Headquarters of Water Injection Project, Yanchang Oil Field Co. Ltd., Yan'an 716000, China; ningtao9226@sxycpc.com (T.N.); 18729325132@163.com (M.X.)
[2] School of Human Settlements and Civil Engineering, Xi'an Jiaotong University, Xi'an 710049, China; hubingtao@stu.xjtu.edu.cn
[3] Mechanical Engineering College, Xi'an Shiyou University, Xi'an 710065, China; 180907@xsyu.edu.cn
[4] School of Chemistry and Chemical Engineering, Shaanxi University of Science and Technology, Xi'an 710021, China; huangcq@sust.edu.cn
* Correspondence: sujunwei@mail.xjtu.edu.cn; Tel.: +86-029-8896-5100

Abstract: Water flooding technology is an important measure to enhance oil recovery in oilfields. Understanding the pore-scale flow mechanism in the water flooding process is of great significance for the optimization of water flooding development schemes. Viscous action and capillarity are crucial factors in the determination of the oil recovery rate of water flooding. In this paper, a direct numerical simulation (DNS) method based on a Navier–Stokes equation and a volume of fluid (VOF) method is employed to investigate the dynamic behavior of the oil–water flow in the pore structure of a low-permeability sandstone reservoir in depth, and the influencing mechanism of viscous action and capillarity on the oil–water flow is explored. The results show that the inhomogeneity variation of viscous action resulted from the viscosity difference of oil and water, and the complex pore-scale oil–water two-phase flow dynamic behaviors exhibited by capillarity play a decisive role in determining the spatial sweep region and the final oil recovery rate. The larger the viscosity ratio is, the stronger the dynamic inhomogeneity will be as the displacement process proceeds, and the greater the difference in distribution of the volumetric flow rate in different channels, which will lead to the formation of a growing viscous fingering phenomenon, thus lowering the oil recovery rate. Under the same viscosity ratio, the absolute viscosity of the oil and water will also have an essential impact on the oil recovery rate by adjusting the relative importance between viscous action and capillarity. Capillarity is the direct cause of the rapid change of the flow velocity, the flow path diversion, and the formation of residual oil in the pore space. Furthermore, influenced by the wettability of the channel and the pore structure's characteristics, the pore-scale behaviors of capillary force—including the capillary barrier induced by the abrupt change of pore channel positions, the inhibiting effect of capillary imbibition on the flow of parallel channels, and the blockage effect induced by the newly formed oil–water interface—play a vital role in determining the pore-scale oil–water flow dynamics, and influence the final oil recovery rate of the water flooding.

Keywords: water flooding; pore scale; enhanced oil recovery; viscosity; capillarity

Citation: Ning, T.; Xi, M.; Hu, B.; Wang, L.; Huang, C.; Su, J. Effect of Viscosity Action and Capillarity on Pore-Scale Oil–Water Flowing Behaviors in a Low-Permeability Sandstone Waterflood. *Energies* **2021**, *14*, 8200. https://doi.org/10.3390/en14248200

Academic Editor: Marcin Kremieniewski

Received: 5 October 2021
Accepted: 19 November 2021
Published: 7 December 2021

1. Introduction

Crude oil, a typical conventional strategical energy resource, plays a significant role in the development of human industrial civilization and economic prosperity. Crude oil is extracted from the production well due to the natural energy and pressure of the reservoirs in the primary oil recovery process. Water flooding technology is widely utilized to further improve the efficiency of oil production, which will ultimately lead to the high ratio of water to oil of producing wells, thus decreasing the oil production efficiency [1]. It is widely acknowledged that ~60% of the original crude oil is still trapped in the reservoirs after water flooding [2]. The crude oil extraction from oil reservoirs is characterized by a typical multiphase flow in porous media, and the flow conditions, the properties of the injected fluid and the pore structure characteristics have a huge impact on the multiphase flow in porous media. The macroscopic characteristics of the water flooding development of sandstone reservoirs are the comprehensive results of oil–water two-phase flow within a large number of microscopic pore channels. The investigation of the flow dynamic characteristic oil–water two-phase flow within porous media at the pore-scale in depth is of great significance to understand the macroscopic phenomenon of the water flooding development process. The pore-scale investigation of multiphase flow behavior and the dynamic process within reservoir rock contributes to clarify the underlying dynamic mechanisms of certain macroscopic water flooding phenomena [3–7].

In the study of pore-scale flow, the gravity effect is generally relatively small, and can be ignored. In such a situation, viscous action and capillarity are essential factors in affecting the process of immiscible displacements in porous media [8]. The current enhanced oil recovery approaches are adopted to improve the oil recovery rate by adjusting the displacement front through the regulation of the viscous effect or capillary effect. For instance, the polymer flooding technique is employed to enhance the oil recovery rate mainly by increasing the viscosity of the displacement fluid, decreasing the viscosity ratio of oil to water, and increasing the sweep region of the displacement fluid [9]. On the other hand, the surfactant flooding technique is used to mobilize the residual oil trapped by the capillary force in the pore space by reducing the interfacial tension of oil and water through the addition of surfactants to the displacement fluid [10]. The increasing injection rate technique is used to mobilize the residual oil by increasing the relative magnitude of the viscous action and capillary force. Particle flooding techniques are used to increase the resistance of the water bearing channel with polymer particles (increasing the effective viscosity of the flooding agents) to make the fluid divert to the oil-bearing channel, so as to improve the oil recovery [11]. Thus, the investigation of the effect of the viscosity action and capillary on the two-phase flow is essentially important in order to enhance oil recovery.

The effects of viscous action and capillary on two-phase flow in the porous media have been widely studied [12]. In terms of the viscous effect on the flow, Homsy systematically reviewed the viscous fingering in porous media in 1987, and pointed out that the root cause of the viscous fingering formation is due to the viscosity difference of two phases [12–23]. The two-phase displacement process within a two-dimensional porous medium was investigated by Cottin, and the variation process of capillary fingering to viscous fingering was obtained [13]. Tsuji investigated the displacement process in three-porous media under different capillary numbers and viscosity ratios, obtained the displacement diagram of three-dimensional porous media, and analyzed the macroscopic characteristics of different displacement modes [14]. Hu investigated the effect of the interplay between disorder and the contact angle on the pore-scale flow behaviors, and obtained a quasi-static immiscible displacement phase diagram in disordered porous media. [15]. Singh reviewed the capillary-dominated flow behaviors in the porous media [16]. Yang investigated the effect of the wettability and capillarity on the remaining oil distribution in 3D porous media [17]. Guo studied the effect of the viscosity and interfacial tension on the distribution of the residual oil distribution [18]. Influenced by the complexity of the pore structure, the morphology of the oil–water interface will change abruptly during the advancing process, thus causing complex flow behaviors induced by the abrupt change of interfacial tension. Specific pore-scale flow behaviors—including the Haines jumps [19], snap-off effect [20–22] and the capillary barrier [11,23]—have been investigated by many researchers at the infancy stage.

Although extensive investigations of the effects of viscous action and capillarity on the flow have been conducted in previous works, these studies mainly focused on the analysis and characterization of the flow behaviors from the observed phenomena, and there are few studies concerning the underlying dynamic mechanism behind these phenomena or the effects of certain phenomena (Haines jumps, the snap-off effect or the capillary barrier, for instance) on the immiscible fluid displacement processes in the open literature.

The pore structure characteristics of a low-permeability sandstone reservoir are obviously different from those of medium- and high-permeability sandstone reservoirs. Compared with medium- and high-permeability sandstone, a low-permeability sandstone reservoir has a smaller pore size and a larger pore–throat ratio. The capillary effect plays a more obvious role in the process of pore-scale flow. The pore space of a low-permeability sandstone reservoir has more unique and complex pore-scale flow characteristics (such as the capillary valve effect). Limited by the chip fabrication process and the accuracy of displacement equipment, there are few studies on pore-scale oil–water movement in low-permeability sandstone reservoirs. Pore-scale direct numerical simulation technology is generally an important technology to study low-permeability sandstone reservoirs. In this paper, the water flooding process in low-permeability sandstone oil reservoirs is numerically simulated using the Navier–Stokes equation coupled with the volume of fluid (VOF) method to investigate the effect of viscous action and capillarity on the pore-scale oil–water flow process from the point of view of two-phase dynamics. The dynamic processes of several specific phenomena or events are analyzed, which will deepen the established understanding of water flooding in low-permeability sandstone oil reservoirs. The numerical model of an oil–water two-phase flow is presented first, then an in-depth analysis of the viscous and capillary effects of the two-phase flow process is performed; results and discussions concerning the numerical simulation of the water flooding process in low-permeability sandstone oil reservoirs follow, and finally the conclusions are drawn in the final part.

2. Dynamic Model and Solution Algorithm for Oil–Water Two-Phase Pore-Scale Flow

The macroscopic characteristics exhibited in the water flooding process of sandstone reservoirs are a concentrated reflection of microscopic oil–water flow in a large number of pore channels. An in-depth investigation on the dynamic process of oil–water two-phase flow in the pore space, as well as the relevant oil–water two-phase flow characteristics, is of great significance for the clarification of the underlying dynamic mechanisms of the water flooding process. In this section, the dynamic equations describing the oil–water two-phase flow in the pore space are presented firstly, followed by its solution algorithm.

In this paper, Navier–Stokes equations are used to describe the oil–water two-phase flow in an Eulerian framework, the VOF (Volume of Fluid) method is used to track the spatial distribution of the oil and water, and the contact angle is employed for the characterization of the wettability of the reservoir rock [24].

2.1. Pore-Scale Dynamic Model for Oil–Water Movement in Porous Media

2.1.1. Mass Conservation Equation

The differential form of the mass equation for an incompressible oil–water two-phase flow is given by

$$\nabla \cdot \mathbf{u} = 0, \tag{1}$$

where $\mathbf{u}$ is average velocity of the oil phase and water phase, $\mathrm{m \cdot s^{-1}}$.

2.1.2. Momentum Conservation Equation

The differential form of the momentum conservation equation for the oil–water two-phase flow is given by

$$\frac{\partial \rho \mathbf{u}}{\partial t} + \nabla \cdot (\rho \mathbf{u} \mathbf{u}) - \nabla \cdot (\mu \boldsymbol{\tau}) = -\nabla p + \rho \mathbf{g} + \mathbf{F}_{\sigma}, \tag{2}$$

where ρ is average density of the oil phase and water phase, $\mathrm{kg \cdot m^{-3}}$; μ is average dynamic viscosity of the two phases, $\mathrm{Pa \cdot s}$; p is dynamic pressure, Pa; $\mathbf{g}$ is acceleration of gravity,

$m{\cdot}s^{-2}$; $\mathbf{F}_{\sigma}$ is the interfacial tension (IFT) between oil and water, $kg{\cdot}m^{-2}{\cdot}s^{-2}$; and $\boldsymbol{\tau}$ is the rate of strain tensor, s^{-1}, which is given as follows:

$$\boldsymbol{\tau} = \left(\nabla \mathbf{u} + (\nabla \mathbf{u})^{T}\right), \tag{3}$$

2.1.3. Oil–Water Interfacial Tension

The last term, $\mathbf{F}_{\sigma}$, which appeared on the right-hand side of Equation (1) represents the interfacial tension between oil and water, which can be expressed as follows:

$$\mathbf{F}_{\sigma} = \sigma \delta_{s} k \mathbf{n}, \tag{4}$$

where σ is the surface tension coefficient, $N{\cdot}m^{-1}$; δ_s is the area of the oil–water interface per unit volume, m^{-1}; k is the curvature of the oil–water interface, m^{-1}; and $\mathbf{n}$ is the unit-normal vector of the interface.

The area of the oil–water interface per unit volume δ_s is given by

$$\delta_{s} = |\nabla \alpha|, \tag{5}$$

where α is the volume fraction of the water phase.

The unit-normal vector of the interface $\mathbf{n}$ is given by

$$\mathbf{n} = \frac{\nabla \alpha}{|\nabla \alpha|}, \tag{6}$$

The curvature of the oil–water interface k is given by

$$k = \nabla \cdot \mathbf{n}, \tag{7}$$

It should be stressed that the interfacial tension given in Equation (4) is a force on the oil–water interface with an area of δ_s. It is different from the capillary force in a tube, which is the total force of the interaction action. It is also different from the capillary pressure function for the porous media. The macroscopic capillary pressure function of the porous media is the result of the interface tension in the microscopic view. According to whether the interface moves or not, the macro impact of the interfacial tension can be divided into two parts: the moving part is described by the macro capillary pressure function (because the capillary force is directly related to the two-phase velocity described in the black oil model for the macroscopic description of the oil–water flow in the porous media). The binding effect of the interfacial tension on oil causes the oil not to move, which is usually described by the endpoint of the relative permeability curve in the macroscopic model. With the model given in Equation (4), the effect of the interface tension of the static oil–water interface (such as the capillary valve effect) and the dynamic oil–water interface (such as the osmotic effect and the capillary inhibition effect) on the pore-scale oil–water motion is taken into account.

2.1.4. Oil–Water Volume Fraction

The VOF method is used to track the spatial distribution of the oil–water two-phase flow. The interface is determined by solving the equation for the volume fraction of water phase α in each cell. When the value of α is unity, the cell is fully occupied by the water phase; when the value of α is zero, the cell is fully occupied by the oil phase; when the value of α is in the range of 0 to 1, the cell contains the free interface. The equation for the volume fraction of the water phase is expressed as follows:

$$\frac{\partial \alpha}{\partial t} + \nabla \cdot (\alpha \mathbf{u}) = 0, \tag{8}$$

With Equation (8), the distribution of water and oil in the pore space can be obtained. The homogeneity of the two-phase flow caused by pore-scale flow behaviors (emulsification or snap-off, for instance) can be obtained easily.

2.1.5. Wettability

The contact angle is commonly employed to characterize the wettability of the reservoir rock. The wettability of the pore wall can be classified into three regimes based on different water contact angles: hydrophilicity (water-wet), intermediate-wettability, and hydrophobicity (oil-wet). The wettability of the pore walls is of great importance in order to accurately simulate the oil–water two-phase flow and predict the distribution of the residual oil. The contact angle of the rock is affected by the surface roughness, the composition of the rock, and the thickness of the water film. It is difficult to accurately determine the contact angle of realistic reservoir rock considering the mixed wettability and the variation of the contact angle with spatial locations. The wettability regimes of water-wettability, intermediate-wettability and oil-wettability are considered by adjusting the value of the contact angle in this paper.

In order to model the wettability of the rock, the contact angle is imposed as a boundary condition, as illustrated in Figure 1. The unit vector normal to the interface needs to be modified as the following from

$$\mathbf{n} = \mathbf{n}_w \cos\theta + \mathbf{s}_w \sin\theta, \tag{9}$$

where $\mathbf{n}_w$ is the unit vector normal to the wall; $\mathbf{s}_w$ is the unit vector perpendicular to the contact line, tangential to and pointing into the wetting–solid interface surface; and θ is the contact angle, radian.

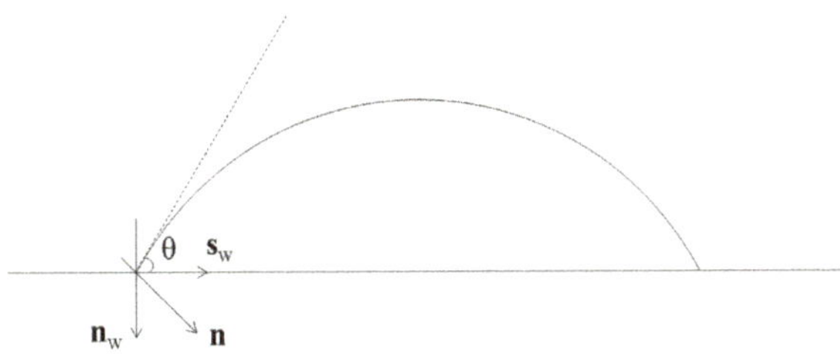

Figure 1. Illustration of the implementation of the contact angle boundary condition.

2.1.6. Averaging Properties of the Oil–Water Flow

In Equation (2), ρ and μ represent the average density and dynamic viscosity of the water phase and the oil phase, respectively, which can be calculated respectively by Equations (10) and (11):

$$\rho = \alpha\rho_w + (1-\alpha)\rho_o, \tag{10}$$

$$\mu = \alpha\mu_w + (1-\alpha)\mu_o, \tag{11}$$

where ρ_w is the density of the water phase, kg·m^{-3}; ρ_o is the density of the oil phase, kg·m^{-3}; μ_w is the dynamic viscosity of the water phase, Pa·s; and μ_o is the dynamic viscosity of the oil phase, Pa·s.

2.2. Solution Method and Procedure

2.2.1. Solution Method

The PISO (Pressure-Implicit with Splitting of Operators) algorithm is used to decouple the continuity and momentum equation [25]. In order to separate the pressure difference caused by surface tension from the actual pressure, Equation (2) can be written as

$$a_p\mathbf{u} = A_H - \nabla p_d - \nabla p_c \tag{12}$$

where a_p is the diagonal coefficient in coefficient matrix discretized from Equation (2). p_c is the pressure gradient caused by surface tension, and A_H can be written as

$$A_H = \sum_N a_N \mathbf{u} + b \tag{13}$$

where a_N is the implicit contribution coefficient of neighbor cells to the cell concerned, and b includes all of the explicit discretization contributions except pressure.

Both sides of Equation (13) are divided by a_p, and we obtain a new equation as follows:

$$\mathbf{u} = \frac{A_H}{a_p} - \frac{\nabla p_d}{a_p} - \frac{\nabla p_c}{a_p} \tag{14}$$

Equation (14) should satisfy the continuity Equation (1). As such, Equation (14) also can be transformed to the following form:

$$\nabla \cdot \left(\frac{1}{a_p} \nabla p_d \right) = \nabla \cdot \left(\frac{A_H}{a_p} \right) - \nabla \cdot \left(\frac{1}{a_p} \nabla p_c \right) \tag{15}$$

Equation (15) is the pressure equation derived from the PISO algorithm. Solving the equation can obtain a new pressure and substitute it into Equation (14) to update the velocity. The finite volume method based on a collocated grid is used in the discretization process in this paper. While updating the body-center velocity, it is necessary to update the flow rate on the interface of the element in order to discretize the convection term next time. The surface flow rate is updated using the interpolation on the surface of Equation (14):

$$\varphi_f = \nabla \cdot \left(\frac{A_H}{a_p} \right)_f S_f - \left(\frac{\nabla p_d}{a_p} \right)_f S_f - \left(\frac{\nabla p_c}{a_p} \right)_f S_f \tag{16}$$

It should be noted that it is necessary to calculate the pressure gradient caused by the surface tension when updating the unit interface flow rate by solving Equation (15) or Equation (16).

2.2.2. Solution Procedure

The arbitrary polyhedral finite volume method in OpenFOAM [26,27] is used to discretize the pore-scale dynamic equations. The following procedure is used to solve the pore-scale models:

Step 1: Solve the momentum conservation equation (Equation (2)) to predict the velocity.

Step 2: Solve the pressure equation (Equation (15)) with the velocity predicted in Step 1.

Step 3: Correct the volume flux at the cell faces using Equation (16) and the velocity at the cell centers using Equation (14), respectively.

Step 4: Repeat Step 2 and Step 3 until the convergence of pressure and velocity.

Step 5: Solve the water phase volume fraction (Equation (8)).

Step 6: Calculate the oil–water interfacial force using Equation (4).

Step 7: Update the average density and average viscosity using Equations (10) and (11).

Step 8: Go to Step 1 for the next time step.

With Steps 1–8 we can obtain the pore-scale distribution of the pressure, velocity, oil fraction, water fraction. A more elaborate description of the solution techniques and model validation can be found in our previous work [6].

3. Viscous Force and Capillary Force in a Capillary Tube

From the dynamic equation of the oil–water two-phase flow (Equation (2)), it can be seen that the oil–water two-phase flow within the pore space is influenced by factors including the applied pressure difference, the viscous effect, the interfacial effect and the gravity effect. The relative magnitudes of the various effects affect the flow behavior of the oil–water two-phase flow within the pore space, and ultimately determine the oil recovery rate of the water flooding method. In general, the water flooding process is a horizontal displacement, and the following section only focuses on the viscous effect and interfacial effect.

3.1. Viscous Force

Oil and water are viscous fluids, and the oil–water two-phase flow within the pore space of the sandstone reservoir is subject to the viscous force. Most of the energy in the process of oil recovery by water flooding is consumed by the viscous effect [28]. The strength of the two-phase viscous effect can be described by the viscous force. Figure 2 demonstrates the oil and water distribution in a capillary tube; the radius of the capillary tube is R, the length of the water side is L_1, the length of the oil side is L_2, and the velocity of the flow is u. According to the Hagen–Poiseuille formula, the viscous force of the fluid can be expressed as

$$F_{vis} = \frac{8(\mu_w L_1 + \mu_o L_2)u}{R^2}, \quad (17)$$

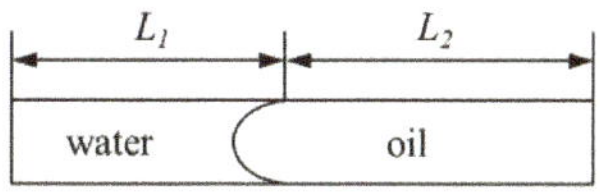

Figure 2. Schematic diagram of the oil and water distribution in a capillary tube.

Equation (17) can be further written in the following form:

$$F_{vis} = \frac{8(\mu_w \alpha_w + \mu_o(1 - \alpha_w))Lu}{R^2}, \quad (18)$$

where L is the length of the capillary tube, and is defined as the sum of L_1 and L_2.

It can be seen from Equation (17) or Equation (18) that the viscous force increases with the increase of the flow rate and the decrease of the capillary tube radius, while the viscous force decreases with the increase of the water content (the viscosity of water is generally lower than that of oil).

It can be concluded from Equation (18) that, with the displacement, the oil in the capillary is gradually replaced by water, and the viscous resistance of the capillary decreases gradually. If the oil–water viscosity ratio is 5, the water in the capillary changes from oil to water, and the resistance becomes 0.2 times that of the original. In the actual water flooding process, with the same pore structure characteristics, the flow resistance in the water-bearing area is 0.2 times that in the oil-bearing area. This inhomogeneity is caused by fluid phase distribution, which is called dynamic inhomogeneity in this paper.

Equation (18) can be further written as the following form:

$$F_{vis} = \frac{8(\mu_w \alpha_w + \mu_o(1 - \alpha_w))LQ}{\pi R^4}, \quad (19)$$

where Q is the volumetric flow rate of the capillary tube, $m^3 \cdot s^{-1}$.

It can be concluded from Equation (19) that the viscous resistance increases 10,000 times when the capillary radius is reduced by a factor of 10 under the same injection rate and the same water content of the capillary tube. The injection pressure of a low-permeability sandstone reservoir is much higher than that of the high-permeability sandstone reservoir under the same injection flow rate.

3.2. Capillary Force

Because the oil and water phases are immiscible, there is an interface between the oil and water phases in the process of flowing in the pore channel. Because the molecular forces on both sides of the interface are not equal, an additional effect is exerted on the flow of oil and water by the interface. The magnitude of the additional force can be expressed by the interfacial tension. Figure 2 shows the capillary force in a straight pipe (the resultant force of local interfacial tension), which can be expressed as

$$p_c = \frac{\sigma \cos(\theta)}{R}, \tag{20}$$

The following conclusions can be drawn from Equation (20):

(1) When $\theta < 90^{\circ}$, the wettability is water-wet, the pressure difference $\Delta p > 0$, the capillary force is a type of driving force, and the direction is consistent with the direction of the oil–water flow.
(2) When $\theta = 90^{\circ}$, the wettability is intermediate-wet, the pressure difference $\Delta p = 0$, and the value of the capillary force is zero.
(3) When $\theta > 90^{\circ}$, the wettability is oil-wet, the pressure difference $\Delta p < 0$, the direction of the capillary force is opposite to the direction of the water flooding, and the capillary force is a type of resistance.
(4) The magnitude of the capillary force is inversely proportional to the radius of the capillary tube: the smaller the radius, the larger the capillary force.

The pore radii at the different locations of the pore channels in real sandstone are different, and there are pores and throats. There are abrupt changes of the pore radius at the connection of the pore and throat channels. Figure 3 shows the pore throat structure of a typical porous medium, as well as the oil and water distribution at different locations. θ is the three-phase contact angle of oil, water and a solid wall, and β is the opening angle of the pore. When the oil–water interface is in the throat channel, the direction of the capillary force (pointing to the concave side of the liquid surface) is the same as the direction of the oil–water movement under water-wet conditions, and the capillary force is a type of driving force, as shown in Figure 3a. When the water flows out of the throat channel, the angle between the oil–water interface and the pore wall after equilibrium is θ. When $\theta + \beta > \pi/2$, the oil–water two-phase interface will be reversed, at which time the capillary force will point to the water side and impede the movement of the oil–water front. That is, affected by the pore structure, even under water-wet conditions, the capillary force formed between the oil–water interface and the pore wall may become resistance under certain conditions.

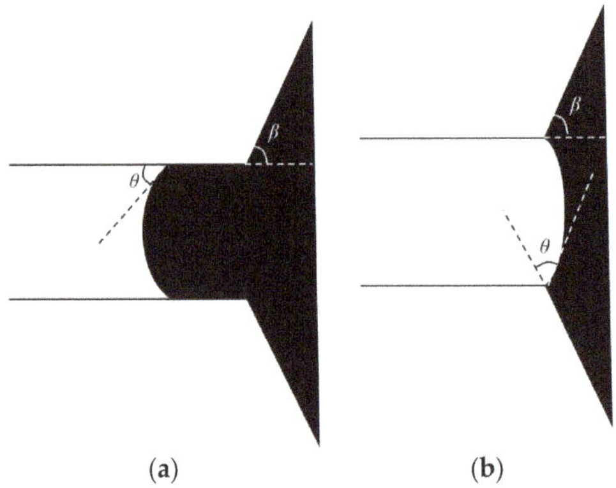

Figure 3. Oil and water distribution at different locations: (**a**) interface at the throat channel; (**b**) interface at the outlet section.

Figure 4 shows the morphology of the oil–water interface at different locations, and the capillary forces at the corresponding locations. In Figure 4, *a–h* indicates the morphology of the oil–water interface, and A–D is the position of the contact line. In the process of the water's movement from the throat to the pore space, the oil–water interface undergoes three processes.

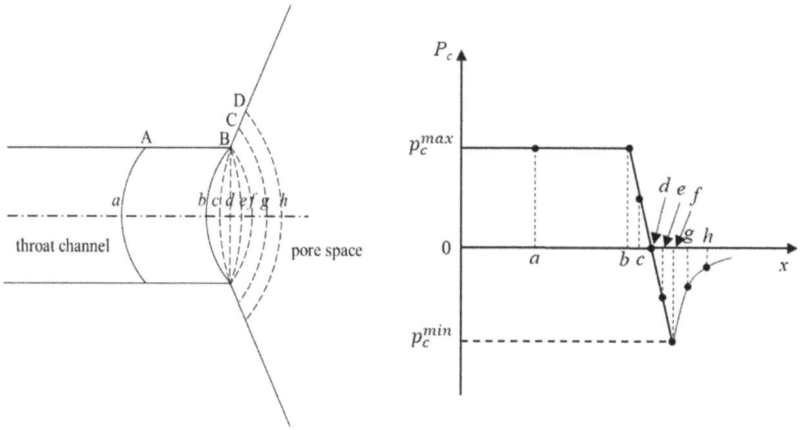

Figure 4. Schematic diagram of the oil–water interface morphology and capillary force variation along the axis: (**a**) the oil–water interface morphology at different locations; (**b**) the capillary force at different locations.

(1) The oil–water interface advances in the throat channel (A→B); the morphology of the oil–water interface is shown as *a* or *b*. At this stage, the capillary force is a kind of driving force, and its value is given as

$$p_c^{max} = \frac{\sigma\cos(\theta)}{R}, \tag{21}$$

(2) After the oil-water interface advances to the B position, the oil–water interface only deforms without moving forward (the three-phase contact line stays at B position) until the equilibrium wetting angle θ between the interface of deformation and the water-side pore wall is reached (the oil–water morphology is shown as *b*). In this process, the capillary force formed on the wall of the oil–water pore first changes from p_c^{max} to 0 (the interface morphology is shown as *d*); after that, the oil–water interface reverses and the capillary force becomes negative until the negative extreme value of p_c^{min} is reached.

$$p_c^{min} = \frac{\sigma\cos(min(\theta+\beta,\pi))}{R}, \tag{22}$$

(3) Thereafter, the contact line moves further forward (B→D), the equivalent radius of the channel gradually increases, and the capillary force gradually decreases, at which time the morphology of the oil–water interface is shown in position *g* and *h*. The change of the capillary force in this stage is shown in Figure 4b (*f*→*h*).

As shown in Figure 4, when $\theta + \beta > \pi/2$, the oil–water interface reverses during the advancing process, the capillary force presents resistance under water-wet conditions, and there is a negative maximum value of p_c^{min}. When the driving force of the fluid is insufficient to overcome this maximum resistance, the oil–water interface will stop advancing, and increasing and decreasing the driving force in this process will only lead to the deformation of the interface. When the driving force of the fluid can overcome the maximum resistance, the value of the resistance will decrease, and the oil–water interface will still advance even at a low driving pressure. Therefore, the driving pressure must be above a certain threshold to induce fluid motion, and once this threshold value is reached, the displacement force required for fluid motion will decrease. For the sake of discussion, conveniently, the phenomenon is called the capillary barrier, and this threshold value is named capillary barrier pressure. When the driving force exceeds the opening pressure of the capillary barrier pressure, the channel will be opened, and the fluid can move with little driving pressure; when the driving force is lower than the opening pressure of the capillary barrier pressure, the channel will be blocked by the capillary force, the fluid cannot move in the channel, and the fluid becomes stagnant.

The motion behavior of oil–water at the pore-scale can be determined by the relative magnitude of the viscous force and capillary force, and the viscosity ratio of oil–water [15]. The relative magnitude of viscous force and capillary force can generally be expressed by the capillary number:

$$Ca = \frac{u\mu_w}{\sigma}, \tag{23}$$

where u is the velocity of the fluid, μ_w is the viscosity of the fluid, and σ is the interfacial tension coefficient. The comparison of the motion characteristics of oil and water under different capillary number can generally be realized by changing one of the three variables in the above equation.

4. Physical and Numerical Conditions

4.1. Physical Model

The core was sampled from the Chang 2 stratum, Xing 5009 well at the Xingzichuan Oil Production Plant in the Yanchang Oilfield, at a depth of 896 m. The stratum is a light gray oil-bearing fine sandstone. The sampled core was scanned via CT technology, and the obtained images were segmented to obtain the pore structure image of the core shown in Figure 5. The gray parts in the figure are pore channels. The pore size distribution is shown in Figure 6.

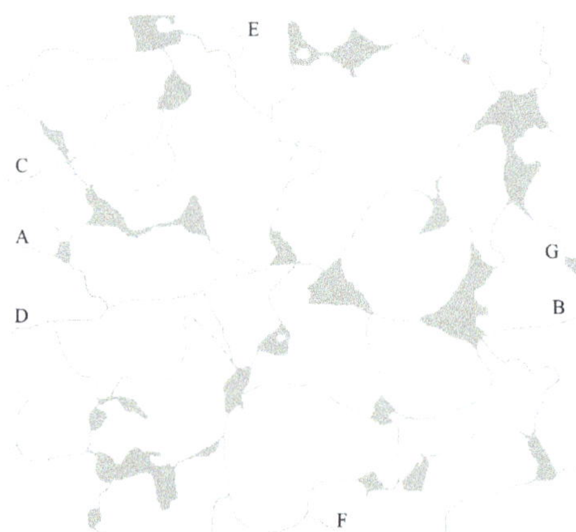

Figure 5. The pore structure of the core. A–G represent the inlet or outlet at different cases.

4.2. Numerical Boundary Conditions

The numerical boundary conditions are given in Table 1. The Dirichlet boundary condition was employed for the velocity at the inlet and the wall, the pressure at the outlet, and the water volume fraction at the inlet. The Neumann boundary condition was employed for the velocity at the outlet, the pressure at the inlet and the wall, and the water volume fraction at the outlet. The constant contact angle boundary condition was used for the water-oil-solid contact line on the wall. The Gamma Scheme presented in work [29] was used to discretize the convection term, and the Crank–Nicolson scheme was for the time term [30]. The residual errors of the different physical quantities were set to 10^{-6}. The courant number in the simulation was set to 0.1, and the time step was adjusted adaptively [6].

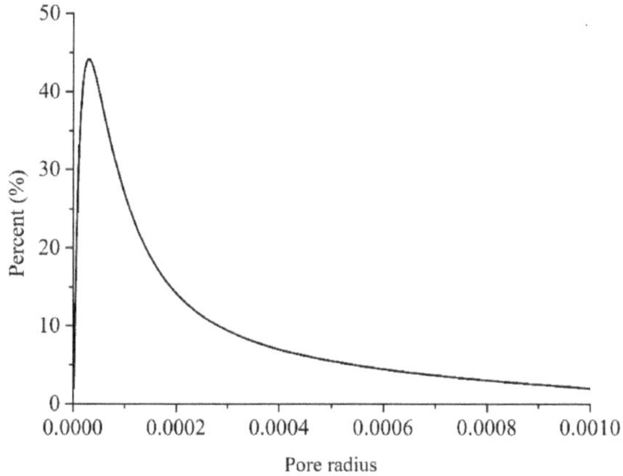

Figure 6. Pore size distribution.

Table 1. Numerical boundary conditions.

Physical Quantity	Boundaries		
	Inlet	Outlet	Wall
velocity	fixed value	zero gradient	no slip
pressure	zero gradient	fixed value	zero gradient
water volume fraction	fixed value	zero gradient	Constant contact angle

5. Results and Discussion

5.1. Oil–Water Two-Phase Flow Characteristics under Different Water Injection Rates

The physical model used for the simulation is shown in Figure 5, where A is the inlet; B is the outlet; C, D, E, F and G are closed and set as walls; and the other boundaries are walls. The pore space of the core is filled with oil at the initial moment, and then water is injected into the pore space at 0.005 m·s^{-1}, 0.01 m·s^{-1}, 0.015 m·s^{-1}, 0.02 m·s^{-1}, and 0.025 m·s^{-1} from inlet A, respectively. The viscosity ratio of oil to water is 10, the wetting angle is 45° (under this wettability condition, the capillary may show resistance or driving force, and the displacement process with this contact includes more extensive flow behaviors), and the tension coefficient at the oil–water interface is 0.07 kg·m^{-2}.

Figure 7 shows the variation of the final oil recovery rate with the injection rate under continuous water injection conditions. It can be seen from Figure 7 that the highest recovery is achieved when the injection rate is 0.01 m·s^{-1}. The capillary number Ca is 2.14 × 10^{-4}. Both too-low and too-high injection velocities fail to achieve the highest oil recovery rate. In order to clarify the underlying reason behind this phenomenon, in-depth analyses of the flow characteristics of the oil and water phases in the pore space at three injection rates of 0.005 m·s^{-1} (Ca = 7.14 × 10^{-5}), 0.01 m·s^{-1} (Ca = 1.43 × 10^{-4}) and 0.03 m·s^{-1} (Ca = 4.29 × 10^{-4}) were conducted, respectively.

Figure 8 demonstrates the spatial distribution of oil and water at different times at an injection velocity of 0.005 m·s^{-1} (Ca = 7.14 × 10^{-5}). As shown in the figure, *a*–*n* are the positions in the pore structure where capillary barriers may occur when the oil–water interface moves to these locations. Table 2 shows the occurring time of capillary barriers at these locations, the time of restart, the blockage duration, and the blockage ratio (the ratio of blockage time to total displacement time). The oil and water distribution at 0.75 s are shown in Figure 8a, and the oil–water interface moves to position *a*, as shown in the figure. This position is at the junction of the throat and the pore. When the oil and water move to this position, the capillary barrier phenomenon will be observed. When the driving pressure is insufficient, the capillary barrier will prevent the oil–water interface from further movement, thus leading to the stagnation of the oil–water interface. As can be seen from Table 2, the capillary pressure barrier pressure here was not overcome until the end of the displacement process. As the displacement proceeds, the oil–water interface is prevented from advancing at position *b* at 0.97 s, and the oil–water movement stops. At this time, the oil–water interface in the upstream channel at position *c* was still advancing until 1.15 s when the oil–water interface was blocked at position *c*. The blockage at position *c* caused the pressure at the inlet to rise, thus restarting the original blockage at the oil–water interface at position *b*. At this time, the distribution of oil and water is shown in Figure 8b. Then, the oil and water continue to move until 1.31 s and the oil–water interface moves to position *d*, where the blockage occurs again. At this time, the oil–water interface in the upstream channels at positions *e* and *f* is still moving; the oil–water interface is blocked at position *e* at 1.41 s (as shown in Figure 8c), and the oil–water interface in the upstream channels at f moves rapidly, and is blocked at position *f* at 1.53 s. The motion of the fluid in all of the channels is blocked by the capillary barrier, resulting in a rise in the upstream pressure; the capillary barrier at position *d* is overcome, and the oil–water interface is restarted at 1.53 s. When the oil–water interface enters the pore after passing through position *d*, the oil–water interface is concave to the water side influenced by the pore structure, showing a resistance state. The increase of the upstream pressure results in the overcoming of the capillary barrier pressure at position *e* at 1.75 s. After the breakthrough, the oil and water distribution at 1.92 s are shown in Figure 8d. Then, the oil and water move further downwards after a brief blockage at 2.81 s at position g, and move to the lower right outlet of position *g*. Although the channel on the upper side of *g* is perpendicular to the displacement direction and the oil–water interface enters the smaller channels, the oil and water can still flow upward by imbibition under the low displacement speed. The final oil and water distribution are shown in Figure 8f.

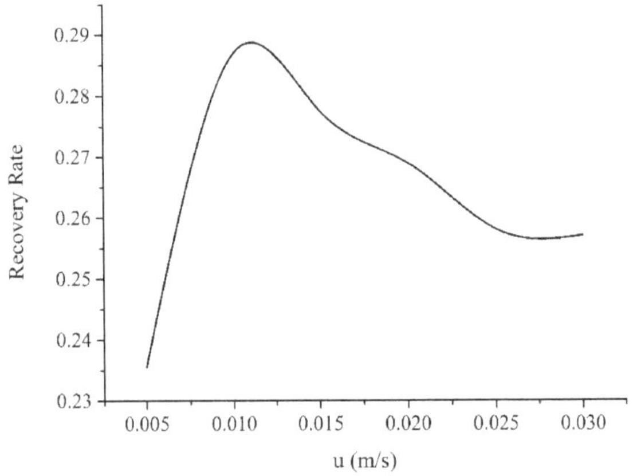

Figure 7. Variation of the final oil recovery rate with the injection velocity.

From the abovementioned oil–water flow process, we can find that the capillary barrier has a significant effect on the oil–water flow dynamics under the low capillary number. When the capillary pressure barrier is formed, the oil–water interface will block the channel where it is located, resulting in a change of the pressure transmission path in space and prompting the fluid flow to divert. When the capillary barrier is formed in the main displacement direction—for example, at position *b* or *d*—the capillary barrier will prevent the oil–water front from advancing, thus causing lateral sweeping. In this case, the capillary barrier formed at position *b* will cause the fluid to advance rapidly in the upstream channel of position *c*. When the capillary barrier is formed at position *d*, the velocity of the fluid will accelerate in the upstream channel of position *e* and the upstream channel of position *f*. When the capillary barrier is formed in the lateral direction, the stagnant oil–water interface will prevent the further advancing of the oil and water flow, thus resulting in the formation of residual oil. For instance, the formation of capillary barriers at positions a and c prevented the further movement of oil and water, resulting in the formation of residual oil on the upper right side of positions a and c.

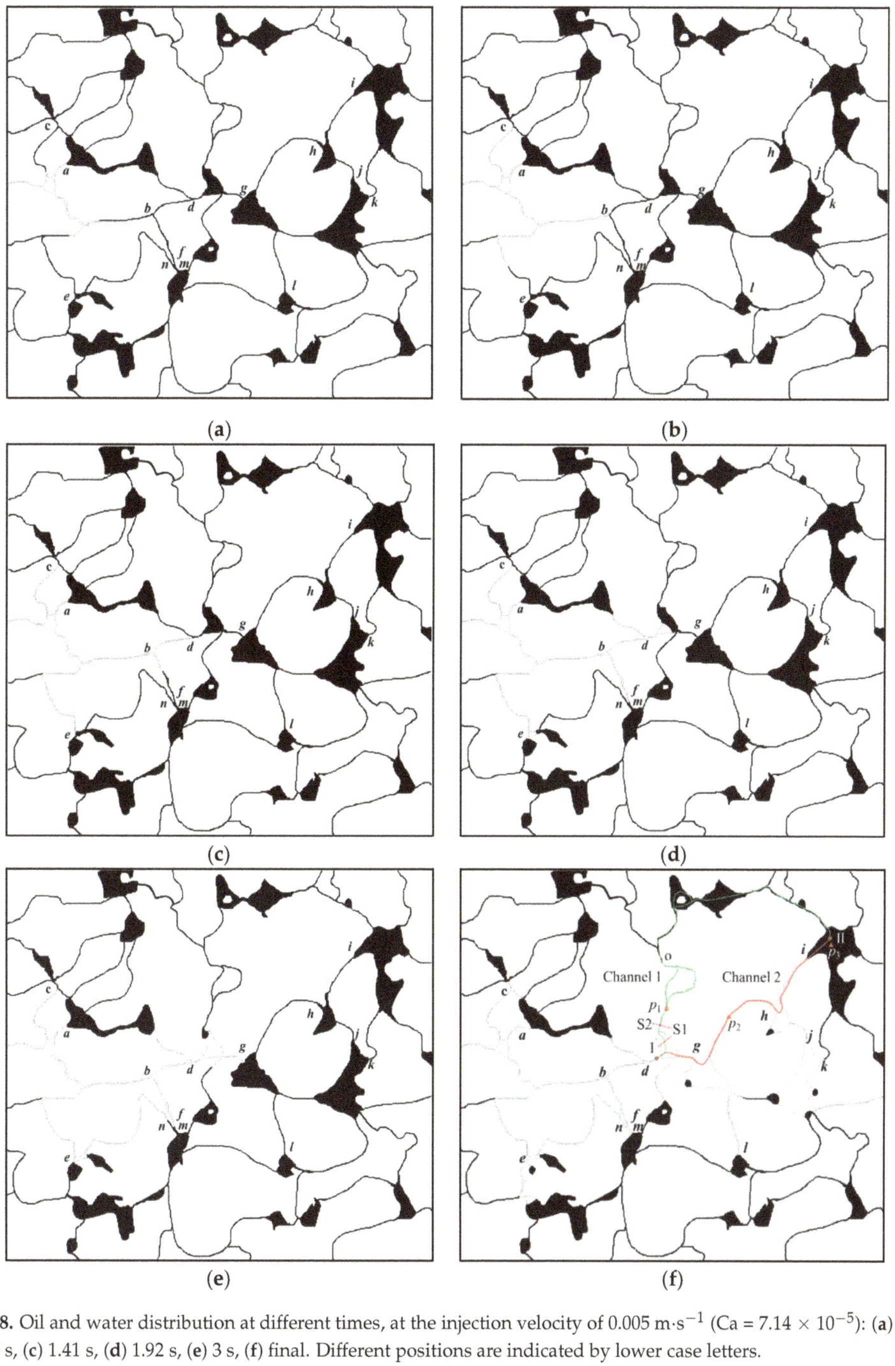

Figure 8. Oil and water distribution at different times, at the injection velocity of 0.005 m·s^{-1} (Ca = 7.14 × 10^{-5}): (**a**) 0.75 s, (**b**) 1.15 s, (**c**) 1.41 s, (**d**) 1.92 s, (**e**) 3 s, (**f**) final. Different positions are indicated by lower case letters.

Table 2. Blocking and restart information at different positions during the injection process, at the injection velocity of 0.005 m·s^{-1} (Ca = 7.14 × 10^{-5}).

Position	Blockage Time(s)	Restart Time(s)	Blockage Duration(s)	Blocking Ratio
a	0.750	∞	∞	∞
b	0.970	1.150	0.180	0.0193
c	1.150	∞	∞	∞
d	1.310	1.530	0.220	0.0236
e	1.410	1.750	0.340	0.0364
f	1.530	3.120	1.590	0.1704
g	2.810	2.880	0.070	0.0075
h	*	*	*	*
i	*	*	*	*
j	*	*	*	*
k	*	*	*	*
l	8.100	∞	∞	∞
m	3.200	∞	∞	∞
n	3.050	3.120	0.070	0.0075

∞ The movement of the oil–water interface at this position has not been resumed at the end of the displacement process. * The oil–water interface has not moved to this position.

Figure 9 shows the oil–water distribution at different times at the injection velocity of 0.01 m·s^{-1} (Ca = 1.43 × 10^{-4}). Also, the blockage and restart information at different locations during the injection process is given in Table 3. The oil–water sweep process before the breakthrough of the capillary barrier at position *d* is almost the same as the oil–water sweep process at the injection velocity of 0.005 m·s^{-1}. After the advancing of the oil–water interface through position *d*, the oil–water interface enters the pore space from the throat channels and presents a resistance state. Influenced by the capillary resistance at position *d*, the capillary barrier pressure at position *c* is overcome at 0.9 s. The oil–water interface advances further until a new capillary barrier appears. Due to the high injection velocity relative to the previous example, the oil–water interface advancing will not form an effective pressure barrier at position *g*, and it advances further forward through this position. The oil–water interface passing through position *g* first enters the channel between *g* and *h* and moves rapidly to position *h*. Position *h* is at the point from the throat to the pore space, and the oil–water interface presents a resistance state after passing through position *h* (as shown in Figure 9b). This resistance effect displaces the fluid to flow along the branch above the pore space in the middle of *dg*. After passing through the pore space downstream of position *h*, the oil–water interface moves further forward, and the capillary pressure barrier is formed at positions *i* and *j*. The blocking phenomenon observed at positions *i* and *j* promotes the transport of oil in the pore space below position *g* toward the outlet, as shown in Figure 9c. The blockage at position *j* is broken at 3.6 s. After the oil–water interface enters the pore space, a new oil–water interface is formed at position *k*, and this oil–water interface prevents the oil from flowing from the throat channel into the pore space. Meanwhile, the interface in the pore channel above the pore space between *d* and *g* stops advancing at position *o*.

It can be observed that there are differences in the sweep process and the final oil–water distribution at the two different injection velocities by comparing Figure 8 with Figure 9. As shown in Figure 8f, the sweep amount of channel 1 is more than that of channel 2 under low-velocity injection conditions; Figure 9d shows that the sweep amount of channel 1 is less than that of channel 2 under high-velocity injection conditions. This is largely due to the difference in the time of the oil–water interface entering the two channels, and an inhibition effect on the flow in the other channel is exerted by the oil–water interface entering the present channel.

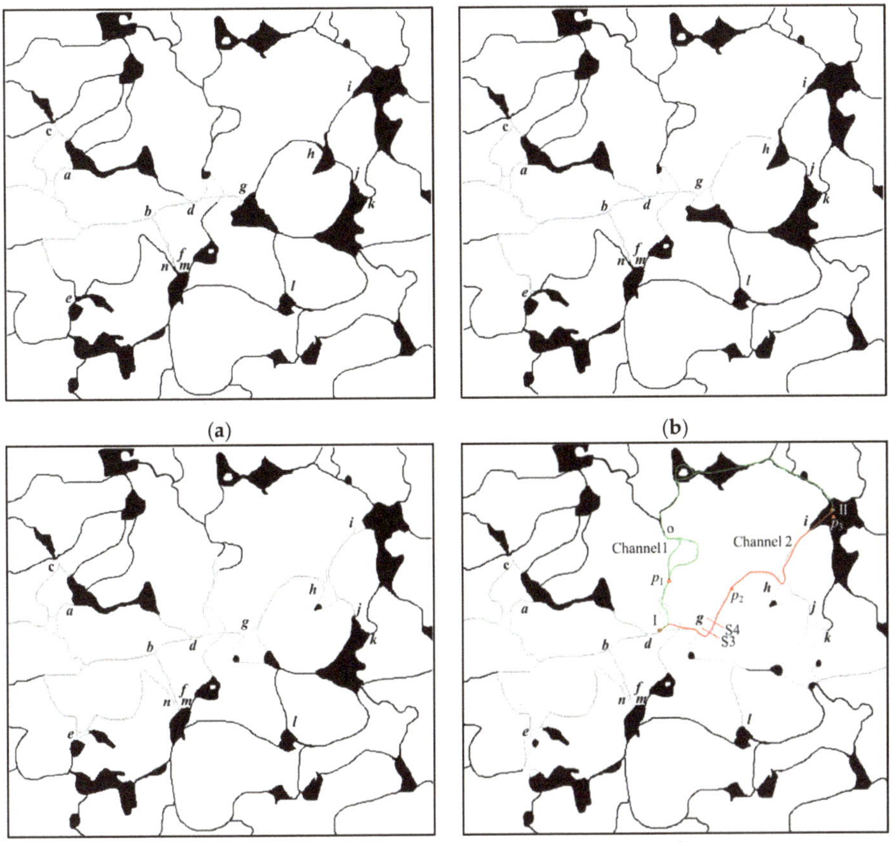

Figure 9. Oil and water distribution at different times, at the injection velocity of 0.01 m·s^{-1} (Ca = 1.43 × 10^{-4}): (**a**) 1.2 s, (**b**) 1.9 s, (**c**) 3.3 s, (**d**) final. Different positions are indicated by lower case letters.

Table 3. Blockage and restart information at different locations during the injection process, at the injection velocity of 0.01 m·s^{-1} (Ca = 1.43 × 10^{-4}).

Position	Blockage Time(s)	Start Time(s)	Blockage Duration(s)	Blockage Ratio
a	0.340	∞	∞	∞
b	0.510	0.540	0.030	0.00528
c	0.540	0.900	0.360	0.0634
d	0.630	0.700	0.070	0.0123
e	1.820	1.970	0.150	0.0264
f	0.700	3.160	2.460	0.433
g	1.180	1.180	0	0
h	1.850	1.850	0	0
i	2.450	∞	∞	∞
j	3.150	3.600	0.450	0.0792
k	4.260	∞	∞	∞
l	5.490	∞	∞	∞
m	3.280	∞	∞	∞
N	3.160	3.160	0	0

∞ The movement of the oil–water interface at this position has not been resumed at the end of the displacement process.

The aforementioned inhibition effect is illustrated in Figure 10. As shown in Figure 10, channel a and channel b are connected with channel c, respectively, and the fluid is driven from left to right. Water first flows into channel b, and an oil–water interface is formed in channel b. The oil–water interface in channel b changes the local pressure state, such that the pressure of the fluid pointed to the concave surface is higher than that of the other side. The pressure on the oil side in channel b is higher than the pressure on the water side under water-wet conditions, resulting in a higher pressure at point C, which reduces the pressure difference between the AC and inhibits the fluid flow in channel a. This is called the inhibiting effect of imbibition on flow in parallel channels.

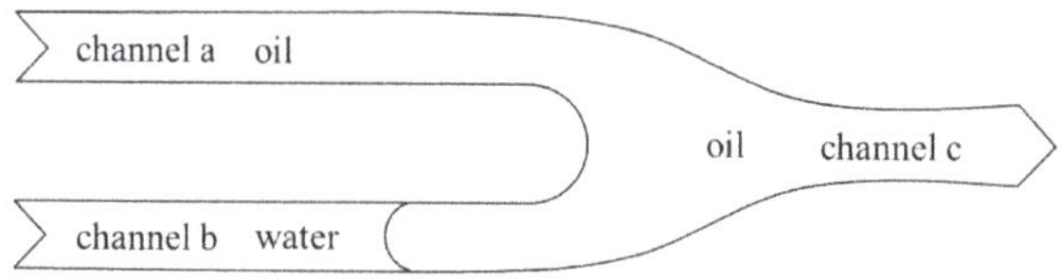

Figure 10. Schematic diagram of the capillary inhibition effect between the parallel channels. Different flow paths are distinguished by a–c.

In order to further illustrate the existence of the inhibition effect induced by the capillarity, the pressure changes at point $p1$, $p2$ and $p3$ in Figures 8f and 9d when the oil–water interface enters channel 1 or channel 2 were monitored. Table 4 shows the pressure at each point and the relevant pressure difference for the following four scenarios:

Table 4. Pressure at each point for different scenarios.

Pressure (Pa)	Scenario I	Scenario II	Scenario III	Scenario IV
p_{p1}	10,242.2	15,495.5	24,403.8	23,604.7
p_{p2}	6895.57	7199.75	22,444	24,639
p_{p3}	6876.09	7430.86	19,975.5	20,088.5
p_{p1}–p_{p3}	3366.11	8064.64	4428.3	3516.2
p_{p2}–p_{p3}	19.48	−231.11	2468.5	4550.5

Scenario I: the injection velocity is 0.005 $m \cdot s^{-1}$, water first enters channel 1 as shown in Figure 8f, and the oil–water interface moves to position S1 shown in Figure 8f.

Scenario II: the injection velocity is 0.005 $m \cdot s^{-1}$, water first enters channel 1 as shown in Figure 8f, and the oil–water interface moves to position S2 shown in Figure 8f.

Scenario III: the injection velocity is 0.01 $m \cdot s^{-1}$, water first enters channel 2 as shown in Figure 9d, and the oil–water interface moves to position S3 shown in Figure 9d.

Scenario IV: the injection velocity is 0.01 $m \cdot s^{-1}$, the water first enters the channel 2 shown in Figure 9d, and the oil–water interface moves to position S4 shown in Figure 9d.

It can be seen from Table 4 that the extent of the pressure increase at point $p1$ is much greater than that at point $p2$ when the water first flows into channel 1. The pressure difference in channel 1 increases greatly, while the pressure difference of the fluid in channel 2 reduces significantly. That is, the flow in channel 2 connected in parallel with channel 1 is inhibited to a large extent after the formation of the oil–water interface in channel 1. The pressure at point $p1$ decreases, while the pressure at points $p2$ and $p3$ increases when the water first flows into channel 2. The pressure difference in channel 2 increases, while the pressure difference in channel 1 decreases. That is, the flow of fluid in channel 1, which is connected in parallel with channel 2, is inhibited after the formation of the oil–water interface in channel 2. When the displacement force of the fluid is not sufficient to overcome the inhibiting effect of the capillarity, the oil–water two-phase flow stops, or a backflow (reverse imbibition) phenomenon is observed, thus reducing the oil recovery rate. Therefore, the inhibiting effect is stronger than the driving effect at an ultra-low injection velocity, which results in a relatively lower oil recovery rate.

A schematic conceptual model of two parallel channels—as shown in Figure 11—was constructed in this paper to further analyze the disturbing effect of capillary forces on the flow in parallel channels from the quantitative perspective.

$$q_1 = \frac{R_2}{R_1 + R_2} q - \frac{p_{c2}}{R_1 + R_2} + \frac{p_{c1}}{R_1 + R_2}, \tag{24}$$

$$q_2 = \frac{R_1}{R_1 + R_2} q - \frac{p_{c1}}{R_1 + R_2} + \frac{p_{c2}}{R_1 + R_2}, \tag{25}$$

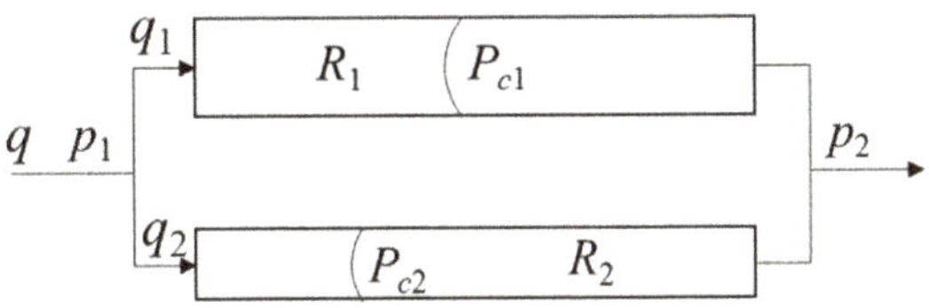

Figure 11. Schematic diagram of the parallel channels.

Figure 11 shows the schematic diagram of two parallel channels. As shown in Figure 11, q is the total flow rate of the two channels; q_1 and q_2 are the flow rates of channel 1 and channel 2, respectively; R_1 and R_2 are the viscous resistance of the two channels; and P_{c1} and P_{c2} are the capillary forces induced by the oil–water interface formed in the channels. The average flow rates of each channel expressed in Equations (24) and (25) are obtained by analyzing the flow in two parallel channels. It can be seen from the equations that the oil-water interface formed in a capillary tube will have an effect on the flow in its parallel channels. For instance, the capillary force induced by the formed oil–water interface in the channel (the capillary pressure is denoted as P_{c1}) will facilitate the increase of the fluid flow rate in channel 1 and the decrease of the fluid flow rate in channel 2. A similar effect is observed at the oil–water interface formed in channel 2. To summarize, the capillary force induced by the oil–water interface formed in a channel acting as a kind of driving force will inevitably impede the two-phase flow in its parallel channel, and vice versa. The magnitude of the impeding and promoting effect is closely related to the ratio of the capillary force to viscous resistance, and the ratio is the distributing amount between parallel-connected channels.

The oil and water distribution at different times at the injection velocity of 0.03 $m \cdot s^{-1}$ is shown in Figure 12. The capillary barrier may be formed along the path *a*–*n*, and the occurring time of the capillary barriers at these locations, the time of restart, the blockage duration, and the blockage ratio are shown in Table 5. As shown in Table 5, a persistent blockage is formed at positions *a* and *m*, while a transient blockages are observed at positions *f* and *n*. Blockage phenomena are not observed at other locations. The water phase advances along the direction of the main displacement under the pressure of the fluid, and a new oil–water interface is formed continuously. The newly formed oil–water interface will impede the lateral oil–water movement, which is referred to as the capillary blockage effect.

The blockage effect of the capillary force is illustrated in Figure 13. The displacement direction is from left to right, and the water first breaks through along channel b and channel c at the initial stage. A new oil–water interface is formed in channel a when the oil–water interface passes through position c. The capillary force induced by the oil–water interface under water-wet conditions will impede the further advancing of oil in channel a. This effect will result in the surplus of the oil in the lateral direction.

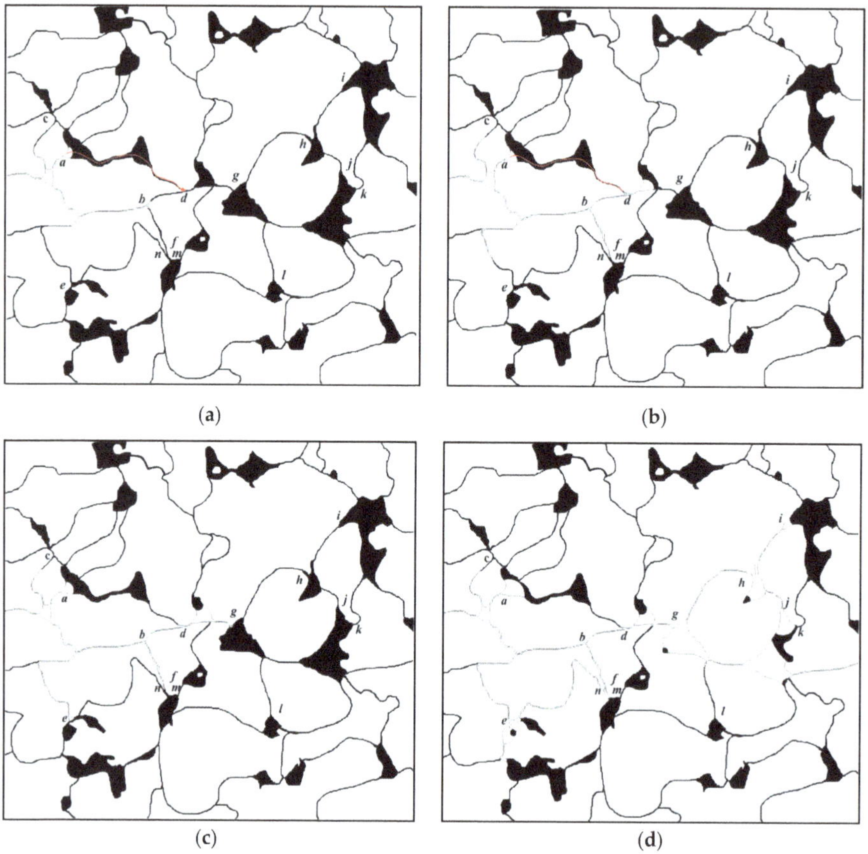

Figure 12. Oil and water distribution at different times, at the injection velocity of 0.03 m·s^{-1} (Ca = 4.29 × 10^{-4}): (**a**) 0.19 s, (**b**) 0.31 s, (**c**) 0.39 s, (**d**) final. Different positions are indicated by lower case letters.

Table 5. Blockage and restart information at different locations during the injection process, at the injection velocity of 0.03 m·s^{-1} (Ca = 4.29 × 10^{-4}).

Position	Blockage Time(s)	Restart Time(s)	Blockage Duration(s)	Blockage Ratio
a	0.210	∞	∞	∞
b	0.185	0.185	0	0
c	*	*	*	*
d	0.220	0.220	0	0
e	0.360	0.360	0	0
f	0.290	1.300	1.010	0.5940
g	0.390	0.390	0	0
h	0.700	0.700	0	0
i	0.980	0.980	0	0
j	1.100	1.100	0	0
k	*	*	*	*
l	*	*	*	*
m	1.350	∞	∞	∞
n	1.150	1.300	0.150	0.0882

∞ The movement of the oil–water interface at this position has not been resumed at the end of the displacement process. * The oil–water interface has not moved to this position.

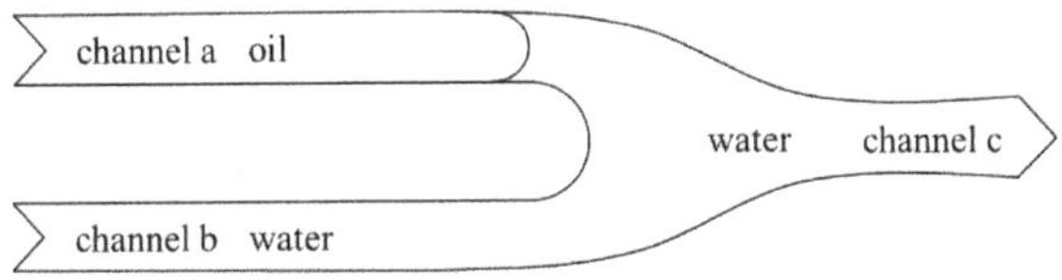

Figure 13. Schematic diagram of the blocking effect in the capillary tube. Different flow paths are distinguished by a–c.

As shown in Figure 12a, the oil in the pore space downstream of position *a* can enter the main channel through the *a–d* channel (colored in red) and flow downstream before the oil–water interface reaches position *d*. Once the water flows past position *d*, an oil–water interface will be formed at one end of position *d* in the *a–d* channel. The formed oil–water interface will impede the movement of oil and water in the *a–d* channel, which will reduce its movement speed or block the channel. A larger distance needs to be bypassed before the oil in the pore space downstream of position *a* enters the main channel, which reduces the speed of the oil movement in the lateral channel. The lateral blockage effect gradually becomes stronger as new oil–water interfaces continue to form downstream, and the distance the oil needs to bypass is greater, eventually causing the oil to fail to flow into the main channel from the lateral direction, which results in a great deal of residual oil being trapped in the pore space. Therefore, the faster the injection rate, the quicker the occurrence of the blocking effect induced by the newly formed oil–water interface. The failure of the oil in the lateral direction to enter the main channel in a timely manner will lead to the lower oil recovery rate. Too high an injection rate will facilitate the rapid formation of water-bearing channels in the main displacement direction, and will reduce the oil recovery rate.

5.2. *Oil–Water Two-Phase Flow Characteristics under Different Viscosity Ratios*

The variation of the final oil recovery rate with the oil–water viscosity ratio at an injection rate of 0.01 $m \cdot s^{-1}$ ($Ca = 1.43 \times 10^{-4}$) is shown in Figure 14. The water viscosity is set to a fixed value of 10^{-3} $kg \cdot m^{-1} \cdot s^{-1}$ in each simulation process, while the oil viscosity is constantly changing. It can be seen from Figure 15 that the oil recovery rate decreases with the increase of the oil–water viscosity ratio when the viscosity ratio is less than 7, while the increase of the viscosity ratio has little impact on the oil recovery rate when the viscosity ratio is greater than 7.

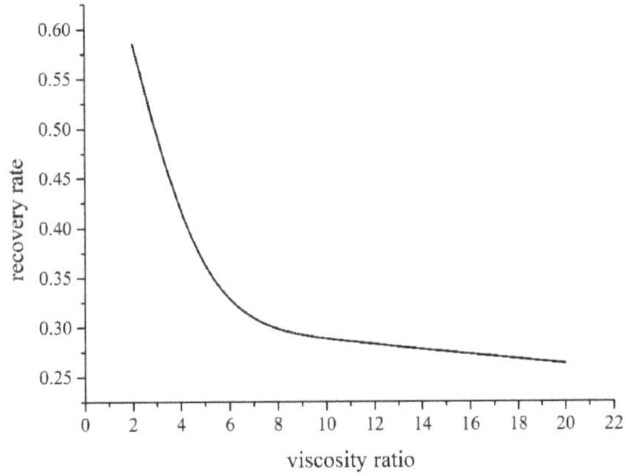

Figure 14. Variation of the final oil recovery rate with the oil–water viscosity ratios.

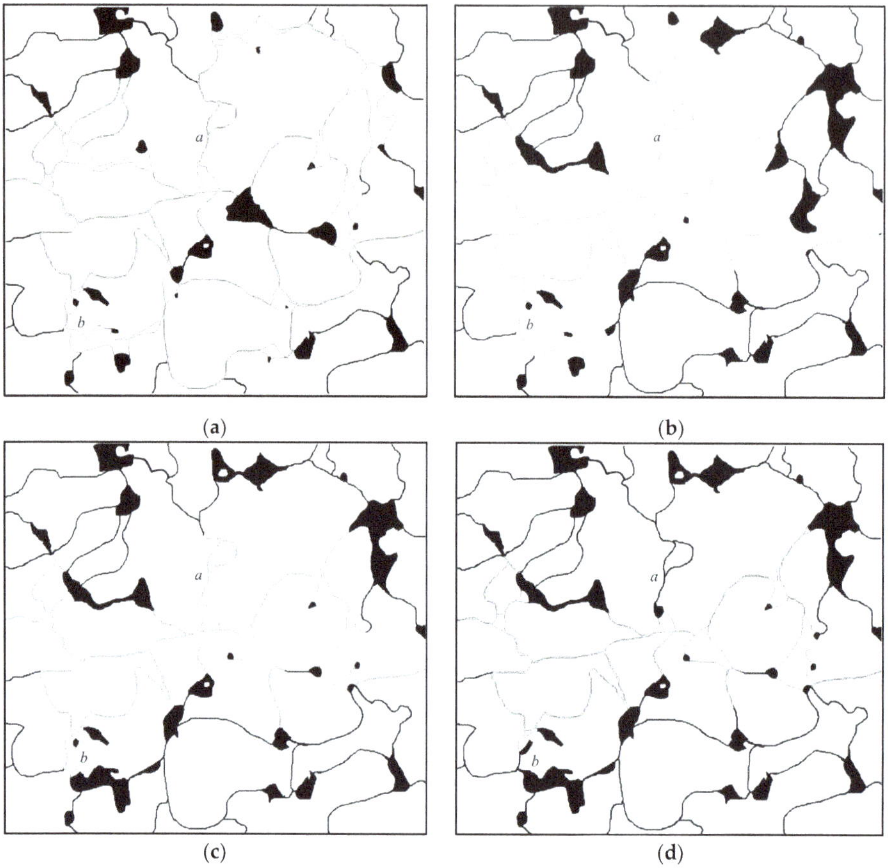

Figure 15. The final oil and water distribution at an injection rate of 0.01 $m \cdot s^{-1}$ with different viscosity ratios: (**a**) at the oil–water viscosity ratio of 2; (**b**) at the oil–water viscosity ratio of 5; (**c**) at the oil–water viscosity ratio of 10; (**d**) at the oil–water viscosity ratio of 20. Different positions are indicated by lower case letters.

Figure 15 shows the final oil and water distribution at an injection rate of 0.01 $m \cdot s^{-1}$ ($Ca = 1.43 \times 10^{-4}$) with different oil–water viscosity ratios. It can be seen from Figure 15a that the sweep space of water flooding is larger when the oil–water viscosity ratio is small. Multiple paths are formed from the inlet to the outlet, resulting in a higher final oil recovery rate. The degree of sweep of the water flooding on both sides of the main channel gradually decreases with the increase of the oil–water viscosity ratio (as shown in Figure 15b–d). The degree of sweep of the water flooding in channel *a* and channel *b* gradually decreases with the increase of the oil–water viscosity ratio.

This change is due to the dynamic inhomogeneity induced by the change of phase distribution during the displacement process, and the dynamic inhomogeneity can be elucidated using the parallel-connected channels shown in Figure 11. The capillary barrier effect is not taken into account in the parallel-connected channels, i.e., $P_{c1} = 0$ and $P_{c2} = 0$. Besides this, assuming that $R_1 < R_2$, according to Equations (24) and (25), the volumetric flow rate in channel 1 is greater than that in channel 2. The oil in the two parallel channels is gradually displaced by water, and the average viscosity of the fluid in the channels decreases as water advances within the two channels. The flow rate in channel 1 is greater and the resistance of the cannel decreases faster than that of channel 2, thus further increasing the flow rate of channel 1. The velocity of the fluid advancing in the channel is higher than that of channel 2, which leads to the viscous fingering in the channel. The larger the oil–water viscosity ratio, the stronger the dynamic inhomogeneity as the displacement process proceeds. Meanwhile, the phenomenon of viscous fingering is more obvious and the degree of sweep of the displacement is lower.

It can be known from Equations (24) and (25) that the flow rate of the two channels is influenced by the viscosity of oil and water, and the capillary effect. The effect of capillarity on the volumetric flow rate of the branch channels is closely related to the ratio of the capillary force to the sum of viscous resistance $R_1 + R_2$ of the two channels. As the displacement process proceeds, the oil in channel 1 and channel 2 is gradually displaced by water, R_1 and R_2 decrease, and the effect of capillarity on the distribution of the volumetric flow rate gradually intensifies. Providing that the oil–water viscosity ratio stays constant in all cases, the capillary effect plays a significant role in the oil displacement process when the absolute value of the viscosity is lower. Figure 16 shows the final oil and water distribution at different absolute viscosities, with a viscosity ratio of 20 at an injection rate of 0.01 m·s^{-1} (Ca = 1.43 × 10^{-4}). In this case, the oil–water viscosity ratio is the same, while the absolute viscosity is different. It can be seen from the figure that the oil in channel *a* is not displaced when the absolute oil–water viscosity is larger, and vice versa. Therefore, under the premise of the same oil–water viscosity ratio, the smaller the absolute viscosity of oil and water, the greater the role of the capillary effect in the flow. The capillary fingering is easier to form, and the lateral displacement of oil will be observed.

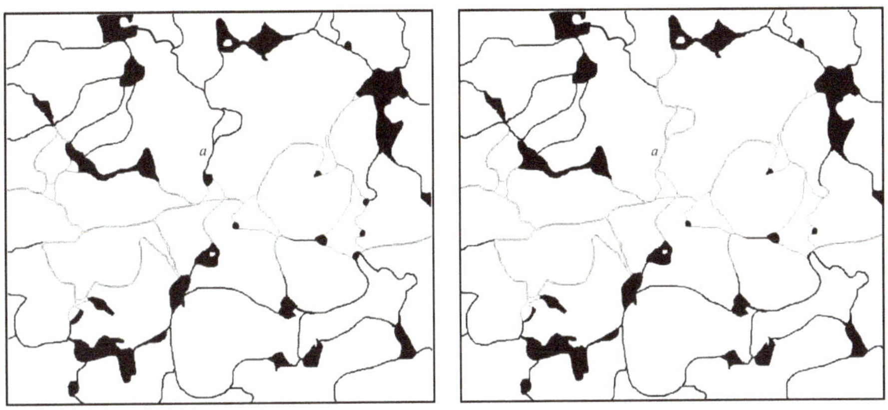

(**a**) (**b**)

Figure 16. The final oil and water distribution at different absolute viscosities, with a viscosity ratio of 20: (**a**) the effective viscosity of water is 0.001 kg·m^{-1}·s^{-1} and oil is 0.02 kg·m^{-1}·s^{-1}; (**b**) the effective viscosity of water is 5 × 10^{-4} kg·m^{-1}·s^{-1} and oil is 0.01 kg·m^{-1}·s^{-1}. Different positions are indicated by lower case letters.

5.3. *Characteristics of the Oil–Water Flow under Different Adjustment Strategies*

The adjustment of the injection direction, turning the oil well to the injection well, and increasing the water injection rate are important measures in the process of oil recovery, and are adopted after the initial water flooding. The spatial distribution of the oil and water is more complex compared with that of the pre-water flooding period. The numerical simulation of water flooding process was conducted using the physical model of the core shown in Figure 5 in order to create a more complex residual oil distribution before the pre-adjustment. Ports A, C and D are injection ports; port G is the outlet; and ports B, E and F are closed and set as walls. The injection velocity is 5 × 10^{-3} m·s^{-1} (Ca = 7.14 × 10^{-5}), the viscosity of oil is 0.02 kg·m^{-1}·s^{-1}, the viscosity of water is 0.001 kg·m^{-1}·s^{-1}, and the oil–water interfacial tension is 0.07 kg·m^{-2}. The final distribution of the residual oil is shown in Figure 17 after continuous water flooding until no oil is displaced from the outlet.

Figure 17. Oil and water distribution before adjustment.

5.3.1. Adjustment of the Injection and Extraction Direction

The oil and water distribution at different times after adjusting the injection and extraction direction (injection from port E and outflow from port F at an injection rate of 5×10^{-3} m·s^{-1}) are shown in Figure 18. A water-bearing channel is formed in the core after continuous water flooding before adjustment, and the remaining oil is observed in the upper and lower parts of the porous structure. There exists a capillary barrier at positions *a*, *b*, *c*, *d*, and *e* where the oil–water interface is stagnant. The displacement pressure is insufficient to overcome the capillary force induced by the capillary barrier and the oil–water interface stops advancing, which results in the remaining of oil in the pore space. The remaining oil on the upper side of the water-bearing channel moves downward under the pressure of displacement after the adjustment of the injection and extraction direction. As shown in Figure 18a, the oil in position *h* moves downward under the displacement pressure, and is split into two streams of fluid—*f* and *g*—at the bifurcation of the pore channel. The capillary force in channel *h* is a kind of driving force (oil displaced by water) due to the water-wet characteristic of the pore structure, while the capillary force (water displaced by oil) induced by the two-phase interface in channels *f* and *g* is a kind of resistance. The capillary force is relatively strong under low-velocity injection conditions, and the capillary resistance in channels *f* and *g* determines the distribution of the volumetric flow rate in the two channels. The final distribution of the volumetric flow rate in channels *f* and *g* is almost identical due to the similar radius of the two pore channels, as shown Figure 18b. As the displacement process proceeds, the flow stream in channel *f* is further divided into two streams along channels f_1 and f_2, respectively, and the amount of oil distributed in each channel is similar, as shown in Figure 18c. The oil distributed in channel f_2 is eventually displaced downstream, forming oil droplets in the pore space adhering to the pore wall (as shown in Figure 18d). Although there is a pressure difference upstream and downstream of channel f_1, as shown in Figure 18d, the magnitude of the capillary force at two ends of the remaining oil is different, and the difference of the capillary forces at two ends of the remaining oil balances the pressure difference between the two ends of the oil column, which leads to the remaining of the oil.

For the fluid below the water-bearing channel, the capillary barriers formed at positions *a*, *b*, *c*, *d*, and *e* have an inhibiting effect on the fluid flow (as shown in Figure 18a), while the capillary force is a kind of driving force at position *i* when the oil–water interface advances within the throat channel. Therefore, the interface at position *i* moves first in the displacement process (comparing the changes of Figure 18c with Figure 18b). The decrease of the capillary force at the interface of position *i* leads to increase of the displacement pressure upstream when water moves to the pore body. The oil–water interface at positions

a, b, c, d, and e resumes moving (as shown in Figure 18c). The channels where the interfaces at position a and position i are formed have low resistance, and the volumetric flow rate of each channel is larger. Eventually, two new oil–water interfaces at positions a_1 and a_2 are formed when the fluid flows downstream from the pore space at position a and then flows out of channel F. The capillary forces induced by the two newly formed interfaces will impede the movement of oil and water under water-wet conditions, causing the oil in the lower right corner to reach an equilibrium state.

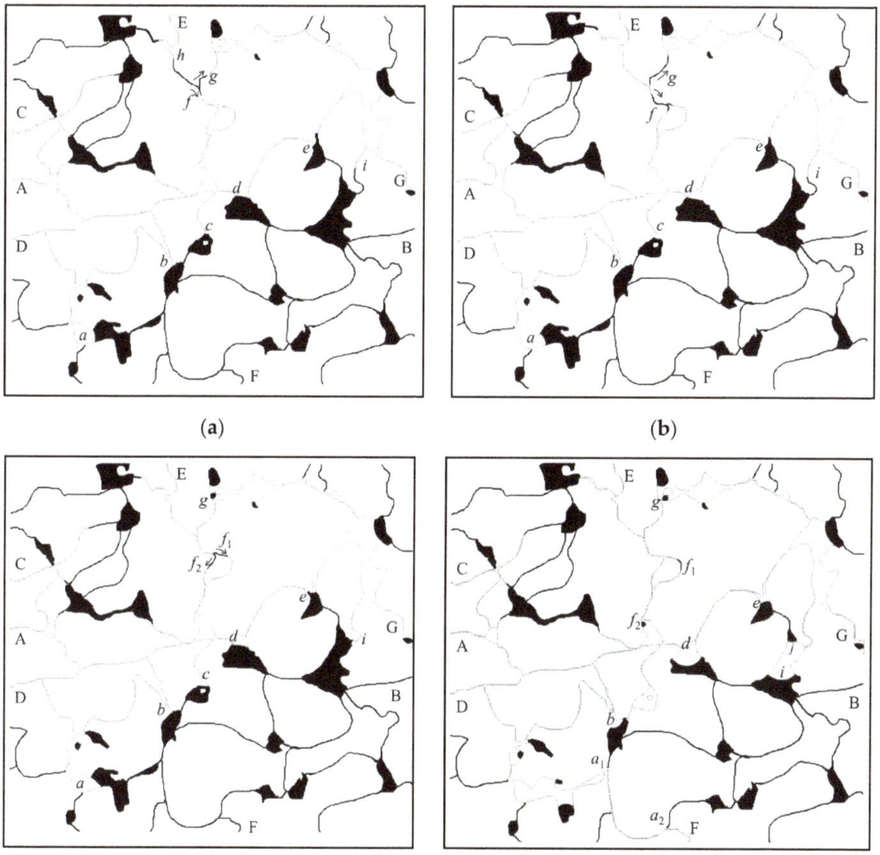

Figure 18. Changes of oil and water distribution after adjusting the injection and extraction direction: (**a**) 0.05 s after adjustment; (**b**) 0.1 s after adjustment, (**c**) 0.15 s, and (**d**) final. Different positions are indicated by lower case letters.

5.3.2. Turning the Extraction Well to the Injection Well

A water-bearing channel was formed between ports A and G in the pre-water-flooding process. No oil will be displaced from port G if water is continuously injected at the same rate. Turning the extraction-well to the injection-well method can be used to further improve the oil recovery rate. Figure 19 shows the oil and water distribution and flow path during the pre-displacement and post-displacement period, at an injection rate of 0.005 $m \cdot s^{-1}$. Water is injected from ports A and G, and F is the outlet. As shown in Figure 19, the light blue overlay area and light red overlay area are the control regions of injection port A and injection port G, respectively. The cyan line is the flow path of fluid flowing from port A, and the red line is the flow path of fluid flowing from port B. Points *a*–*n* are the intersection points of the flow paths.

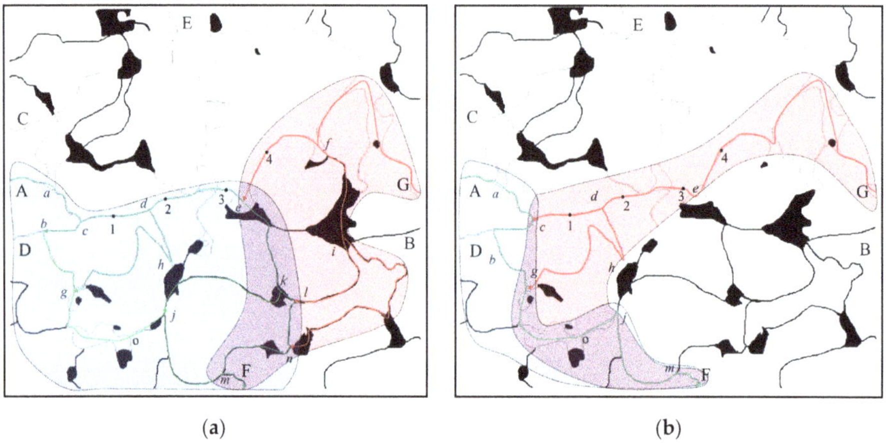

Figure 19. Flow paths and the residual oil distribution in different periods under simultaneous continuous injection conditions at port A and port G: (**a**) oil and water distribution during the pre-displacement, (**b**) final residual oil distribution.

Figure 19a shows the oil and water distribution in the mobilization process of the remaining oil during the pre-displacement. It can be seen from the figure that the control area of port A is much larger than that of injection port G. The fluid in port A plays a significant role in the mobilization of the oil in the channel along the path *bgjm*F, as well as the oil in the channel along path *eklnm*F and the channel along path *ekjm*F. The oil gradually flows out through outlet F as the displacement process proceeds, and the remaining oil in the channel along the path *bgjm*F is gradually displaced by water. A new oil–water interface is formed when the water front advances through the intersection point *j*. The remaining oil flowing through the *kj* channel to the *jm* channel will be prevented from moving by the newly formed oil–water interface. The movement of the oil in the *kj* channel will stop when the upstream displacement pressure is insufficient to overcome the resistance. The newly formed oil–water interface will prevent the migration of oil in channel *nm* to channel *m*F when the water front advances through the intersection point *m*. The final residual oil distribution and the flowing path is shown in Figure 19b.

In order to further clarify the change of the flowing path in the displacement process, flow velocities at the four selected points (points 1–4) shown in the figure were monitored. The velocity variation with time at these four points is shown in Figure 20. The velocity value is positive if the fluid flows from left to right, and the velocity value is negative if the fluid flows from right to left. It can be seen from the figure that the velocity of each point has been changing as the injection process proceeds. Small-scale fluctuations appeared in the local region, and dramatic changes of the velocity with time were observed. In addition, the velocity of points 1, 2, and 3 is reversed between 1 s and 1.5 s.

In order to account for the aforementioned velocity reversal, the forces exerted on the two-phase fluid within the pore space were analyzed in detail. The fluid flow within the pore space mainly resulted from the viscous action and the capillarity:

(1) Viscous action: The viscous action allows the fluid to flow at a more uniform velocity, and the viscosity will not purely cause large velocity fluctuations during the flow process. The viscous effect determines the magnitude of the volumetric flow rate in different channels when there is no oil–water interface formed. For instance, the volumetric flow rate of channels along the paths *iln* and *in* at intersection point *i* are different at the initial displacement stage.

(2) Capillarity: (a) The magnitude of the capillary force is influenced by the size of the pore channel, and changes with the positions of the oil–water interface, thus causing the acceleration or deceleration of the fluid in local regions. As shown in Figure 20, the velocity of point 2 at time I fluctuates dramatically (flow reversal), which is mainly due to the abrupt decrease of the pore channel when the oil–water interface moves to position *o*, as shown in Figure 19. When the fluid flows from the pore space to the throat channels, the capillary force will accelerate the flow rate of the local fluid, which is a type of driving force, resulting in the velocity decrease of point 2 first, and then the velocity acceleration of point 2 in the reverse direction. (b) Influenced by the abrupt change of the pore channel and the wettability conditions of the wall, the capillary force may show resistance or driving force characteristics. The oil–water interface stops moving if the displacement pressure is insufficient to overcome the capillary resistance, prompting the flow path to change. As shown in Figure 20 at time II (1.12 s), the large velocity fluctuations of point 2 and the velocity reversal of point 3 can be observed. The oil–water interface stops moving when the displacement pressure is insufficient to overcome the capillary resistance induced by the newly formed interface at position *j*, and the original flow path *kj* is blocked, resulting in the reversal of the flow path *ed*. Therefore, the main reason for the fluid reversal at points 1, 2, and 3 is the capillary blockage effect.

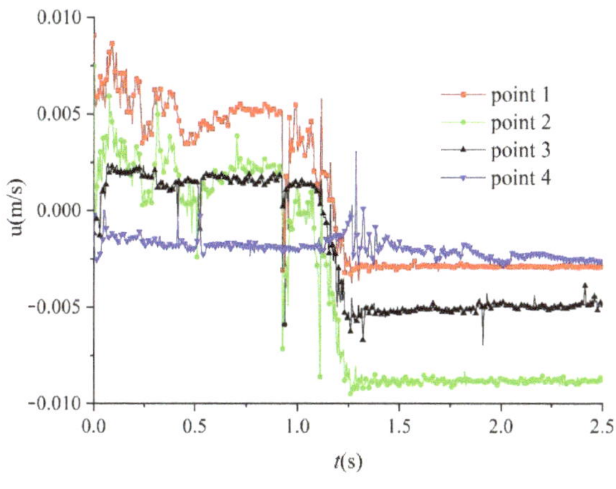

Figure 20. Velocity variation of different detection points with time.

Figure 21 shows the variation of the inlet pressure and the oil saturation of the core with time. It can be seen from Figure 21a that the oil saturation of the core gradually decreases with time, and the oil saturation of the core is hardly changed after 1.25 s. The pressure at the left and right injection points almost remains constant before the formation of the oil–water interface at the intersection *j* (1.12 s) when injecting water from the left and right points simultaneously, and the oil saturation in the core remains almost constant with time during this period. The pressure variation is significantly different from that under single-point injection conditions. The pressure difference at the inlet and outlet gradually decreases with the decrease of the oil saturation in the core under single-point injection conditions due to the lower viscosity of water compared with that of oil. The precondition of the drawn conclusion is that the channels dominated by the injection port almost remain unchanged. However, the system has the effect of stabilizing the pressure when high water-bearing channels are formed between the two ports by injecting water into the core from the two ports, and the process of stabilizing the pressure is achieved by adjusting the flowing paths.

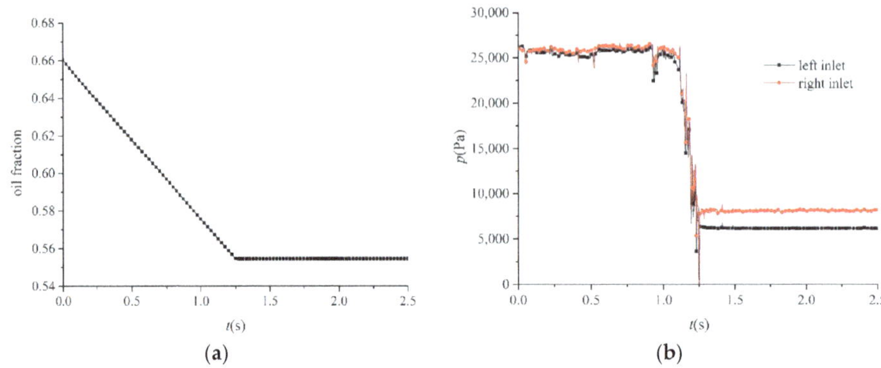

Figure 21. Variation of the inlet pressure and oil saturation with time: (**a**) oil saturation in the core sample with time; (**b**) pressure at the injection port with time.

5.3.3. Increasing the Water Injection Rate

Figure 22a shows the distribution and flow paths of the final residual oil after increasing the injection velocities. The black color in the figure is the remaining oil, and the red, light blue and cyan colors are the flow paths. Water is injected into the pore space at a rate of 0.005 m/s from ports A, D and C, and flows out from port G. As shown in the figure, there are clusters of residual oil in area I and II, and the rest of the residual oil is trapped in dead-end pore space. For the remaining oil in area I, the interfaces of the oil and water are located at positions *a*, *b*, *c*, *d* and *e*, respectively. Positions *a*, *b*, and *c* are located upstream, and *d* and *e* are located downstream. The capillary barrier formed due to the abrupt change of the pore radius at positions *a*, *b*, and *c* impede the further movement of the oil–water interface. The oil–water interface has an inhibiting effect on the movement of the oil–water two-phase flow when the oil–water interface at positions *d* and *e* move downstream (water displaced by oil). *a*′, *b*′, *c*′, *d*′, and *e*′ are points on the flow path connected to the channel where points *a*, *b*, *c*, *d*, and *e* are located, respectively. There is no movement of water in channels *a*′*a*, *b*′*b*, *c*′*c*, *d*′*d* or *e*′*e* due to the barrier effect of the oil–water interface, and due to there being no pressure drop in the channels. The oil and water movement will stop if the displacement pressure on the flow path is insufficient to overcome the capillary barrier pressure and the resistance of the water displacement by the oil. For instance, the oil in the channel between points *a* and *e* will not be mobilized when the pressure difference between points *a*′ and *e*′ is insufficient to overcome the capillary resistance (water displaced by oil) formed at points *a* and *e*. The dual resistance effect of the capillary barrier (point a) and the capillary blockage effect (point e) is the main reason for the formation of residual oil under water-wet conditions. Under water flooding conditions, increasing the volumetric flow rate of channels can mobilize the oil in channels *a*–*e*. The oil will flow if the pressure difference between points *a*′ and *e*′ on the flow path is sufficient to overcome the capillary resistance formed at points *a* and *e*. The formation process of residual oil in region II is similar to the formation process of residual oil in region I.

Figure 22b shows the final residual oil distribution after water flooding at an injection rate of 0.005 $m \cdot s^{-1}$ (as shown in Figure 22a) with 0.01 $m \cdot s^{-1}$ liquid extraction. From the perspective of the distribution and flow path characteristics, the mobilization effect of the residual oil is not obvious, and no flowing path is formed. It can be known from the oil–water interface position that interface *f* moves forward a little. Accordingly, interface *l* slightly moves forward along the elongated throat channel. Finally, the oil–water interface reaches a new equilibrium and the oil stops moving. The relationship between the injection rate and the mobilized amount of the remaining oil used is not always linear. Only when the injection rate reaches a certain threshold value and the pressure difference formed at the different oil–water interfaces of the same remaining oil is higher than the dual resistance of the capillary barrier and capillary blockage, will the oil be mobilized.

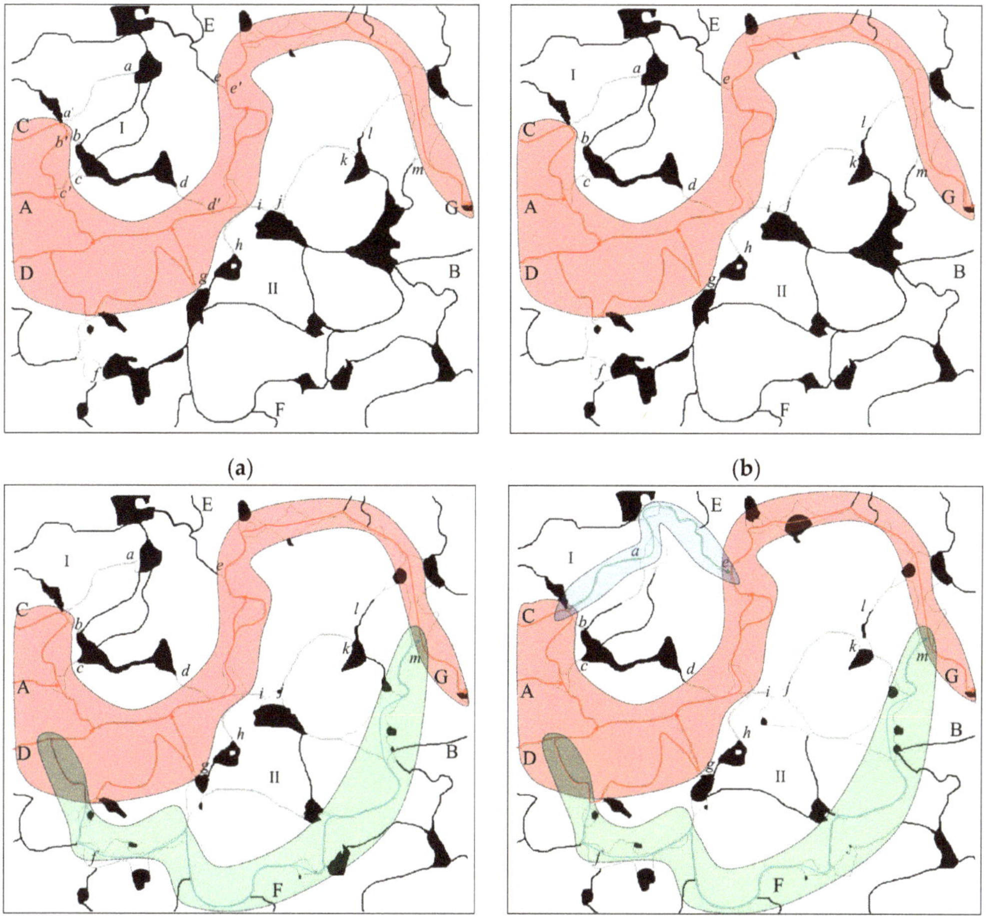

(**a**) (**b**) (**c**) (**d**)

Figure 22. Final residual oil distribution and flow path characteristics after increasing the water injection rate: (**a**) residual oil after water flooding, at an injection velocity of 0.005 m·s^{-1}; (**b**) final residual oil at injection velocity of 0.01 m·s^{-1} based on the results at a velocity of 0.005 m·s^{-1}; (**c**) final residual oil at an injection velocity of 0.015 m·s^{-1} based on the results at a velocity of 0.005 m·s^{-1}; (**d**) final residual oil at injection velocity of 0.02 m·s^{-1} based on the results at a velocity of 0.005 m·s^{-1}. Different positions are indicated by lower case letters.

Figure 22c,d shows the residual oil at injection rates of 0.015 m·s^{-1} and 0.02 m·s^{-1} after the initial water injection at the rate of 0.005 m·s^{-1}, respectively. The remaining oil is mobilized to some extent, and new flow paths are formed when increasing the injection rate of the water. For instance, the residual oil in region II is mobilized and a new flow path is formed (light blue) at the injection rate of 0.015 m·s^{-1}, while the residual oil in region I and II is mobilized and two new flowing paths are formed (cyan and light blue) when the injection rate is increased to 0.02 m·s^{-1}. Once a flow channel fully filled with water is formed between the main inlet and the flow outlet, increasing the water injection rate is an important measure to improve the oil recovery rate.

6. Conclusions

In this paper, the water flooding process in low-permeability sandstone oil reservoirs was numerically simulated using a Navier–Stokes equation coupled with the VOF method in order to systematically investigate the effects of viscous action and capillarity on the oil–water two-phase flow dynamics of water flooding. By studying the oil–water two-phase flow dynamics under different injection rates, the viscosity ratios of oil to water, and different adjustment strategies after water flooding, the following conclusions can be drawn:

(1) The larger the viscosity ratio is, the stronger the dynamic inhomogeneity will be as the displacement process proceeds, and the greater the difference in the distribution of volumetric flow rate in different channels, which will lead to the formation of a growing viscous fingering phenomenon, thus lowering the oil recovery rate. Under the same viscosity ratio, the absolute viscosity of the oil and water will also have an essential impact on the oil recovery rate by adjusting the relative value of viscous action and capillarity. The change of the dynamic inhomogeneity induced by the viscous effect is a process of gradual change, and does not cause abrupt changes of the fluid velocity in the pore space. In the case of unidirectional displacement, the flow path does not change, the oil with high viscosity is gradually replaced by the oil with low viscosity in the pore channels, and the pressure difference between the inlet and the outlet decreases continuously as the displacement process proceeds. The two-way displacement strategy promotes the fluid to flow along the vertical direction of the displacement. The control regions of the injection with two ports constantly change as the displacement proceeds, and thereby the process of stabilizing the pressure is achieved by adjusting the flow paths.

(2) Pore-scale phenomena induced by the capillary effect have a crucial impact on the pore-scale flow dynamics. The capillary barrier in the main displacement direction causes the flow of the fluid in lateral direction, and the capillary barrier in the lateral direction will impede the further flow of the fluid. Reducing the angle between the displacement direction and the direction of the capillary barrier pressure induced by adjusting the displacement direction can further improve the sweep area of the displacement fluid. The capillary imbibition will accelerate the fluid in the channel, and has an inhibiting effect on the fluid in parallel channels. The capillary blocking effect induced by the newly formed interface at the pore intersections can result in the sudden change of the fluid flow in the pore space.

There is an optimal displacement velocity for a specific pore structure. The capillary barrier in the lateral direction is difficult to break through at too low a displacement velocity, and the inhibiting effect of channels with percolation on parallel channels will be intensified, which affects the final oil recovery rate. Too high a displacement velocity will prompt the rapid formation of a new oil–water interface, and will form blockage in the measurement channel, thus reducing the final oil recovery rate. The dual effect of the capillary barrier and capillary blockage is the direct cause of the residual oil formation, and the main reason that residual oil can be mobilized only under a certain dynamic condition. The production of the remaining oil can only be completed under certain power conditions. Increasing the injection rate step by step is an important measure to improve oil recovery.

Author Contributions: Conceptualization, J.S. and T.N.; methodology, M.X. and J.S.; investigation, M.X. and B.H.; writing—original draft preparation, T.N. and B.H.; writing—review and editing, L.W. and C.H. All authors have read and agreed to the published version of the manuscript.

Funding: This research was funded by the National Natural Science Foundation of China, grant number: 12072256; the Natural Science Basic Research Plan in Shaanxi Province of China, grant number: 2021JM-406; the Excellent Youth Foundation of SINOPEC (P20009); and the Key Research and Development Program of Shaanxi (Program No. 2021GXLH-Z-071).

Institutional Review Board Statement: Not applicable.

Informed Consent Statement: Not applicable.

Data Availability Statement: Some or all data, models, or code that support the findings of this study are available from the corresponding author.

Conflicts of Interest: The authors declare no conflict of interest.

References

1. Chai, R.; Liu, Y.; He, Y.; Liu, Q.; Xue, L. Dynamic behaviors and mechanisms of fluid-fluid interaction in low salinity waterflooding of carbonate reservoirs. *J. Pet. Sci. Eng.* **2021**, 109256. [CrossRef]
2. Zallaghi, M.; Khaz'Ali, A.R. Experimental and modeling study of enhanced oil recovery from carbonate reservoirs with smart water and surfactant injection. *Fuel* **2021**, *304*, 121516. [CrossRef]
3. Su, J.; Wang, L.; Gu, Z.; Zhang, Y.; Chen, C. Advances in Pore-Scale Simulation of Oil Reservoirs. *Energies* **2018**, *11*, 1132. [CrossRef]
4. Chai, G.L.; Wang, L.; Gu, Z.L.; Yu, C.; Zhnag, Y.; Shu, Q.; Su, J. A consistent sharp interface fictious domain method for moving boundary problems with arbitrarily polyhedral mesh. *Int. J. Numer. Methods Fluids* **2021**, *93*, 2065–2088. [CrossRef]
5. Su, J.; Chai, G.; Wang, L.; Yu, J.; Cao, W.; Gu, Z.; Chen, C.; Meng, W. Direct numerical simulation of particle pore-scale transport through three-dimensional porous media with arbitrarily polyhedral mesh. *Powder Technol.* **2020**, *367*, 576–596. [CrossRef]
6. Su, J.; Chai, G.; Wang, L.; Cao, W.; Yu, J.; Gu, Z.; Chen, C. Direct numerical simulation of pore scale particle-water-oil transport in porous media. *J. Pet. Sci. Eng.* **2019**, *180*, 159–175. [CrossRef]
7. Su, J.; Chai, G.; Wang, L.; Cao, W.; Gu, Z.; Chen, C.; Xu, X.Y. Pore-scale direct numerical simulation of particle transport in porous media. *Chem. Eng. Sci.* **2019**, *199*, 613–627. [CrossRef]
8. Lenormand, R.; Touboul, E.; Zarcone, C. Numerical models and experiments on immiscible displacements in porous media. *J. Fluid Mech.* **1988**, *189*, 165–187. [CrossRef]
9. Swpu, P. Review on chemical enhanced oil recovery using polymer flooding: Fundamentals, experimental and numerical simulation. *Petroleum* **2020**, *6*, 115–122.
10. Kamal, M.S.; Hussein, I.A.; Sultan, A.S. Review on Surfactant Flooding: Phase Behavior, Retention, IFT, and Field Applications. *Energy Fuels* **2017**, *31*, 7701–7720. [CrossRef]
11. Liu, X.; Wang, L.; Wang, J.; Su, J. Pore-Scale Simulation of Particle Flooding for Enhancing Oil Recovery. *Energies* **2021**, *14*, 2305. [CrossRef]
12. Homsy, G.M. Viscous fingering in porous media. *Annu. Rev. Fluid Mech.* **1987**, *19*, 271–311. [CrossRef]
13. Cottin, C.; Bodiguel, H.; Colin, A. Drainage in two-dimensional porous media: From capillary fingering to viscous flow. *Phys. Rev. E* **2010**, *82*, 046315. [CrossRef] [PubMed]
14. Tsuji, T.; Jiang, F.; Christensen, K. Characterization of immiscible fluid displacement processes with various capillary numbers and viscosity ratios in 3D natural sandstone. *Adv. Water Resour.* **2016**, *95*, 3–15. [CrossRef]
15. Hu, R.; Lan, T.; Wei, G.-J.; Chen, Y.-F. Phase diagram of quasi-static immiscible displacement in disordered porous media. *J. Fluid Mech.* **2019**, *875*, 448–475. [CrossRef]
16. Singh, K.; Jung, M.; Brinkmann, M.; Seemann, R. Capillary-Dominated Fluid Displacement in Porous Media. *Annu. Rev. Fluid Mech.* **2019**, *51*, 429–449. [CrossRef]
17. Yang, Y.; Cai, S.; Yao, J.; Zhong, J.; Zhang, K.; Song, W.; Zhang, L.; Sun, H.; Lisitsa, V. Pore-scale simulation of remaining oil distribution in 3D porous media affected by wettability and capillarity based on volume of fluid method. *Int. J. Multiph. Flow* **2021**, *143*, 103746. [CrossRef]
18. Guo, Y.; Zhang, L.; Zhu, G.; Yao, J.; Sun, H.; Song, W.; Yang, Y.; Zhao, J. A Pore-Scale Investigation of Residual Oil Distributions and Enhanced Oil Recovery Methods. *Energies* **2019**, *12*, 3732. [CrossRef]
19. Berg, S.; Ott, H.; Klapp, S.A.; Schwing, A.; Neiteler, R.; Brussee, N.; Makurat, A.; Leu, L.; Enzmann, F.; Schwarz, J.-O.; et al. Real-time 3D imaging of Haines jumps in porous media flow. *Proc. Natl. Acad. Sci. USA* **2013**, *110*, 3755–3759. [CrossRef]
20. Roof, J. Snap-off of oil droplets in water-wet pores. *Soc. Pet. Eng. J.* **1970**, *10*, 85–90. [CrossRef]
21. Li, Z.; Gu, Z.; Li, R.; Wang, C.; Chen, C.; Yu, C.; Zhang, Y.; Shu, Q.; Su, J. Investigation on droplet dynamic snap-off process in a short, abrupt constriction. *Chem. Eng. Sci.* **2021**, *235*, 116496. [CrossRef]
22. Cha, L.M.; Xie, C.Y.; Feng, Q.H.; Balhoff, M. Geometric criteria for the snap-off of a non-wetting droplet in pore-throat channels with rectangular cross-sections. *Water Resour. Res.* **2021**, *57*. [CrossRef]
23. Rabbani, H.S.; Zhao, B.; Juanes, R.; Shokri, N. Pore geometry control of apparent wetting in porous media. *Sci. Rep.* **2018**, *8*, 1–8. [CrossRef]
24. Ubbink, O. *Numerical Prediction of Two Fluid Systems with Sharp Interfaces*; University of London: London, UK, 1997.
25. Issa, R.I. Solution of the implicitly discretised fluid flow equations by operator-splitting. *J. Comput. Phys.* **1986**, *62*, 40–65. [CrossRef]
26. OpenCFD, OpenFOAM. Available online: www.openfoam.com (accessed on 13 September 2021).
27. Wang, L.; Tian, W.-X.; Zhao, X.-Y.; Huang, C.-Q. Numerical simulation of the effects of canopy properties on airflow and pollutant dispersion in street canyons. *Indoor Built Environ.* **2021**. [CrossRef]

28. Raeini, A.Q.; Blunt, M.J.; Bijeljic, B. Direct simulations of two-phase flow on micro-CT images of porous media and upscaling of pore-scale forces. *Adv. Water Resour.* **2014**, *74*, 116–126. [CrossRef]
29. Jasak, H.; Weller, H.G.; Gosman, A.D. High resolution NVD differencing scheme for arbitrarily unstructured meshes. *Int. J. Numer. Methods Fluids* **1999**, *31*, 431–449. [CrossRef]
30. Crank, J.; Phyllis, N. A practical method for numerical evaluation of solutions of partial differential equations of the heat-conduction type. *Math. Proc. Camb. Philos. Soc.* **1947**, *43*, 431–449. [CrossRef]

energies

MDPI

Article

Influence of the Addition of Silica Nanoparticles on the Compressive Strength of Cement Slurries under Elevated Temperature Condition

Anna Pikłowska [1], Jan Ziaja [1] and Marcin Kremieniewski [2,*]

1 Department of Drilling and Geoengineering, Faculty of Drilling, Oil and Gas, AGH University of Science and Technology, 30-059 Kraków, Poland; annapiklowska@gmail.com (A.P.); ziaja@agh.edu.pl (J.Z.)
2 Oil and Gas Institute—National Research Institute, 31-503 Krakow, Poland
* Correspondence: kremieniewski@inig.pl

Abstract: Drilling ever deeper, and thus in increasingly difficult conditions, is associated with restrictive requirements that must be met by cement slurries. This implies the need to use advanced, innovative measures that will significantly improve the performance parameters of the cement slurry and cement stone. Due to its unique properties, an admixture of nanosilica improves the properties of the cement stone and allows for appropriate zone insulation. The article presents the results of strength tests of cement stone samples with the addition of silica nanoparticles deposited in an environment of increased temperature of 90 °C. In all three cases of modification with an admixture of nanosilica (type 1, 2 and 3, concentration 0.5%, 1% and 5%), the cement stone shows an improvement in mechanical properties, which is manifested by an increase in compressive strength. The most homogeneous results of strength measurements are for cement slurries with an admixture of type 3 nanosilica (the highest average strength: 132–149% in relation to the base sample). They show the smallest stretch marks and deviations from the average. The highest average increase in strength is for the sample with the addition of 1% nanosilica (on average 124% in relation to the base sample). This amount causes the greatest increase in strength with no significant deterioration of rheological parameters.

Keywords: well cementing; improved borehole sealing; nanosilica; cement stone; cement slurry; mechanical parameters

Citation: Pikłowska, A.; Ziaja, J.; Kremieniewski, M. Influence of the Addition of Silica Nanoparticles on the Compressive Strength of Cement Slurries under Elevated Temperature Condition. *Energies* **2021**, *14*, 5493. https://doi.org/10.3390/en14175493

Academic Editors: Pål Østebø Andersen and George Avgouropoulos

Received: 30 July 2021
Accepted: 28 August 2021
Published: 3 September 2021

1. Introduction

In the petroleum industry, the purpose of cementing in boreholes is primarily to create a resistant and durable barrier against fluid migration. Migration takes place between zones in rock formations. It is important to obtain a permanent connection of the cement sheath in the annular space between the borehole wall and the casing column. Additionally, the cement stabilizes the casing and protects it against corrosion and shock loads. The task of the cement sheath is also to seal the mud escape zone and other sections of the hole where drilling is difficult. Cement used in drilling will hydrate under different conditions than the cement used in conventional construction. The high temperature and pressure at great depths are a major obstacle here. The presence of highly mineralized reservoir brines and hydrogen sulfide, which adversely affect the durability of the cement sheath, is also important [1,2]. The increase in global energy consumption and the growing demand for fossil fuels as the dominant energy source significantly accelerated the development of new technologies in hydrocarbon production [3,4]. Drilling deeper and deeper creates more and more difficult conditions for the cement slurry and increases the requirements [5,6]. Cement slurries must be able to cope with them, therefore it implies the need to use advanced, innovative measures that will improve the functional parameters of cement slurry and cement stone [7–10]. Such activities allow for obtaining appropriate zone insulation.

For this purpose, the use of nanotechnology is very beneficial, which has become one of the paths of development and progress in recent years [11,12]. It allows you to design, manufacture, study and apply structures at the level of single atoms and particles with nanometric sizes—in which at least one of the size dimensions is smaller than 100 nm [13]. Thanks to the rapid development of this field, a new class of materials, such as nanoparticles, is widely studied in order to provide simpler and more economical hydrocarbon exploration and production processes, especially in difficult reservoir conditions [14–16]. In the last dozen or so years, huge efforts have been made to develop methods of producing nanometric-sized materials. Nanosized particles can be prepared by physical, chemical or biological methods, depending on the nature of the materials and the contribution of chemical reactions [17]. The synthesis of nanoparticles can be classified as either "top-down" or "bottom-up". In a synthetic top-down method, nanoparticles are typically made by comminuting with a variety of physical and/or chemical treatments. There are mechanical milling, ball milling, chemical etching, thermal ablation, laser ablation, explosion and spraying. In the bottom-up approach, nanoparticles are formed by smaller entities such as atoms, molecules, and smaller particles as building blocks that are combined to form the final product. Typically, this method relies on chemical or biological mechanisms. Examples are chemical or electrochemical precipitation, vapor deposition, atomic or molecular condensation, sol-gel processes, spray pyrolysis, laser pyrolysis, aerosol processes, and chemical or biological reduction. Due to their large surface area, high aspect ratio, small size, low density, large surface area and interesting physicochemical properties, they have a strong influence on the mechanical properties of cement materials [18,19]. Research on the use of nanomaterials as additives to cement slurries began relatively recently [20]. The literature [11,21–23] provides examples of the use of nanoparticles as well as nano-SiO_2 and nano-Fe_2O_3 additives improving the strength parameters of the cement matrix. In addition to improving strength, the addition of nano-Fe_2O_3 improves the ability to monitor stresses in the cement stone [24–26]. Increasingly, scientists are also using less popular nanoxides, such as zinc and titanium nanoxides [21,26–28]. Some authors also conducted research on the influence of carbon nanotubes on the mechanical parameters of the hardened cement slurry, however, the research in this area is ambiguous [8].

Nanosilica in Cement Slurries

One of the important additives in cement is silica, which is used in a certain amount to counteract the loss of strength. The use of nanosilica in cement resulting from the Stober method (a variant of the sol-gel method) improves the properties of hardened cement. Due to the extremely small size, high sphericity and relatively high quality, nanoparticles are preferred because their abrasive action is negligible with a lower impact of kinetic energy [1,9]. When nanosilica is used, the mechanism of its action is multifaceted. First of all, the addition of nanosilica increases the efficiency of C-S-H gel formation by promoting pozzolanic reactions. It influences the improvement of the microfilling effect of the cement matrix space. Silica nanoparticles act as nucleation sites for hydration products, and due to their high surface energy, cement hydration products are deposited on them. On the nano-SiO_2 surface, the hydration products grow into conglomerates containing nanoparticles as nuclei. The addition of nanosilica controls crystallization, limiting the amount of formation of large portlandite crystals. The synergistic effect of all these mechanisms results in the formation of a compact, durable and impermeable microstructure [18,19,28]. With the right composition, a higher packing density results in lower water requirements for the mixture and contributes to increased strength due to reduced capillary porosity (Figure 1).

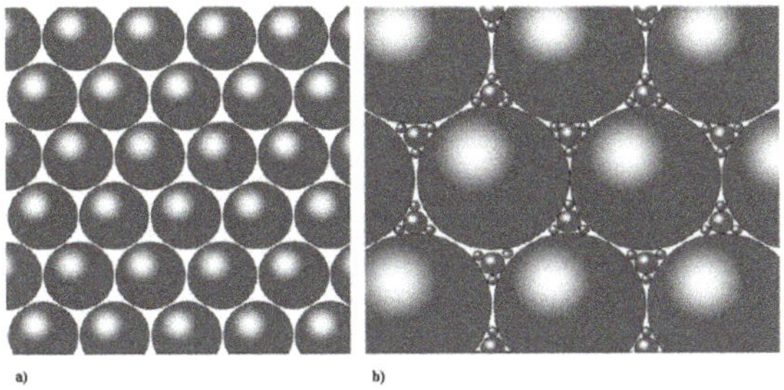

Figure 1. (**a**) Packing density of a single size system. (**b**) Increased packing density of a system using an additive with different particle sizes Reprinted with permission from ref. [19]. Copyright 2013 Woodhead Publishing: Sawston, UK.

The nanoparticles have a large specific surface area, thus ensuring high chemical reactivity [29,30]. As a result, the nano-SiO_2 present in the cement slurry additionally supports and accelerates the hydration process (Figure 2) [31]. Therefore, it seems advisable to add nanosilica particles to cement mixtures in order to obtain a better quality cement slurry [32,33].

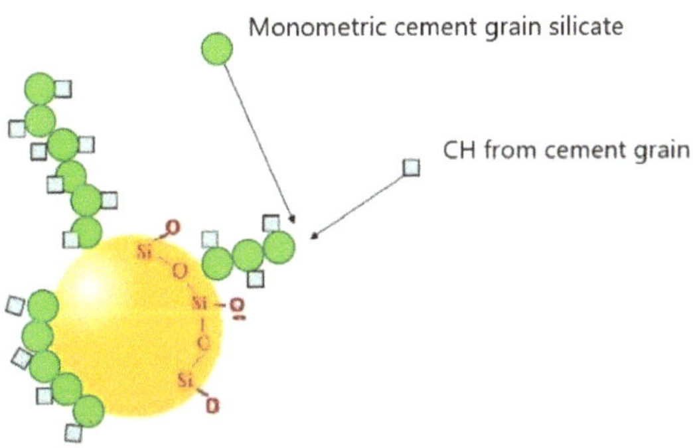

Figure 2. Mechanism of action of nanosilica. Monomeric silicates dissolved from the clinker phases are attracted by colloidal silica. The formed "silicate polymer" is stabilized with calcium hydroxide [manufacturer's materials, www.akzonobel.com (accessed on 6 March 2021)].

Based on the research, it is stated that the smaller the size of nanosilica particles, and therefore the larger the specific surface area, the higher the compressive strength of the tested samples [34,35]. However, the analysis of the results of research on nanosilica shows large discrepancies in the results. The reasons for these discrepancies are the differences in the methods and conditions under which the experiments are conducted. The molar ratio of the reactants, the type of silica precursor, the shape (nanospheres, fibers) and the type of nanoparticles, as well as the duration of the reaction in the sol-gel method and the degree of particle dispersion in the cement matrix are important. An important issue is the stability of the obtained mixtures due to the content of nanoparticles, as it depends to a degree on their size (Figure 3). Therefore, this article presents the effect of nanosilica admixtures of various sizes and concentrations on the mechanical properties of cement slurry matured under elevated temperature conditions.

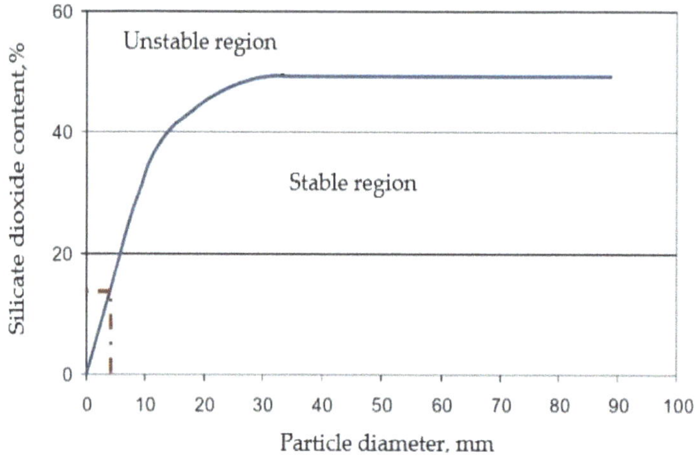

Figure 3. The dependence of the possible concentration of silica nanoparticles on the particle size. Reprinted with permission from ref. [9]. Copyright 2018 Egypt. J. Pet.

Patil and Deshpande [7] described the reports on the positive effect of nanosilica on the properties of cement slurries deposited in the borehole environment. They document a case where nanosilica was used in cementitious compositions to obtain high early strength. The addition of 0.2 gal/sk of nanosilica, the rate of strength development increased from 172 to 460 psi/hr. The ultimate strength of cement composition was found to be three times that of the control and control plus microsilica. At the same time, Santra, Boul, and Pang [16] confirmed the positive effect of nanosilica, indicating that improvements in properties attributed to the incorporation of nanosilica into cement-based materials include increased early and final compressive strength. The nanosilica synthesized by Qalandari, Aghajanpour, and Khatibi [29] significantly increased the uniaxial compressive strength of the cement system (an increase of 16.59% with an addition of only 0.25% by weight). El-Gamal [2] claims that the addition of 1% nanosilica to oilwell cement leads to a marked increase in the compressive strength values at all ages of hydration, as compared to those of the neat paste (after 1, 3, 7 and 14 days). The positive effect of nanoparticles is also described by Dębińska [15,21]. One of the groups of nanoparticles she studies is nanosilica. She notes that the slurry with the addition of nanosilica, tested at 90 °C, was characterized by good rheological parameters, and the cement stone obtained from it was characterized by high values of early compressive strength (18 MPa after 24 h, 21 MPa after 48 h) and strength after 28 days reaching about 28 MPa. It is worth noting that the cement stones obtained from slurries with the addition of nanosilica were characterized by significantly higher values of compressive strength compared to the base slurry. Therefore, she concludes that these slurries can be the basis for the development of recipes for which high mechanical strength is required.

2. Materials and Methods

2.1. Materials

2.1.1. Cement

Class G German cement (Dyckerhoff) was used for the tests. The specific surface area (measured by the Blaine method) was 3462 cm^2/g. The alkali content was kept at 0.89%, chlorides at 0.061% and sulfates at 3.23%. The loss on ignition was 3.03%, while the insoluble residue was 0.54%. The stability of the volume was determined at the level of 0.44 mm.

2.1.2. Silica Nanoparticles

Three types of water dispersion of nanosilica particles were used for the tests. The additives differ in properties, mainly in the size of the particles.

Silica 1 is an alkaline, aqueous dispersion of colloidal silica that contains about 50% by weight solids. The silica dispersion is stabilized with sodium and the amorphous silica particles have a negative surface charge. SiO_2 particles have a smooth, spherical shape and a wide size distribution, on average around 32 nm. It is physically a white liquid, slightly more viscous than water.

Silica 2 is a transparent, aqueous suspension of silicon oxide nanopowder particles that contains 25% by weight solids. SiO_2 particles have a spherical shape, particle size: 30 nm (Figure 4).

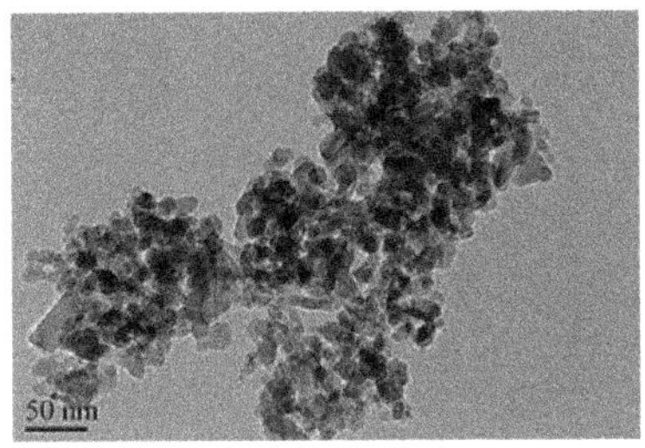

Figure 4. Type 2 nanosilica [manufacturer's materials, www.us-nano.com (accessed on 14 March 2021)].

Silica 3 is a transparent, aqueous suspension of silicon oxide nanopowder particles that contains 25% by weight solids. SiO_2 particles have a spherical shape, particle size: 5–35 nm.

2.2. Preparation of Samples

In order to eliminate the influence of many variables on the result of the experiment, the water-cement ratio (w/c) of all tested cement slurries was equal to 0.5. The slurry was made by dosing the appropriate amount of tap water and cement. Before mixing them, additives regulating the technological parameters of the slurry were introduced into the mixing water, i.e., defoamer, fluidizing agent, delaying setting and regulating filtration. Selected types of nanosilica in concentrations (0.5%, 1%, 5%) were introduced into the solution prepared in this way, followed by cement. The cement slurry prepared in this way was mixed for 30 min with a rotational speed of 160 rpm. The adopted mixing conditions corresponded to the slurry preparation technology in borehole conditions. Due to the form of nanosilica addition (water dispersion), the calculated amount of water resulting from the assumed w/c ratio was reduced by appropriate values. For comparison, a reference cement slurry was also prepared (without the addition of silica). The cement slurry was poured into 4 × 4 × 4 cm cubic molds and placed in an autoclave at 90 °C and 80% humidity for 24 h. Then, for the next seven days, the samples were placed in water at 90 °C (Figure 5). Detailed methodology chart is in Table 1. Compressive strength tests were performed from samples seasoned in this way.

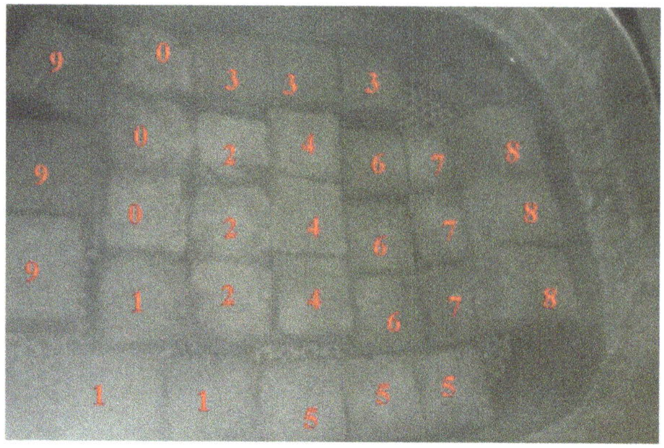

Figure 5. Autoclave samples (photo by the author).

Table 1. Detailed methodology chart.

w/c		**0.5**
Concentrations of nanosilica		0%, 0.5%, 1%, 5%
Conditions for first 24 h	humidity	80%
	temperature	90 °C
Conditions for first 24 h	humidity	100%
	temperature	90 °C

2.3. Experimental Procedures

The mechanical strength was determined using a testing machine (hydraulic press) model E183 PN 100 Matest. It was designed to determine the bending and compressive strength of cement stone. The test consists in measuring the crushing force needed to destroy the sample. The tested samples were loaded with a speed dependent on the expected compressive strength.

3. Results and Discussion

The results of the compressive strength measurements are presented in the table (Table 2) and the graph (Figure 6). Tables 3 and 4 present the statistical processing of the obtained results.

Table 2. Results of compressive strength of samples seasoned for 7 days (MPa).

System No.	Concentration of Nanosilica	Sample 1	Sample 2	Sample 3	Average Value
0	0.0%	22.8	24.31	18.87	21.99
1	0.5%	13.15	15.34	3.97	10.82
2	1.0%	19.52	27.02	19.62	22.05
3	5.0%	12.31	23.82	23.61	19.91
4	0.5%	26.66	25.6	24.16	25.47
5	1.0%	30.55	23.09	27.4	27.01
6	5.0%	32.11	29.08	25.62	28.94
7	0.5%	27.8	31.21	28.36	29.12
8	1.0%	31.5	36.89	29.88	32.76
9	5.0%	30.19	33.1	31.37	31.55

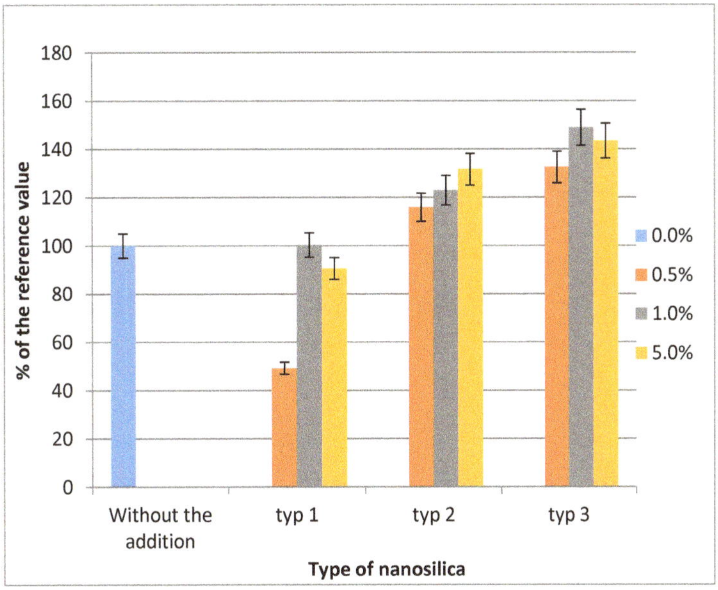

Figure 6. Compressive strength change in relation to the type and amount of nanosilica additive.

Table 3. Selected statistical parameters of the obtained results.

System No.	Max. (MPa)	Min. (MPa)	Range (MPa)	Range as% of the Mean Value (%)	Standard Deviation (MPa)	Standard Deviation as% of the Mean (%)	% of the Reference Value (%)
0	24.31	18.87	5.44	24.73	2.81	12.77	100.00
1	15.34	3.97	11.37	105.08	6.03	55.75	49.20
2	27.02	19.52	7.5	34.01	4.30	19.51	100.27
3	23.82	12.31	11.51	57.80	6.59	33.07	90.54
4	26.66	24.16	2.5	9.81	1.25	4.93	115.82
5	30.55	23.09	7.46	27.62	3.75	13.86	122.83
6	32.11	25.62	6.49	22.43	3.25	11.22	131.57
7	31.21	27.8	3.41	11.71	1.83	6.28	132.42
8	36.89	29.88	7.01	21.40	3.67	11.20	148.94
9	33.1	30.19	2.91	9.22	1.46	4.64	143.47

Table 4. Change in the strength of samples in relation to the unmodified sample.

	No Add-on	Type 1	Type 2	Type 3	Average Value for the Selected Concentration (MPa)	Average as Reference (%)
0.5%	-	10.82	25.47	29.12	21.81	99.15
1.0%	-	22.05	27.01	32.76	27.27	124.01
5.0%	-	19.91	28.94	31.55	26.80	121.86
0%	21.99	-	-	-	-	-
Average for a given type (MPa)	-	17.60	27.14	31.14	-	-
Average as reference (%)	-	80.00	123.41	141.61	-	-

The article presents the effect of the size distribution of silica nanoparticles (types of nanosilica) and their concentration on the mechanical properties of cement slurry (compressive strength) matured under elevated temperature conditions. Compressive strength was chosen as the evaluation parameter. This parameter is the most basic criterion and is crucial for the further drilling of the hole. It is also very important due to the universality of the research method, as well as its connection with structural parameters. The initial hypothesis of the research undertaken assumed the improvement of the strength of samples containing nanosilica, due to the several times' greater specific surface of these nanoparticles. Analyzing the results (Table 4), it can stated that the highest percentage increase in compressive strength (on average by 41.6%) was obtained for type 3 nanosilica with different particle sizes. All types of nanoparticles had the same order of diameters, but the specific surface area varied considerably due to the differences in the particle size distribution. Improvement in strength occurs in the case of slurries with the addition of type 2 nanosilica (on average by 23.4%). However, the strength increase in relation to the reference cement slurry was lower than that of the slurries with the addition of type 3 nanosilica. In the case of type 1 nanosilica addition, the lack of strength increase of the samples was related to their uneven distribution in the cement matrix. This problem also applied to higher concentrations (5%) for type 3 nanosilica. Then the high surface energy contributed to the agglomeration of particles and, due to the too high water demand of nanoparticles, it was problematic to mix the cement even in the initial period. Probably the increase in the concentration of the superplasticizer allowed for a more even distribution of SiO2 nanoparticles in the cement slurry, even with their high content.

Cement slurries with the addition of type 1 nanosilica were the most diverse in terms of strength. In all cases, they showed the greatest difference between the extreme values (range). In the case of 0.5% concentration, the range value was as much as 105% of the mean value, and the standard deviation was 55.75% of the mean. This was due to the much lower strength of one of the samples due to its defect. However, due to the small sample size, it was not decided to reject it earlier. For the concentration of 1% and 5% of nanosilica of this type, the range value was also relatively high, 34.01% and 57.8%, respectively (standard deviations of 19.51% and 33.07% of the mean, respectively). This proves the high heterogeneity in terms of strength. For cement slurries with type 2 nanosilica admixture, the strength measurements were more homogeneous. For increasing concentrations, the range was 9.81%, 27.62% and 22.43% of the mean, respectively, which translates into a smaller standard deviation of 4.93%, 13.86% and 11.22% of the mean. The most homogeneous results of strength measurements were observed for cement slurries with an admixture of type 3 nanosilica. They present the smallest stretch marks and deviations from the average. The range was 11.71%, 21.4% and 9.22% of the mean, respectively, while the standard deviation was 6.28%, 11.2% and 4.64% of the mean. This indicates the group of results most concentrated around the mean. In all three cases there was an increasing trend, i.e., an increase in strength along with an increase in nanosilica content. The analysis of the linear correlation for all three types of admixture shows the greatest dependence for type 2 (Pearson correlation coefficient 0.94, coefficient $R^2 = 0.8771$. In other cases the linear correlation was not very clear (for type 1 admixture: Pearson coefficient equal to 0, 43, $R^2 = 0.1851$) or basically did not occur (for type 3 admixture, Pearson's coefficient was at the level of 0.29, $R^2 = 0.084$). On this basis, it can be concluded that the relationship between the content of the admixture and the increase in strength for groups 1 and 3 is nonlinear (Figures 7–9).

Comparing the obtained results for individual concentrations, it can be stated that the highest average increase in strength occurred for the sample with the addition of 1% nanosilica. This amount caused the greatest increase in strength with no significant deterioration in rheological parameters.

The mechanism responsible for improving the compactness of the microstructure and increasing the strength of cement stones can be presented as follows. When small amounts of nanoparticles are well dispersed in the cement slurry, the cement hydration products begin to deposit on the nanoparticles due to their high surface energy, and during hydration, begin to grow into conglomerates containing nanoparticles as nuclei. The nanoparticles located in the cement slurry will additionally support and accelerate cement hydration due to their high reactivity. This is due to the pozzolanic reaction in the cement and the effect of filling the pores, as a result of which the porosity of the cement system decreased. The use of colloidal silica accelerates the dissolution of C_3S and the rapid formation of the CSH phase in the cement slurry, and prevents the formation of large crystals, such as $Ca(OH)_2$. By achieving an even dispersion of nanoparticles, it is possible to obtain an appropriate microstructure with good dispersed conglomerates characterized by high mechanical strength.

The addition of silica nanoparticles affects the morphology and mineralogy of hydration products and the rate of their formation. The additional amount of CSH gel formed and the density of the structure is a key factor for which the slurry strength and durability depend. Taking all this into account, it can be confirmed with certainty that nanosilica particles can effectively increase the compressive strength of class G cement deposited in the oilwell.

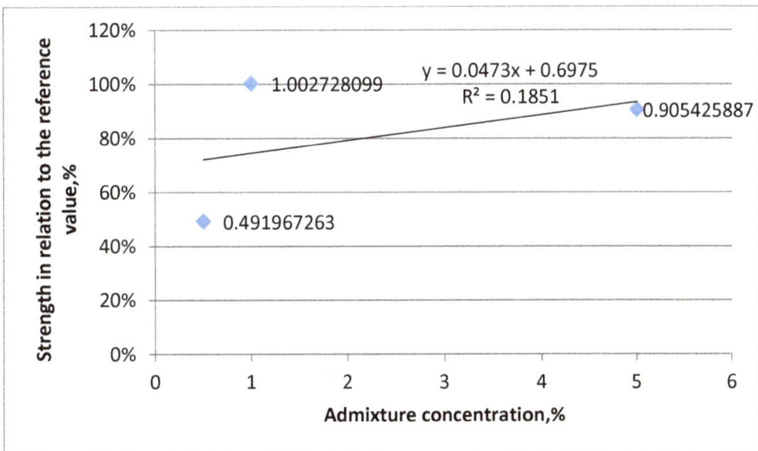

Figure 7. Compressive strength change depending on the type and amount of nanosilica additive—correlation analysis for cement slurry with nanosilica type 1.

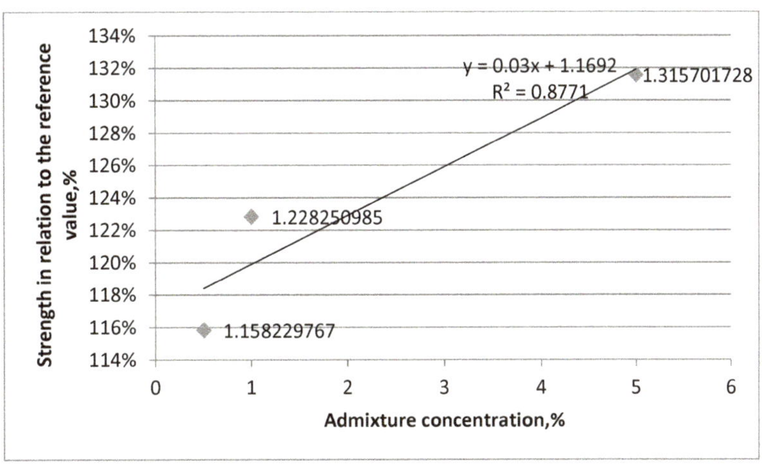

Figure 8. Compressive strength change depending on the type and amount of nanosilica additive—correlation analysis for cement slurry with nanosilica type 2.

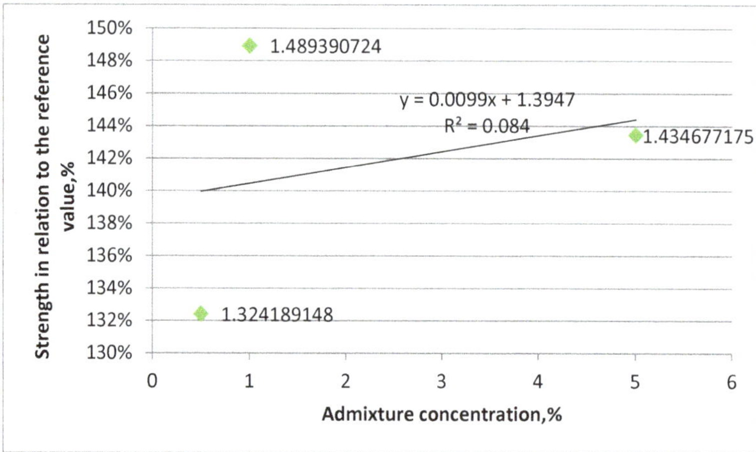

Figure 9. Compressive strength change depending on the type and amount of nanosilica additive—correlation analysis for cement slurry with nanosilica type 3.

4. Conclusions

- The addition of each of the tested types of nanosilica increases the strength, but the increase is different due to the type and concentration of the admixture.
- Type 1 nanosilica admixture causes the smallest increase in strength. This can be explained by the uneven dispersion of the admixture in the cement matrix. There is no linear relationship between the increase in strength and the increasing concentration of nanosilica.
- The admixture of type 2 nanosilica causes a greater increase in strength than type 1, it also has a smaller scatter of results and shows a linear trend between the increase in strength and the increase in concentration.
- Type 3 nanosilica admixture causes the greatest increase in strength, as evidenced by the highest values and the smallest dispersion. It may be the result of a tighter filling of the cement matrix with particles with the smallest and most diverse particles. There is no linear dependence of the increase in strength on the increasing concentration of this type of nanosilica.
- The concentration of 0.5%, a fairly low concentration, resulted in the slightest increase in strength, in one case this increase is even less noticeable due to a defective sample.
- For the concentration of 1%, the addition of nanosilica in such an amount caused the greatest increase in strength and small scatter of results probably due to good dispersion in the cement matrix.
- The concentration of 5%, a relatively high concentration, resulted in a noticeable increase in strength, but it was also the cause of difficulties in the dispersion of nanosilica particles in the cement matrix.

On the basis of the obtained results, we conclude that the optimal amount of nanosilica is a concentration equal to 1% (bwoc). Cement stone modified with the addition of nanosilica shows an improvement in mechanical properties, which is manifested by an increase in compressive strength. However, the underlying problem is the difficulty in properly dispersing the nanoparticles. The next step to confirm the positive effect of nanosilica is the study of the microstructure and permeability of cement stone carried out in order to visualize the condition of the cement matrix of slurries modified with the addition of nanosilica deposited in a borehole-like environment.

Author Contributions: Methodology, A.P.; software, M.K.; formal analysis, A.P. and J.Z.; data curation, A.P. and M.K.; writing—original draft preparation, A.P.; writing—review and editing, J.Z. and M.K.; visualization, J.Z.; project administration, J.Z.; funding acquisition, A.P. and J.Z. All authors have read and agreed to the published version of the manuscript.

Funding: The work was financially supported by Ministry of Science and Higher Education Warsaw (Internal order Oil and Gas Institute—National Research Institute Project No. 0015/KW/21) and by AGH-UST in Kraków—Faculty of Drilling, Oil and Gas, Drilling and Geoengineering Department, project No. 16.16.190.779.

Institutional Review Board Statement: Not applicable.

Informed Consent Statement: Not applicable.

Data Availability Statement: Not applicable.

Acknowledgments: The author thanks the anonymous reviewers for their constructive comments and the editor for handling the paper.

Conflicts of Interest: The author declares no conflict of interest.

References

1. Shadravan, A.; Amani, M. HPHT 101—What Every Engineer or Geoscientist Should Know about High Pressure High Temperature Wells. Presented at the SPE Kuwait International Petroleum Conference and Exhibition, Kuwait City, Kuwait, 10 December 2012. [CrossRef]
2. El-Gamal, S.M.; Hashem, F.S.; Amin, M.S. Influence of carbon nanotubes, nanosilica and nanometakaolin on some morphological-mechanical properties of oil well cement pastes subjected to elevated water curing temperature and regular room air curing temperature. *Constr. Build. Mater.* **2017**, *146*, 531–546. [CrossRef]
3. Murtaza, M.; Rahman, M.K.; Al-Majed, A.A. Effect of Nanoclay on Mechanical and Rheological Properties of Oil Well Cement Slurry under HPHT Environment. Presented at the International Petroleum Technology Conference, Bangkok, Thailand, 12 November 2016. [CrossRef]
4. Li, L.; Yuan, X.; Xu, X.; Li, S.; Wang, L. Vital Role of Nanotechnology and Nanomaterials in the Field of Oilfield Chemistry. Presented at the International Petroleum Technology Conference, Beijing, China, 26 March 2013. [CrossRef]
5. Hoelscher, K.P.; Young, S.; Friedheim, J.; De Stefano, G. Nanotechnology Application in Drilling Fluids. Presented at the Offshore Mediterranean Conference and Exhibition, Ravenna, Italy, 13 March 2013.
6. Kong, X.; Ohadi, M.M. Applications of Micro and Nano Technologies in the Oil and Gas Industry-An Overview of the Recent Progress. Presented at the Abu Dhabi International Petroleum Exhibition and Conference, Abu Dhabi, United Arab Emirates, 1 November 2010. [CrossRef]
7. Patil, R.; Deshpande, A. Use of Nanomaterials in Cementing Applications. Presented at the SPE International Oilfield Nanotechnology Conference and Exhibition, Noordwijk, The Netherlands, 12 June 2012. [CrossRef]
8. Khalil, M.; Jan, B.M.; Tong, C.W.; Berawi, M.A. Advanced nanomaterials in oil and gas industry: Design, application and challenges. *Appl. Energy* **2017**, *191*, 287–310. [CrossRef]
9. Mohammed, A.S. Vipulanandan models to predict the electrical resistivity, rheological properties and compressive stress-strain behavior of oil well cementmodified with silica nanoparticles. *Egypt. J. Pet.* **2018**, *27*, 1265–1273. [CrossRef]
10. Fakoya, M.F.; Shah, S.N. Emergence of nanotechnology in the oil and gas industry: Emphasis on the application of silica nanoparticles. *Petroleum* **2017**, *3*, 391–405. [CrossRef]
11. Dębińska, E. Niekonwencjonalne zaczyny cementowe z dodatkiem nanokrzemionki. *Nafta-Gaz* **2015**, *5*, 290–300.
12. Vryzas, Z.; Mahmoud, O.; Nasr-El-Din, H.A.; Kelessidis, V.C. Development and Testing of Novel Drilling Fluids Using Fe_2O_3 and SiO_2 Nanoparticles for Enhanced Drilling Operations. Presented at the International Petroleum Technology Conference, Doha, Qatar, 6 December 2015. [CrossRef]
13. Cocuzza, M.; Pirri, F.; Rocca, V.; Verga, F. Is the Oil Industry Ready For Nanotechnologies? Presented at the Offshore Mediterranean Conference and Exhibition, Ravenna, Italy, 23–25 March 2011.
14. Gibson, S. Novel Solution to Cement Strength Retrogression. Presented at the SPE/IADC Drilling Conference and Exhibition, Amsterdam, The Netherlands, 1 March 2011. [CrossRef]
15. Murtaza, M.; Rahman, M.K.; Al-Majed, A.A.; Samad, A. Mechanical, Rheological and Microstructural Properties of Saudi Type G Cement Slurry with Silica Flour Used in Saudi Oil Field under HTHP Conditions. Presented at the SPE Saudi Arabia Section Technical Symposium and Exhibition, Al-Khobar, Saudi Arabia, 19 May 2013. [CrossRef]
16. Santra, A.; Boul, P.J.; Pang, X. Influence of Nanomaterials in Oilwell Cement Hydration and Mechanical Properties. Presented at the SPE International Oilfield Nanotechnology Conference and Exhibition, Noordwijk, The Netherlands, 12 June 2012. [CrossRef]
17. Kim, J.J.; Rahman, M.K.; Al-Majed, A.A.; Al-Zahrani, M.M.; Taha, M.M. Nanosilica effects on composition and silicate polymerization in hardened cement paste cured under high temperature and pressure. *Cem. Concr. Compos.* **2013**, *43*, 78–85. [CrossRef]

18. Singh, L.P.; Zhu, W.; Howind, T.; Sharma, U. Quantification and characterization of C-S-H in silica nanoparticles incorporated cementitious system. *Cem. Concr. Compos.* **2017**, *79*, 106–116. [CrossRef]
19. Constantinides, G. Nanoscience and nanoengineering of cement-based materials. In *Nanotechnology in Eco-Efficient Construction*; Woodhead Publishing: Sawston, UK, 2013. [CrossRef]
20. Yuan, B.; Wang, Y.; Yang, Y.; Xie, Y.; Li, Y. Wellbore sealing integrity of nanosilica-latex modified cement in natural gas reservoirs with high H2S contents. *Constr. Build. Mater.* **2018**, *192*, 621–632.
21. Dębińska, E. Wpływ nanotlenków glinu i cynku na parametry świeżego i stwardniałego zaczynu cementowego. *Nafta-Gaz* **2016**, *4*, 251–261. [CrossRef]
22. Chartier, M.A.; Thompson, S.; Bordieanu, M.; Bustamante, G.; Saunders, J.R.; Kaiser, T.M. Performance Characterization and Optimization of Cement Systems for Thermally Stimulated Wells. Presented at the SPE Canada Heavy Oil Technical Conference, Calgary, AB, Canada, 9 June 2015. [CrossRef]
23. Saebom, K.; Chun, H. Use of nanoparticles for oil production applications. *J. Pet. Sci. Eng.* **2019**, *172*, 97–114.
24. Lau, H.C.; Yu, M.; Nguyen, Q.P. Nanotechnology for Oilfield Applications: Challenges and Impact. Presented at the Abu Dhabi International Petroleum Exhibition & Conference, Abu Dhabi, United Arab Emirates, 7 November 2016. [CrossRef]
25. Lau, H.C.; Yu, M.; Nguyen, Q.P. Nanotechnology for oilfield applications: Challenges and impact. *J. Pet. Sci. Eng.* **2017**, *157*, 1160–1169. [CrossRef]
26. Wang, C.; Chen, X.; Wei, X.; Wang, R. Can nanosilica sol prevent oil well cement from strength retrogression under high temperature? *Constr. Build. Mater.* **2017**, *144*, 574–585. [CrossRef]
27. Ghoddousi, P.; Javid, A.A.; Zareechian, M. Physical and chemical effects of siliceous particles at nano, micro, and macro scales on properties of self-consolidating mortar overlays. *Constr. Build. Mater.* **2018**, *189*, 1140–1154. [CrossRef]
28. Murtaza, M.; Mahmoud, M.; Elkatatny, S.; Majed, A.A.; Chen, W.; Jamaluddin, A. Experimental Investigation of the Impact of Modified Nano Clay on the Rheology of Oil Well Cement Slurry. Presented at the International Petroleum Technology Conference, Beijing, China, 22 March 2019. [CrossRef]
29. Qalandari, R.; Aghajanpour, A.; Khatibi, S. A Novel Nanosilica-Based Solution for Enhancing Mechanical and Rheological Properties of Oil Well Cement. Presented at the SPE Asia Pacific Oil and Gas Conference and Exhibition, Brisbane, Australia, 23 October 2018. [CrossRef]
30. Friedheim, J.; Young, S.; De Stefano, G.; Lee, J.; Guo, Q. Nanotechnology for Oilfield Applications—Hype or Reality? Presented at the SPE International Oilfield Nanotechnology Conference and Exhibition, Noordwijk, The Netherlands, 12 June 2012. [CrossRef]
31. Horszczaruk, E.; Mijowska, E.; Cendrowski, K.; Mijowska, S.; Sikora, P. The influence of nanosilica with different morphology on the mechanical properties of cement mortars. *Cem. Wapno Beton* **2013**, *18*, 24–32.
32. Luo, Z.; Li, W.; Wang, K.; Shah, S.P. Research progress in advanced nanomechanical characterization of cement based materials. *Cem. Concr. Compos.* **2018**, *94*, 277–295.
33. Bera, A.; Belhaj, H. Application of nanotechnology by means of nanoparticles and nanodispersions in oil recovery—A comprehensive review. *J. Nat. Gas Sci. Eng.* **2016**, *34*, 1284–1309.
34. Mahmoud, A.A.; Elkatatny, S.; Mahmoud, M. Improving Class G Cement Carbonation Resistance Using Nanoclay Particles for Geologic Carbon Sequestration Applications. Presented at the Abu Dhabi International Petroleum Exhibition & Conference, Abu Dhabi, United Arab Emirates, 12 November 2018. [CrossRef]
35. Murtaza, M.; Rahman, M.K.; Al Majed, A.A.; Tariq, Z.; Mahmoud, M. Scratch Test for Strength and Toughness of Oil Well Cement with Nanoclay as an Additive. Presented at the Abu Dhabi International Petroleum Exhibition & Conference, Abu Dhabi, United Arab Emirates, 11 November 2019. [CrossRef]

energies

MDPI

Article

Increasing the Efficiency of Sealing the Borehole in Terms of Spacer Pumping Time

Marcin Kremieniewski *, Miłosz Kędzierski and Sławomir Błaż

Oil and Gas Institute—National Research Institute, 25A Lubicz Str., 31-503 Krakow, Poland; kedzierski@inig.pl (M.K.); blaz@inig.pl (S.B.)
* Correspondence: marcin.kremieniewski@inig.pl

Abstract: The tightness of a borehole is essential for its long-term durability. For this purpose, the column of the pipe is sealed with cement slurry. After contacting the slurry, mud in the borehole is removed. However, the slurry does not effectively remove the remaining drilling mud. Therefore, the annular space is cleaned with a wash. Effectively cleaning the borehole presents quite a problem, as many variables that affect the stability of the borehole need to be considered. The time of contact between the borehole and the wash is very important. On the one hand, insufficient contact time does not guarantee proper removal of the mud. On the other hand, a long contact time may destroy the wall of the borehole. To address these problems, studies were carried out to assess the effect of the wash contact time on annular space cleaning. When determining the time of washing, a compromise between effective cleaning and the stability of the borehole wall is required. In the research presented in this publication, the simplest wash was used, i.e., water. This choice was based on the objective of observing the influence of the wash time on cleaning, i.e., the preparation of the borehole for cementing. By using water, the physicochemical action of surfactants can be ignored. In order to capture changes in cleaning due to differences in contact time, a control test was performed using a pure sandstone core without mud. The effect of the wash contact time on the cleaning of the annular space was investigated by determining the adhesion of the cement sheath to the rock core. First, mud was formed on the core, and then it was removed. By comparing the obtained adhesion to the reference sample, the effectiveness of the deposit removal was determined. On the basis of this research, the optimal wash contact time was determined.

Keywords: drilling fluids; wash; mud cake; annular space; borehole cleaning; cementing; wash contact time

Citation: Kremieniewski, M.; Kędzierski, M.; Błaż, S. Increasing the Efficiency of Sealing the Borehole in Terms of Spacer Pumping Time. *Energies* **2021**, *14*, 6702. https://doi.org/10.3390/en14206702

Received: 2 September 2021
Accepted: 12 October 2021
Published: 15 October 2021

1. Introduction

To remove drilling fluid and prepare the annular space for the addition of cement slurry, a sequence of drilling fluids is pumped into the borehole. After drilling to the required depth, the drill string is removed from the borehole, and the mud circulation is interrupted. Subsequently, geophysical measurements are carried out, casing pipes are installed, and the mud cycle is repeated [1–4]. When the casing is placed in the bore after the rheological parameters of the mud have been adjusted, further drilling fluids are pumped into the borehole. The following process describes the preparation of the borehole for cementing. After conditioning the well, fluids are pumped into the borehole in the following order: cement slurry, spacer, and displacement fluid (usually drilling fluid). The wash should leave the surface of the pipe and the wall of the borehole moistened, which will improve the bonding of the cement slurry. The wash contact parameters are very important. If wash flow conditions are dense with a turbulent regime, the mud is more effectively removed. However, it is also important to use wash flow conditions that do not cause fractures in the formation, which can occur when the pumped fluid at the bottom of the borehole becomes excessively high due to a high delivery rate. Sufficient contact time between the wash and the annular space is necessary, as this allows the mud to be effectively removed [5–9]. Cleaning the borehole prior to cementing is essential for many reasons [10–12]. Effective removal of the mud cake from the annular space prior to cementing enables the required tightness to be obtained between the rock, the cement sheath, and the casing pipe [13–16]. Tight contacts limit gas release after cementing [17,18]. Achieving the required tightness of the borehole also enables long-term operations while preventing the corrosive environment from adversely affecting the pipe columns [19–21]. Borehole cleaning issues involve such factors as the rheology of drilling fluids, fluid mechanics, the mechanism of action of chemicals, surfactants, and the chemistry of cement during hydration [22,23]. Conducting research to improve borehole cleaning is essential because, despite recent progress in the development of techniques and technology for sealing pipe columns, some cementing treatments are still unsuccessful. Successful sealing of the borehole depends on the correct performance of cementing, but this is only one component of an effective borehole seal [24–26]. Experience shows that the quality and quantity of the remaining mud cake affect the effectiveness of sealing the pipe column [27–30]. Therefore, it is necessary to consider the borehole clearance in specific geological conditions and the washer composition. However, the composition and effect of the wash on the mud cake are only part of the problem. The second important factor is the wash contact conditions. Wash liquids have Newtonian characteristics, which enable the dispersal of mud. The low viscosity of the wash makes it possible to pump it in turbulent conditions, but it requires the use of higher delivery rates, which is a problem in weakly compact geological structures. Injecting the wash at high delivery rates can cause depressurization of the zone near the well. This should be borne in mind when establishing wash contact conditions. However, effective removal of the mud cake is possible with a suitable Reynolds number (Re > 2300). In some cases, mud cake residues will not be completely removed. Then, a second variable must be adjusted, which determines the effectiveness of mud cake removal. This second factor is the wash contact time [25,27,28,31,32], which can be from 2 to 15 min. The literature frequently reports values of 4, 5, or 8 min [23,25,27,33–37]. The optimal wash contact time depends on the delivery rate, mud cake removal characteristics, type of mud cake, borehole conditions, and other parameters. Therefore, the correct preparation of the borehole depends not only on the appropriate selection of the wash contact time but also on the optimal contact conditions and the chemical composition of the wash. This study shows borehole-like conditions with respect to deposit formation on the core surface and deposit removal. In previous studies, a deposit was produced on the core in a viscometer. Then, a core rotation was used and the liquid was under static conditions. However, in this paper other possibilities are presented. The cores are stationary as they are in the bore, while the liquid is in motion. The liquid flows from the bottom of the hole upwards as in the borehole. Thereby, a mud cake is produced and then, when washing is performed, the spacer removes the mud cake. Publications on borehole preparation for annular cleaning are available. However, as far as the precise consideration of the contact time of the washing liquid on the wall of the bore and the surface of casing pipes is concerned, the literature on such items is scarce. Therefore, it has been decided to fill this gap in the literature and conduct research to improve borehole sealing with respect to spacer contact time.

2. Materials and Methods

2.1. Materials

Tap water was used to remove the mud cake. The mud cake was prepared from polymer-potassium mud, the composition of which is given in Table 1.

A mud cake was formed on a core cut from heavy sandstone. The core was 25 mm in diameter and 60 mm in height. The average value of the compressive strength of the core was 8.5 MPa. The core had a uniform, non-cracked structure.

Table 1. Composition of the drilling mud.

Ingredients	Concentration Percent by Weight
Biocide	0.2
Xanthan gum	0.3
Carboxymethylcellulose	2.5
Partially hydrolyzed polyacrylamide (polymeric clay hydration inhibitor)	0.2
KCl (potassium chloride as an ionic inhibitor)	0.5

CEM I 42.5R Portland cement was used to create the cement slurry. It contained 2.66% SO_3 and 0.065% Cl^-. Means were used to adjust the parameters of the cement slurry. Plasticizer PSP 046 is a modified naphthalene lignosulfonate dispersant. Its bulk density is approximately 500 ± 50 kg/m^3, and its pH is between 6.6 and 8.5. A defoaming agent, which is a mixture of unsaturated fatty acid esters and refined hydrocarbons, was added to the cement slurry. An anti-filtration agent was included in the starter and setting accelerator. Latex, which is a dispersion of styrene–butadiene copolymer, was also used to seal the set cement. The sponge contained 10% microcement, which is characterized by grains up to 10 μm and a specific surface area of approximately 1400 m^2/kg. Microcement reduces porosity and increases compressive strength. The percentages of materials used to prepare the cement slurry are summarized in Table 2 [28].

Table 2. Recipe and parameters of the cement slurry used in the adhesion test on the hardened cement–rock contact [27].

Ingredients	Percent by Mass of Cement
Water–cement ratio	0.45
Plasticizer	0.2
Latex	10.0
Stabilizer	1.0
Defoaming agent	0.5
Anti-filtration agent	0.2
Setting accelerator	4.0
Microcement	10.0
Cement CEM I 42.5R	100.0

All components in percent by mass of cement.

2.2. Preparation of the Cement Slurry

A mud cake was formed on the rock cores to determine the effectiveness of borehole cleaning for different wash contact times. Then, the cores were washed and sealed with the cement slurry. The slurry is prepared as follows. The required amount of water is measured with a measuring cylinder. Water is poured into the mixer, which is then turned on and set to a speed of 1600 rpm. Cement slurry parameter-modifying agents are added and mixed for 10 min. Then, fine-grained agents (microcement, microspheres, microsilica, and cement) are poured into the water with the other agents. The agents are mixed for another 20 min. Mixing at low speed is similar to preparing slurry in borehole conditions.

2.3. Experimental Procedures

The properties of the cement paste were tested in accordance with the standard:

- PN-EN ISO 10426-2. Oil and gas industry. Cements and materials for cementing holes. Lot 2: Testing of drilling cements. In these tests, the following were measured: slurry density, filtration, and thickening time.

The adhesion test was carried out according to the standard:

- PN-EN 196-1: 2006 Cement testing methods. Strength marking.

The research methodology was designed to determine the optimal wash contact time to improve the hole preparation method. The research was carried out in a drilling fluid flow simulator constructed for the needs of the tests (Figure 1) [35]. The liquid in the device flows in a closed circuit. The liquid flow rate was 11.2 L/min in all tests. This value allows the contact of the liquid while maintaining the Reynolds number at over 2300. During the tests, the contact times were 4, 6, 8, and 10 min. Cores cut from Ciężkowice sandstone were used for the tests. The core sample is Ciężkowice sandstone with a porosity of about 20%. It is a quartz sandstone that was formed in the period from the Upper Paleocene to the Lower Eocene. It is sandstone, red-gray in color, very compact, with a loose grain skeleton. It has coarse-grained, deep-sea sediments. The permeability of the sandstone core is from 10 to 50 mD, and in some samples it ranges from 80 to 250 mD. The cores were cylindrical with a diameter of 0.025 m and a length of 0.06 m (Figure 2) [27,28]. The cores were placed in a tripod inside a PVC pipe (Figure 3) [27,28]. The drilling fluids flowed in the pipe. The device simulates the annular space of the borehole.

Figure 1. A photo of the drilling fluid flow simulator [27].

Figure 2. A core cut from a sandstone sample [27].

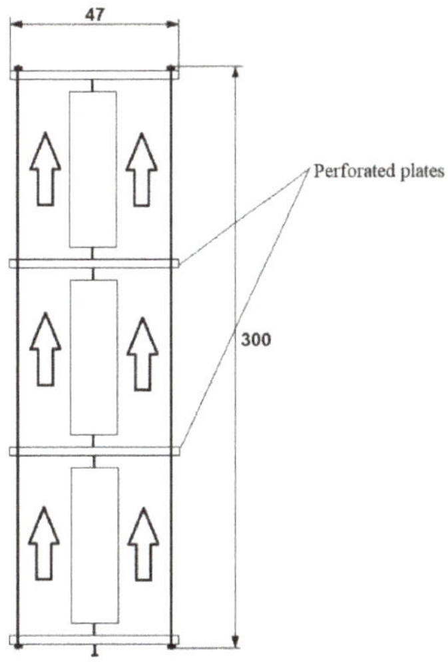

Figure 3. Arrangement of sandstone cores inside the plastic pipe [27].

In the first stage of the research, a mud cake was formed on the sandstone core. For this purpose, mud was pumped into the PVC pipe, which imitates the annular space of the borehole. The 3 cores were washed with the rinse for 1 h. The mud flow rate was determined on the basis of previously performed preliminary tests and amounted to 4.5 L/min. The mud cake was made with a scrubber, the composition of which is given in Table 1. After preparation of the mud cake, the cores were washed with water for a period of 4, 6, 8, or 10 min. In the tests, tap water was used as the standard wash liquid. After washing for the specified contact times, borehole preparation was assessed by testing the cement slurry adhesion to the core after it was cleaned. For this purpose, after forming a mud cake on the sandstone cores and then washing them with water for a specified period of time, the cores were covered with cement slurry (Figure 4). After a fixed setting time of 48 h, an adhesion test was carried out on the contact between the hardened cement slurry and the sandstone core. The test was carried out in a testing machine. By measuring the applied load on the sample, the force required to break the adhesion was determined.

Figure 4. Sandstone core with cement slurry prepared for adhesion tests on the contact surface between the core and cement.

The adhesion expressed in MPa is calculated according to Equation (1) [27]:

$$\sigma_p = \frac{\mathrm{P}}{\mathrm{s}} \cdot 10^3 [\mathrm{MPa}] \tag{1}$$

where σ_p is adhesion at the point of contact between the bonded cement and the rock core (MPa), P is the force required to break the connection between the bonded cement and the rock core (kN), and s is the contact area of the core with the cement slurry (m^2) Equations (2) and (3).

The testing machine monitor shows the pressure force (P), and the outer surface area is derived from the circumference of the core and the height of the cement bound to the core. The dimensions needed for the calculations are shown in Figure 5 [27,28].

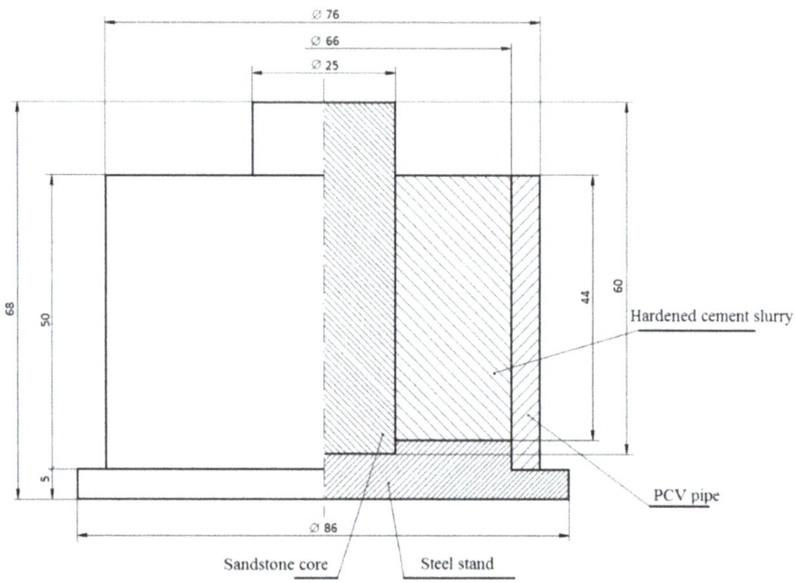

Figure 5. Diagram of a mold with a core and hardened cement slurry [27].

Explanation:

Core diameter (d) = 25 mm = 0.025 m, and the height of the cemented part of the core (h) = 44 mm = 0.044 m.

$$s = \pi \times d \times h \tag{2}$$

$$s = \pi \times 0.025 \times 0.044 = 0.003456 \left[m^2\right] \tag{3}$$

The adhesion of the bonded cement to the core is determined according to Equation (4):

$$\sigma_p = \frac{P}{3.456}\ [\text{MPa}] \tag{4}$$

It is also necessary to set benchmarks. For this purpose, adhesion measurements were taken for a "clean" core without a mud cake and for a core with a mud cake but without washing. The results in Table 3 are the maximum and minimum baseline adhesion. The obtained test results were compared to these values. Sandstone cores were flooded with a cement slurry of the same composition and parameters (Table 2).

Table 3. Baseline adhesion between hardened cement slurry and rock [27].

	The Force Required to Break the Adhesion (N)	Adhesion of Cement to the Sandstone Core (MPa)
Maximum control value	8.2	2.37
Minimum control value	2.1	0.61

3. Results and Discussion

The adhesion of the bonded cement to the cleaned rock core after the removal of the prepared mud cake ranged from 1.07 MPa to 1.77 MPa. Comparing these values with the minimum baseline adhesion of 0.61 MPa reveals that the percentage increase in the adhesion value ranged from 75% (for 4 min of contact time) to 189% (for 10 min of wash contact), as shown in Table 4. These values are higher than those obtained using CEM I 32.5R class cement, which was tested in [27]. Comparing the obtained adhesion to the maximum baseline adhesion of 2.37 MPa (Table 3) shows that the results ranged from 55% lower (contact time of 4 min) to 26% lower (10 min of wash contact). Figure 6 shows a graphical summary of the results of the mud cake removal efficiency in terms of the contact time of the rinse compared to the maximum and minimum baseline values. This study supplements the work in [27]; in the previous study, another type of cement slurry based on CEM I 32.5R cement was used, and in this study, a cement slurry based on CEM I 42.5R cement was used.

Table 4. Adhesion between the bound cement and the rock for different wash contact times.

Contact Time (min)	2 min	4 min	6 min	8 min	10 min	12 min
Adhesion to the sandstone core (MPa)		1.07	1.51	1.62	1.77	
Percentage change in the adhesion in relation to the maximum baseline adhesion (%)	−68% ± 0.5%	−55% ± 0.5%	−37% ± 0.5%	−32% ± 0.5%	−26% ± 0.5%	−24% ± 0.5%
Percentage improvement in mud cake removal efficiency (%)	+32% ± 0.5%	+45% ± 0.5%	+63% ± 0.5%	+68% ± 0.5%	+74% ± 0.5%	+76% ± 0.5%
Percentage change in the adhesion compared to the minimum baseline adhesion (%)	+23% ± 0.5%	+75% ± 0.5%	+146% ± 0.5%	+165% ± 0.5%	+189% ± 0.5%	+195% ± 0.5%

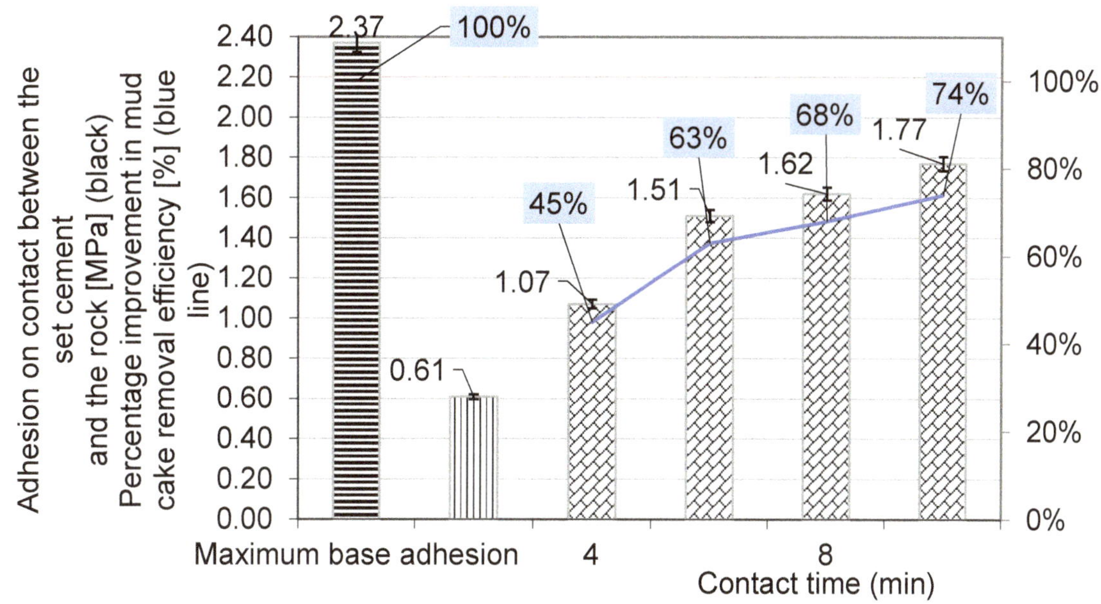

Figure 6. Values of adhesion of the hardened cement slurry to the core and the percentage improvement in mud cake removal (adhesion 2.37 = 100% mud cake removal). Results after different washing times.

In this study, an increase in the wash contact time was observed to increase the adhesion of the hardened cement slurry to the sandstone core. This results in an improvement in mud cake removal from the formation. The contact time of the wash strongly influences the efficiency of mud cake removal in the annular space. With the shortest contact time, which was 4 min, the adhesion increased by 75%. Extending the contact time of the wash by 2 min (6 min contact) increased the adhesion by another 71% compared to a wash contact time of 4 min. The improvement over the baseline was 146% for a 6 min wash contact time. An adhesion value of 2.37 MPa corresponds to 100% removal of the mud cake, as the core is clean, and the mud cake is eliminated. The core was only dipped in water to bring the measurement conditions closer to those of the core after rinsing. Based on the percentage reduction in adhesion compared to the maximum baseline value (2.37 MPa), the percentage improvement in the cleaning efficiency of the annular space was obtained. Analysis of the results shows that the improvement in mud cake removal efficiency was 45% after 4 min of wash contact and 74% after 10 min of contact. Figures 7–10 show the mud cake removal efficiency after different wash contact times.

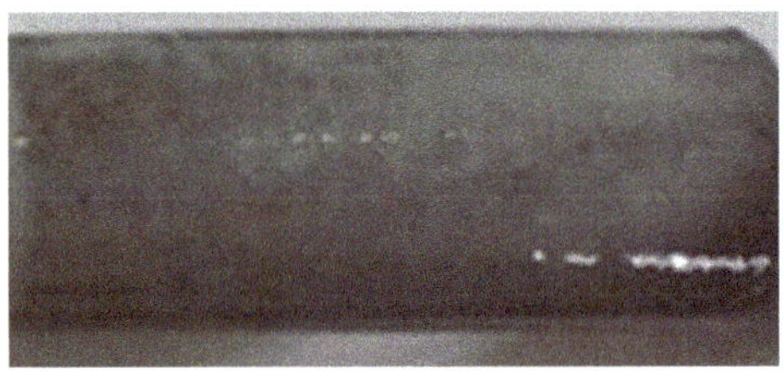

Figure 7. Wash contact time of 4 min.

Figure 8. Wash contact time of 6 min.

Figure 9. Wash contact time of 8 min.

Figure 10. Wash contact time of 10 min.

When optimizing the obtained values, it is assumed that the best wash contact time is 6 min, which corresponds to a 63% mud cake removal efficiency. This wash contact time is optimal because if the wash is in contact with rock formations for a longer duration, it will remove weakly bonded structures. This 6 min time was considered the best on the basis of the adhesion value obtained only. This is the best value under laboratory conditions. However, bear in mind that the experiments optimize the contact time and do not allow for an assessment of the mechanical aspects. It is also important to choose a suitable wash composition. Then, the optimal contact time and wash flow rate allow for the efficient removal of the mud cake. This action contributes to the improvement of hole preparation by increasing the tightness of the contacts between the casing pipe, cement sheath, and borehole wall. The conducted research confirms the beneficial effect of an appropriate wash contact time on the preparation of the hole prior to cementing by increasing the efficiency of mud cake removal. This was observed for a mud cake made of polymer-potassium mud. Mud cakes are physicochemically diverse. The mud cake composition depends on the type of rock being drilled, the type and parameters of the mud, and borehole conditions. Therefore, each borehole should be considered individually. A considerable limitation in this type of research is the inability to generate bore pressure and liquid temperature in the borehole. However, despite these limitations, the tests carried out make it possible to determine to a very good degree the influence of the contact time of the washing liquid on the effectiveness of the washing mud cake removal.

4. Conclusions

1. On the basis of the obtained test results, the efficiency of mud cake removal largely depends on the wash contact time in the annular space.
2. The effectiveness of mud cake removal from a borehole is mainly determined by the type of flow, i.e., the amount of wash contact.
3. Although increasing the wash contact time improves the mud cake removal efficiency, it may damage the weakly compact zone; therefore, the required wash contact time and delivery rate for a given borehole must not be exceeded.
4. The tested wash contact times resulted in adhesion values ranging from 1.07 MPa to 1.77 MPa (adhesion of the sample with the mud cake was 0.61 MPa, while the baseline sample without the mud cake had an adhesion of 2.37 MPa).
5. The obtained test results show that the adhesion improved by 75–189% relative to the minimum baseline value for a cement slurry based on class G drilling cement.
6. Comparing the obtained mud cake removal values to the maximum value of the baseline adhesion (removal of 100% of the mud cake), the improvement in the mud cake removal efficiency ranged from 45% after 4 min of wash contact to 74% after 10 min of wash contact.
7. The analysis of the obtained test results shows that the optimal wash contact time was 6 min, which resulted in 63% efficiency in removing the mud cake.
8. Tests on the effectiveness of mud cake removal under laboratory conditions do not completely replicate actual borehole conditions, but the constructed drilling fluid flow simulator allows borehole conditions to be reproduced to a large extent.

Author Contributions: Data curation, M.K. (Marcin Kremieniewski) and M.K. (Miłosz Kędzierski); investigation, M.K. (Marcin Kremieniewski); methodology, M.K. (Marcin Kremieniewski); writing—original draft, M.K. (Marcin Kremieniewski); project administration, M.K. (Miłosz Kędzierski); visualization, M.K. (Miłosz Kędzierski) and S.B.; resources, S.B.; supervision, S.B.; writing—review & editing, S.B. All authors have read and agreed to the published version of the manuscript.

Funding: This work was financially supported by Ministry of Science and Higher Education Warsaw (Internal order Oil and Gas Institute—National Research Institute Project No. 58/KW/17 and 0015/KW/21).

Institutional Review Board Statement: Not applicable.

Informed Consent Statement: Not applicable.

Data Availability Statement: Not applicable.

Acknowledgments: The authors thank the anonymous reviewers for their constructive comments and the editor for managing the paper.

Conflicts of Interest: The authors declare no conflict of interest.

Nomenclature

kN	Kilonewton
MPa	Megapascal
PCV	Polyvinyl chloride
μm	Micrometer

References

1. Błaż, S. Nowe rodzaje cieczy przemywających osady z płuczki inwersyjnej przed zabiegiem cementowania otworów wiertniczych. *Nafta-Gaz* **2017**, *5*, 302–311. [CrossRef]
2. Jasiński, B. Ocena wpływu cieczy przemywającej na jakość zacementowania rur w otworze wiertniczym po użyciu płuczki glikolowo-potasowej. *Nafta-Gaz* **2016**, *6*, 413–421. [CrossRef]
3. Adari, R.B.; Miska, S.; Kuru, E.; Bern, P.A.; Saasen, A. Selecting drilling fluid properties and flow rates for effective hole cleaning in high-angle and horizontal wells. In Proceedings of the Paper Presented at the SPE Annual Technical Conference and Exhibition, Dallas, TX, USA, 1–4 October 2000.
4. Zima, G. Wpływ właściwości płuczek wiertniczych na jakość cementowania w gazonośnych poziomach miocenu. *Nafta-Gaz* **2014**, *12*, 899–907.
5. Kremieniewski, M. Influence of Graphene Oxide on Rheological Parameters of Cement Slurries. *Energies* **2020**, *13*, 5441. [CrossRef]
6. Huang, S.; Feng, B.; Li, Z.; Tang, S.; Li, J.; Su, D.; Qi, L. Remediation of oil-based mud contaminated cement with talcum in shale gas well primary cementing: Mechanical properties, microstructure, and hydration. *Constr. Build. Mater.* **2021**, *300*, 124047. [CrossRef]
7. Zhang, F.; Miska, S.; Yu, M.; Ozbayoglu, E.; Takach, N.; Osgouei, R.E. Is well clean enough? A fast approach to estimate hole cleaning for directional drilling. In Proceedings of the Paper Presented at the SPE/ICoTA Coiled Tubing & Well Intervention Conference & Exhibition, The Woodlands, TX, USA, 24–25 March 2015.
8. Ahmed, A.; Mahmoud, A.A.; Elkatatny, S.; Chen, W. The effect of weighting materials on oil-well cement properties while drilling deep wells. *Sustainability* **2019**, *11*, 6776. [CrossRef]
9. Kremieniewski, M.; Rzepka, M. *Przyczyny i Skutki Przepływu Gazu w Zacementowanej Przestrzeni Pierścieniowej Otworu Wiertniczego Oraz Metody Zapobiegania Temu Zjawisku*; Nafta-Gaz: Hünenber, Switzerland, 2016; p. 9.
10. Kmieć, M.; Karpiński, B.; Antoszkiewicz, M.; Szkodo, M. Laboratory research on the influence of swelling clay on the quality of borehole cementing and evaluation of clay-cutting wellbore tool prototype. *Appl. Clay Sci.* **2018**, *164*, 13–25. [CrossRef]
11. Kremieniewski, M. *Ocena Skuteczności Oczyszczania Kolumny Rur Okładzinowych Przed Cementowaniem na Podstawie Badań Przy Użyciu Wiskozymetru Obrotowego*; Nafta-Gaz: Hünenber, Switzerland, 2018; Volume 9, pp. 59–66.
12. Boyou, N.V.; Ismail, I.; Sulaiman, W.R.W.; Haddad, A.S.; Husein, N.; Hui, H.T.; Nadaraja, K. Experimental investigation of hole cleaning in directional drilling by using nano-enhanced water-based drilling fluids. *J. Pet. Sci. Eng.* **2019**, *176*, 220–231. [CrossRef]
13. Kremieniewski, M.; Kędzierski, M. *Badanie Frakcjonowania Lekkich Materiałów Obniżających Gęstość Jako Wstępnego Parametru Podczas Projektowania Receptury Zaczynu Lekkiego*; Nafta-Gaz: Hünenber, Switzerland, 2019; Volume 12, pp. 35–42.
14. Saasen, A. Hole cleaning during deviated drilling—The effects of pump rate and rheology. In Proceedings of the Paper Presented at the European Petroleum Conference, The Hague, The Netherlands, 20–22 October 1998.
15. Yu, M.; Takach, N.; Nakamura, E.; David, R.; Shariff, M.M. An experimental study of hole cleaning under simulated downhole conditions. In Proceedings of the Paper Presented at the SPE Annual Technical Conference and Exhibition, Anaheim, CA, USA, 11–14 November 2007.
16. Wiśniowski, R.; Skrzypaszek, K.; Małachowski, T. Selection of a suitable rheological model for drilling fluid using applied numerical methods. *Energies* **2020**, *13*, 3192. [CrossRef]
17. Kremieniewski, M. *Korelacja Skuteczności Działania Środków Dyspergujących o Różnym Mechanizmie Upłynniania*; Nafta-Gaz: Hünenber, Switzerland, 2020; pp. 816–826.
18. Kateev, R.I. *Casing of Wells in Abnormal Hydrodynamic Conditions of the Development of Oil Fields in Tatarstan*; Nedra: Moscow, Russia, 2005; pp. 65–67.
19. Kremieniewski, M. Ultra-lightweight cement slurry to seal wellbore of poor wellbore stability. *Energies* **2020**, *13*, 3124. [CrossRef]
20. Ryan, D.F.; Browne, S.V.; Burnham, M.P. Mud clean-up in horizontal wells: A major joint industry study. In Proceedings of the Paper Presented at the SPE Annual Technical Conference and Exhibition, Dallas, TX, USA, 22–25 October 1995.

21. Kremieniewski, M. Receptura zaczynu lekkiego do uszczelniania otworów w strefie niskich ciśnień złożowych. *Nafta-Gaz* **2020**, *9*, 577–584. [CrossRef]
22. Sifferman, T.; Becker, T. Hole Cleaning in Full-Scale Inclined Wellbores. *SPE Drill. Eng.* **1992**, *7*, 115–120. [CrossRef]
23. Saasen, A.; Løklingholm, G. The effect of drilling fluid rheological properties on hole cleaning. In Proceedings of the Paper Presented at the IADC/SPE Drilling Conference, Dallas, TX, USA, 11 February 2002.
24. Kremieniewski, M. Recipe of lightweight slurry with high early strength of the resultant cement sheath. *Energies* **2020**, *13*, 1583. [CrossRef]
25. Okrajni, S.; Azar, J. The Effects of Mud Rheology on Annular Hole Cleaning in Directional Wells. *SPE Drill. Eng.* **1986**, *1*, 297–308. [CrossRef]
26. Sanchez, R.A.; Azar, J.J.; Bassal, A.A.; Martins, A.L. The effect of drillpipe rotation on hole cleaning during directional well drilling. In Proceedings of the Paper Presented at the SPE/IADC Drilling Conference, Amsterdam, The Netherlands, 4–6 March 1997.
27. Li, Z.; Su, G.; Zheng, L. Enhancing filter cake removal by engineering parameter optimization for clean development of fossil hydrogen energy: A numerical simulation. *Int. J. Hydrogen Energy* **2021**, *46*, 12784–12800. [CrossRef]
28. Kremieniewski, M. Hybrid Washer Fluid for Primary Cementing. *Energies* **2021**, *14*, 1295. [CrossRef]
29. Kremieniewski, M. Badania środków poprawiających stabilność sedymentacyjną zaczynów cementowych. *Nafta-Gaz* **2020**, *6*, 387–395. [CrossRef]
30. Hirpa, M.M.; Arnipally, S.K.; Kuru, E. Effect of the particle size on the near-wall turbulence characteristics of the polymer fluid flow and the critical velocity required for particle removal from the sand bed deposited in horizontal wells. *Energies* **2020**, *13*, 3172. [CrossRef]
31. Valluri, S.G.; Miska, S.; Yu, M.; Ahmed, R.M.; Takach, N. Experimental study of effective hole cleaning using sweeps in horizontal wellbores. In Proceedings of the Paper Presented at the SPE Annual Technical Conference and Exhibition, San Antonio, TX, USA, 24–27 September 2006.
32. Bilgesu, H.; Nekkhil, I.M.; Ameri, S. Understanding the effect of drilling parameters on hole cleaning in horizontal and deviated wellbores using computational fluid dynamics. In Proceedings of the Paper Presented at the Eastern Regional Meeting, Lexington, KY, USA, 17–19 October 2007.
33. Kremieniewski, M.; Kędzierski, M. Badania nad opracowaniem hybrydowej cieczy buforowej. *Nafta-Gaz* **2020**, *76*, 517–526. [CrossRef]
34. Zamora, M.; Jefferson, D.T.; Powell, J.W. Hole-cleaning study of polymer-based drilling fluids. In Proceedings of the Paper Presented at the SPE Annual Technical Conference and Exhibition, Houston, TX, USA, 3–6 October 1993.
35. Sayindla, S.; Lund, B.; Ytrehus, J.D.; Saasen, A. Hole-cleaning performance comparison of oil-based and water-based drilling fluids. *J. Pet. Sci. Eng.* **2017**, *159*, 49–57. [CrossRef]
36. Li, J.; Walker, S.A. Sensitivity Analysis of Hole Cleaning Parameters in Directional Wells. *SPE J.* **2001**, *6*, 356–363. [CrossRef]
37. Bizhani, M.; Corredor, F.E.R.; Kuru, E. Quantitative Evaluation of Critical Conditions Required for Effective Hole Cleaning in Coiled-Tubing Drilling of Horizontal Wells. *SPE Drill. Complet.* **2016**, *31*, 188–199. [CrossRef]

energies

MDPI

Article

The Influence of Temperature on Degradation of Oil and Gas Tubing Made of L80-1 Steel

Dariusz Bęben

Oil and Gas Institute—National Research Institute, 25A Lubicz Str., 31-503 Krakow, Poland; beben@inig.pl

Abstract: Corrosion in the oil and gas industry is very common due to the simultaneous action of a chemically active environment, temperature, and other non-chemical factors, for example, mechanical erosion by friction, and for these reasons corrosion is a very complex process. Corrosion at higher temperatures is an important aspect when extracting natural gas from a field with high temperatures (120 °C in the Lubiatow deposit and 180 °C in the gas well in Kutno). Water in the reservoir is often in the form of steam, with a pressure of about 25 MPa; as a result of its extraction, it cools down, which causes condensation. Condensed water in contact with the acid components of the gas causes corrosion, especially in the presence of aggressive gases, such as CO_2 and H_2S. Therefore, the aim of the work was to conduct research on the influence of water condensation, as a result of temperature changes in gasses containing CO_2 and H_2S on the corrosion of L80-1 steel at the junction of extraction pipes with casing pipes. The tests are carried out at temperatures of 65–95 °C, under a pressure of 7.5 MPa, so in quite aggressive conditions. The duration of the studies was 720 h (within a month). The results of the research allowed an answer to be provided for the question of what influence temperature, gas components, and pressure have on the corrosion of the well construction material. Moreover, the results clearly showed the selection of the material for the well, in order to prevent corrosion in aggressive environments.

Keywords: high temperature on corrosion of mining pipes; water; aggressive natural gas components; L80-1 steel

Citation: Bęben, D. The Influence of Temperature on Degradation of Oil and Gas Tubing Made of L80-1 Steel. *Energies* **2021**, *14*, 6855. https://doi.org/10.3390/en14206855

Academic Editor: Marcin Kremieniewski

Received: 14 September 2021
Accepted: 14 October 2021
Published: 19 October 2021

1. Introduction

Water, in contact with the acid components of gas, causes corrosion in mining pipes. The pH of the reservoir water is affected by components of natural gas, such as CO_2 and H_2S [1–3]. The extension of the service life of pipelines requires engineering knowledge related to the degradation of the materials, the capability of the diagnostics of the corrosion (chemical) processes, and the analysis of numerical measurement data. The simultaneous presence of moisture (as an electrolyte) results in the accelerated corrosion of steel, more frequent pipeline failures, operational equipment damage, and environmental pollution [4–8]. The mechanism of the corrosion caused by the acid components of natural gas has been described and extensively studied. Steel, with the addition of chromium (Cr), is now the most frequently used metal, as it is resistant to the corrosion caused by the presence of CO_2 and H_2S. It was found that steel containing Cr (1 to 5%) increases the corrosion resistance almost ten-fold in comparison to that of carbon steels [3,9]. Therefore, steel with a low Cr content is used the most often. The research was aimed at studying the corrosion of chromium-containing (8%) steel at various temperatures. Despite numerous studies on the corrosion of steel with a low and high Cr content in the corrosive environment, there are no studies on the effect of temperature on the degradation process. The process of local electrochemical corrosion (pitting), which features a point mass loss, is related to the action of a local cell existing at the point of contact of the pipes [10]. The rate of metal dissolution at the point of contact of steels is very high, causing perforation of pipe walls in a very short time, without a bigger mass loss outside the attacked place. The pitting is intensified by the aggressive chemical environment and the effect of high temperature. The presence of aggressive anions in the reservoir water, which feature high molar polarization, facilitates local damage to the passive layer [11]. In areas where the passive layer is the thinnest, a very large potential drop occurs, accelerating the permeation of ions, creating soluble chloride oxides through the oxide layer [12]. The pitting usually occurs on any internal inhomogeneities of the metal (non-metallic inclusions, precipitations, deformations), as well as on external inhomogeneities (edges, scratches, indentations, remnants of scale, deposits, etc.). The literature states that smooth and uniform surfaces are much more resistant to such corrosion [13]. The corrosion occurring in the environment of reservoir extraction is most frequently an example of chemical corrosion. The reaction of metal corrosion is a complex process and depends on the following factors:

1. Adsorption and chemisorption, that is, accumulation of substances originating from the gas on the metal surface, as a result of the formation of surface chemical connections with metal, creating a thin oxidation layer;
2. Origination of oxidation products on the absorption surface of the corrosion layer, and integrated into the crystalline lattice of the scale;
3. Diffusion/flow of metal ions to the formed scale.

The corrosion processes of oxidation depend on the pressure and temperature. Based on the literature, it is known that the corrosion rate of carbon steel under supercritical CO_2, without protective $FeCO_3$, is very high ($\geq$20 mm/y) [14–18]. In certain conditions, the corrosion rate can decrease to low values (<1 mm/y) during long-term exposure, due to the formation of a protective film of $FeCO_3$ [16–19]. A high pressure results in increased origination of pitting, and the corrosion rate may even be 20 times higher, due to a higher real pressure [20]. A layer of scale usually adheres to the metal, and cracks and microfissures exist close to the edge of the metal and scale. The formation of cracks and microfissures causes the start of a further pitting process. Increased pressure in the crack accelerates dissociation, forming a secondary phase of scale. The diffusion of oxidizer into the microfissures results in the formation of a second zone, and accelerates the internal corrosion. The rate of internal corrosion depends on the oxidising inclusions, metal composition, and cyclic temperature changes, and is the reason for the weakening of the mechanical properties, primarily metal plasticity and elasticity. One of the best corrosion protection solutions is now the application of an appropriate metal alloy and corrosion monitoring, and dosing inhibitors or cathodic protection.

To achieve corrosion protection, corrosion inhibitors are most frequently added to the reservoir water. They are already batched at the initial phase of production. During the extraction of natural gas with reservoir water, the investors seldom pay attention to the selection of an effective inhibitor to provide the best corrosion protection with varying pH of the fluid [21]. The temperature and pressure existing in the inter-tube space are also important parameters that affect the progression of the corrosion.

There are many methods of corrosion protection or the reduction in its effect, depending on the corrosion type and the chemical nature of the corrosive agents. The selection of an appropriate material for aggressive environment conditions, and the use of inhibitors to retard it, is the basic method of protection against chemical corrosion. The corrosive action of some agents may be substantially reduced by the use of corrosion inhibitors (retarders). On the metal surface, inhibitors usually form protective layers, which retard the corrosion rate. By definition, a corrosion inhibitor is a chemical agent, which, after application in small amounts, will effectively reduce corrosion. Laboratory tests should be carried out prior to the application of an inhibitor, whose action under conditions of high temperatures and pressures should then be checked. An inhibitor selected in such a way should effectively reduce the progression of the corrosion. A well-chosen inhibitor reduces the corrosion progression by approx. 95% with the use of 0.008% of the agent, and by 90% with the use of 0.004% of the agent [22]. The effectiveness of the agent usually depends on many factors, such as pressure, temperature, flow rate, and the composition, as well as on the number of corrosive agents (water, CO_2, H_2S, NaCl, etc.) [23]. Inhibitors may conduct the following:

1. Creating a passivation layer on the surface via the growth of insoluble metal oxide on the surface. These bonds create a protective barrier coating, which becomes an impermeable layer, and, at the same time, it is very flexible and adheres well to the substrate [24]. Phosphates and chromates are typical examples of such an application;
2. Neutralising ions, which cause corrosion in the environment. Neutralising amines and ammonia are typical components of such an inhibitor. These are inhibitors that are effective in boiler waters and in slightly acid environments;
3. Removing caustic ions from the solution. In-hydrazine and sodium sulphate are typical components of the inhibitor. Such inhibitors remove oxygen dissolved in water.

2. Experimental Procedure

2.1. Methodology

The carried-out laboratory tests were aimed at determining the corrosion type occurring at the contact point of mining pipes and casing, proceeding as a result of temperature and reservoir water action. Laboratory tests on steel were carried out based on the schedule determined for specific solutions, determined temperatures and pressure. The samples of tested steel should be entirely immersed. The total test time was determined depending on the corrosion rate; for example, for samples whose corrosion rate, at an elevated temperature (calculated after 24 h), exceeds 0.3 mm/year, the test time should be at least 7 days, and at the rate below 0.3 mm/year—at least 30 days [25–29].

2.2. Materials

The tests were carried out using reservoir water containing CO_2 and H_2S, extracted from a well. Its composition is presented in Table 1 below.

Table 1. Composition of deposit water.

Variable	Value	Standard Deviation
Temperature [°C]	20.0	1.5
Density [g/cm^3]	1.182	0.001
pH	4.8	1.5
Eh [mV]	117.8	1.12
Carbonates [mg/dm^3]	n.s.	n.s.
Hydrogencarbonates [mg/dm^3]	152	15.6
Chlorides [mg/dm^3]	148.000	38.540
Calcium [mg/dm^3]	29.260	9.230
Magnesium [mg/dm^3]	5.350	1.320
Potassium [mg/dm^3]	590	62
CO_2 [wt. %]	0.6	0.1
H_2S [wt. %]	0.15	0.01

The research was carried out using 16 different batches of water.

Plates of steel L80-1 were used for testing, with the composition as in Table 2, and with the dimensions 50 × 20 × 4 and 10 × 100 × 4 mm, cut out from a tube fragment (Figure 1).

Table 2. Chemical composition (wt. %.) of L80-1 steel, norm ISO 11960 API 5CT.

C [%]	Mn [%]	Si [%]	P [%]	S [%]	Al [%]	Ni [%]	Mb [%]	Cr [%]	V [%]	Fe [%]
0.26	1.25	0.24	0.012	0.004	0.018	0.11	0.23	0.08	0.006	Bal.

Figure 1. Coupons used for testing.

The performed tests should determine the reason for pitting origination at various temperatures and in the reservoir water environment.

To study the corrosion of steel L80-1, depending on the temperature and corrosion time, visual inspection of the corroded samples surface was carried out as well as scanning microscopy and X-ray microanalysis of corrosion products. The view of geometrical structure of the corroded surface of samples was documented on a scanning microscope at diversified magnifications. X-ray microanalysis was carried out on the corroded surface in specific areas, marked with consecutive numbers. Metallographic microsections transverse to the corroded surface were made to determine the corrosion type and its progress depending on the adopted conditions. Upon that basis, the thickness of the layer of corrosion products on non-etched and etched microsections was evaluated, as was the shape of its external and internal boundary depending on the corrosion conditions. To evaluate the chemical composition, a linear qualitative analysis by means of X-ray microanalysis was carried out on the cross section of the corroded layer and steel substrate.

3. Results and Discussion

The research was carried out in a pressurized chamber (7.5 MPa) at 65, 80 and, finally, 95 °C for 720 h (30 days). The facility is shown in Figure 2. The tests were performed on steel L80-1. The tests were carried out in a level gauge type of chamber, which enabled observation of the samples during the test. The chamber was equipped with a manometer, in order to control the pressure during the tests, and a valve to maintain the pre-set pressure. Reservoir water was introduced into the chamber, as well as long samples, which were stacked up to observe the corrosion on the contact between the metals.

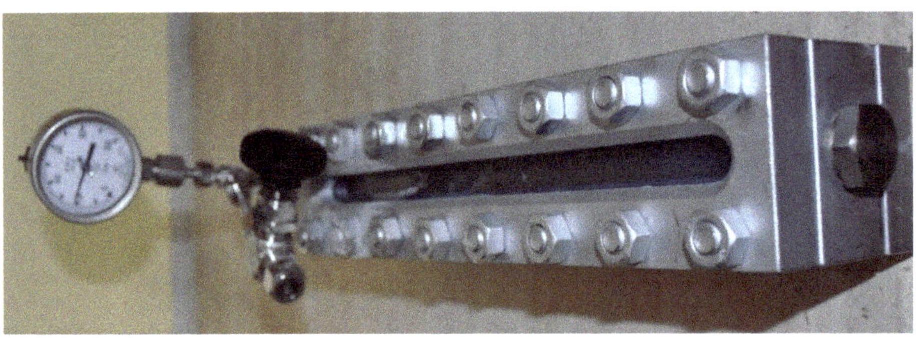

Figure 2. The chamber pressure tests.

As Figure 3 shows, all the samples taken out after the tests had a non-corroded top layer, which proves that a corrosion inhibitor was added to the reservoir water, preventing surface corrosion. At the metal/metal contact interface in Figure 3, a coat of dark brown–black bloom was observed, which proves that passivation occurs at the point of contact between metals. The observed passivation (surface corrosion) is a surface bloom; at a temperature of 65 °C, it is difficult to remove, and at temperatures of 80 and 95 °C, it is poorly adhesive. No cases of pitting were observed upon visual inspection with the naked eye. The assessment of the steel surface was based on the microscopic examinations for the studied temperatures. The carried-out tests were aimed at determining the surface topography, microstructure, and the chemical composition of the top layer, as well as of the microstructure of the corroded samples. The chemical composition and surface topography were studied on an X-ray microanalyzer and a scanning microscope, at characteristic points of the corroded surface of the studied steel samples. The microstructure of the substrate material of long samples and the corrosion layer were studied on a Neophot 2 microscope [28]. Figure 4 shows, at the macro scale, the microstructure of the corroded samples used in this work, the microstructures and chemical composition of the top layer, and the microstructures of the samples pressure-conditioned at 95 °C.

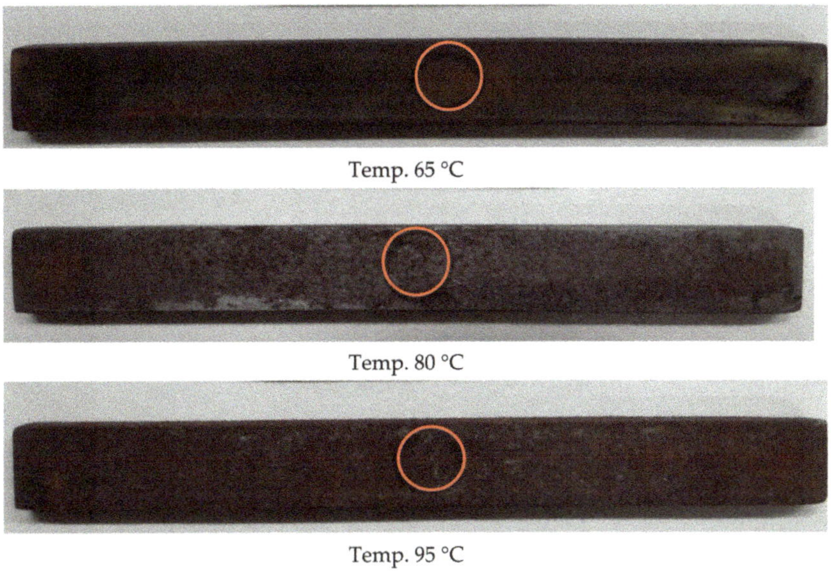

Figure 3. View of a coupon for research with selected places observed by SEM and EDS microanalysis.

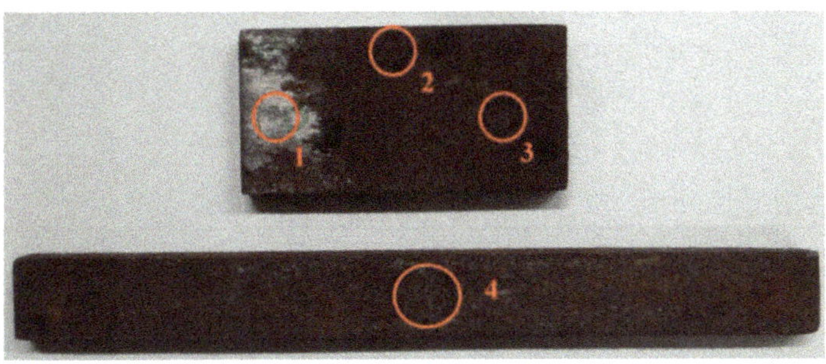

Figure 4. The samples prior to SEM/EDS analysis with indicated analysis points.

The topography of the sample surface (Figure 5) with visible craters is typical of pitting. The chemical composition determined on the surface of the studied sample showed a high oxygen and carbon content, which indicates the presence of iron oxides. As line 2 in Table 3 shows, the smallest steel loss was at this point; in the other points, 1 and 3, a high amount of oxygen was observed, as well as a reduced amount of iron, which suggests the existence of iron oxides (corrosion products).

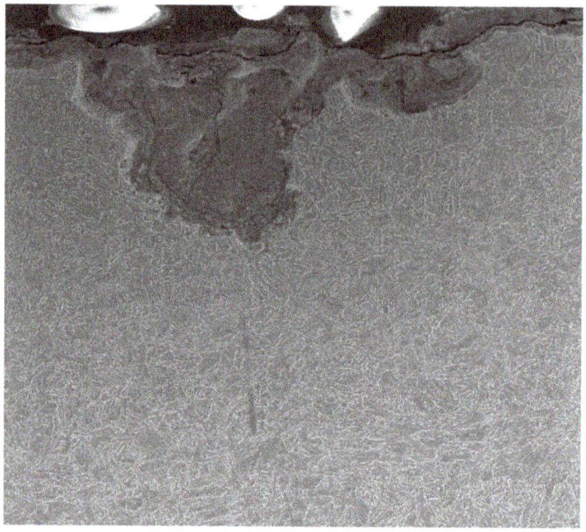

Figure 5. Cross section of L80-1 sample exposed to 95 °C (a long sample).

Table 3. EDS analyses for the sample indicated in Figure 6, the sample L80-1 exposed to 95 °C.

		Element	Wt. %	At. %
1.	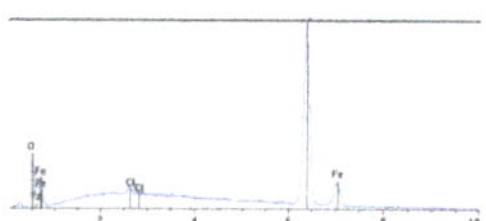	O	9.88	27.57
		Cl	0.79	0.99
		Fe	89.33	71.44
		Total	**100.00**	**100.00**
		Element	**Wt. %**	**At. %**
2.	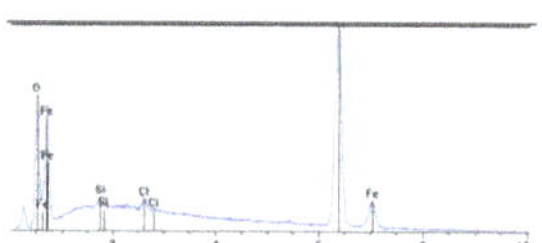	O	18.83	44.38
		Cl	0.72	0.96
		Fe	0.92	0.98
		Si	79.52	53.68
		Total	**100.00**	**100.00**
		Element	**Wt. %**	**At. %**
3.	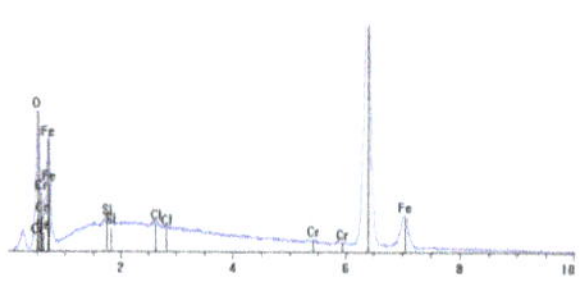	O	16.17	40.00
		Cl	0.46	0.51
		Fe	82.59	58.52
		Si	0.57	0.81
		Cr	0.20	0.15
		Total	**100.00**	**100.00**
		Element	**Wt. %**	**At. %**
4.	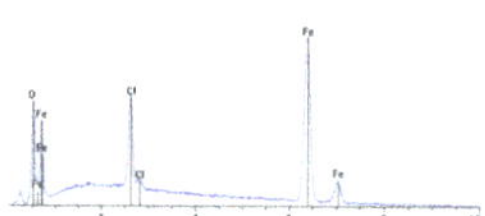	O	17.60	41.31
		Cl	8.51	9.02
		Fe	73.89	49.68
		Total	**100.00**	**100.00**

Figure 6. Cross section of L80-1 sample exposed to 95 °C corrosion scale (a long sample).

On metallographic microsections, transverse to the surface (Figure 6), the microstructure of the steel substrate of a sample fragment and a layer of corrosion products are visible. In the upper part, apart from visible pits, there is a black layer of corrosion products. The layer is porous, uneven, and cracked. The corrosive medium has access to the steel substrate via the unevenness, pores, and cracks. White inclusions are visible in certain places; this is a precipitated salt crystal, which integrated into the corrosion scale (Table 4). The majority of the studied sample is covered with scale, uniform corrosion, and single deep pits, 40 to 65 µm deep, visible under the microscope (Figure 7).

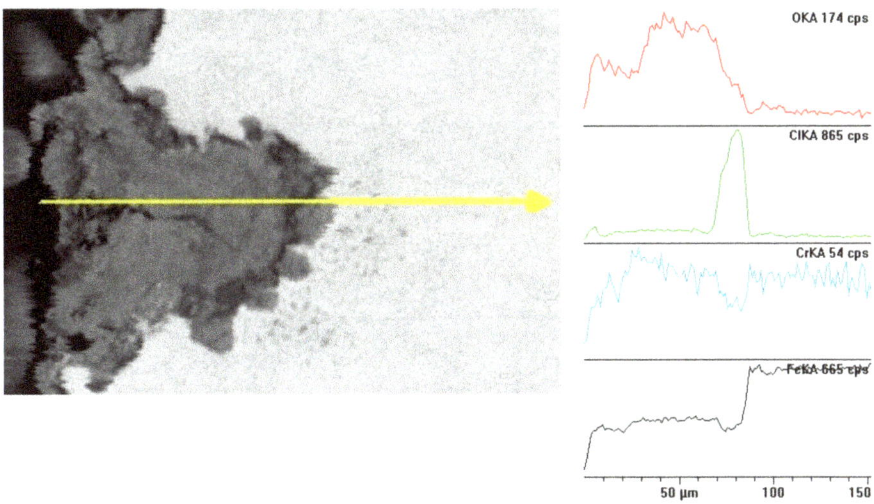

Figure 7. Cross section and linear cross scan of the elements in L80-1 steel at 95 °C.

Table 4. Results of chemical composition studies in the area from Figure 8.

		Element	Wt. %	At. %
5.	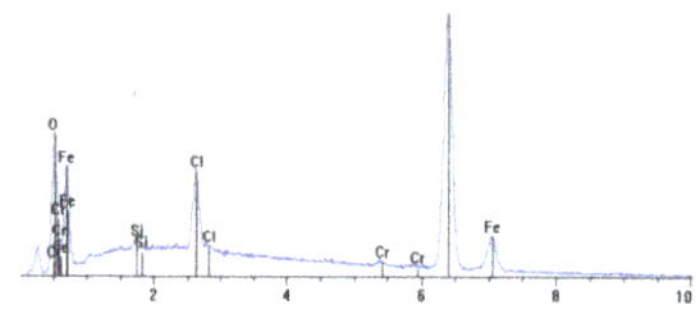	O	14.77	36.73
		Cl	5.18	5.81
		Fe	78.99	56.26
		Si	0.60	0.85
		Cr	0.46	0.35
		Total	**100.00**	**100.00**

Figure 8. Topography studies of surface in Figure 7, microstructure and chemical composition of the top layer as well as the microstructure of the corroded sample, pressure-conditioned at 80 °C.

The metallographic micro-section, transverse to the surface, presented below (Figure 9), shows the microstructure of the steel substrate of a sample fragment and a layer of corrosion products. When comparing it with the samples conditioned at 80 °C, we can notice that the pits are larger and deeper. In the upper part, apart from visible small pits, as compared with 95 °C, there is a dark brown layer of corrosion products. This layer is porous and uneven. The corrosive medium has access to the steel substrate via the unevenness, pores, and cracks. White inclusions are visible in certain places; this is a precipitated salt crystal, which integrated into the corrosion scale. The majority of the studied sample is covered with scale and uniform corrosion, as well as cracks and single deep pits, approx. 25 to 30 µm deep, visible under the microscope; see Figure 10. At a temperature of 95 °C, the material was subjected to deeper destruction than at 80 °C.

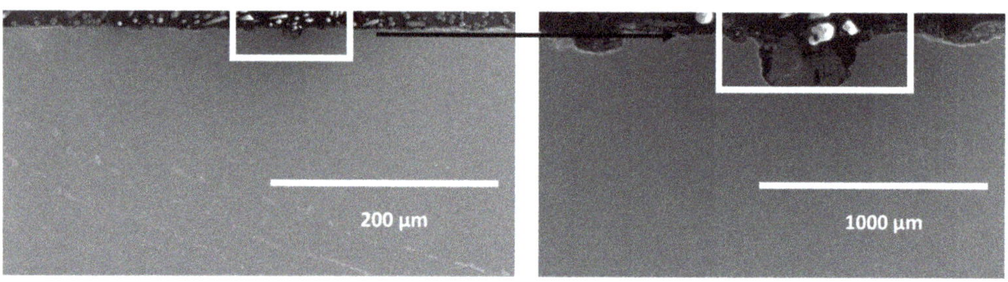

Figure 9. Cross section of L80-1 sample exposed to 80 °C (a long sample).

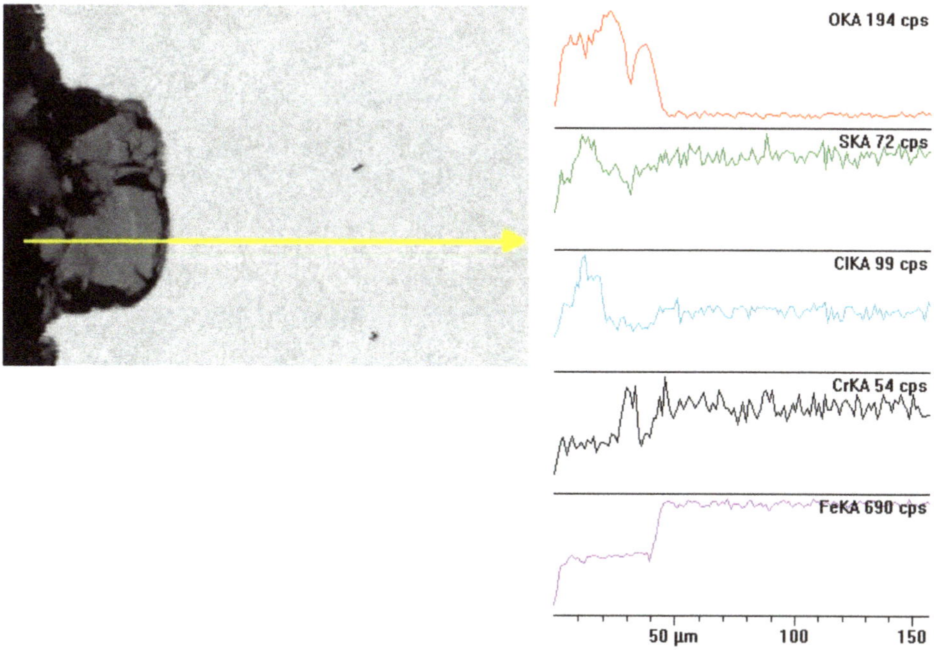

Figure 10. Cross section and linear cross scan of the elements in L80-1 steel at 80 °C.

White inclusions are visible in certain places; this is a precipitated salt crystal (see Figure 11), which integrated into the corrosion scale. The majority of the sample is covered with scale, uniform corrosion, and cracks that are visible under the microscope; there are no pits. Figure 12 shows the metallographic microsection transverse to the surface, with a visible microstructure of the steel substrate of a sample fragment and a layer of corrosion products. When comparing it with the samples studied at 95 and 80 °C, we can notice that there are no bigger pits. The surface is covered with a thin, relatively even black layer of corrosion products. The scale (the top layer of corrosion products) is approx. 10 to 12 µm thick. The surface was not subject to major destruction (see Figure 13), as compared with previously studied samples.

Figure 11. Macrostructure of the corroded sample at 65 °C.

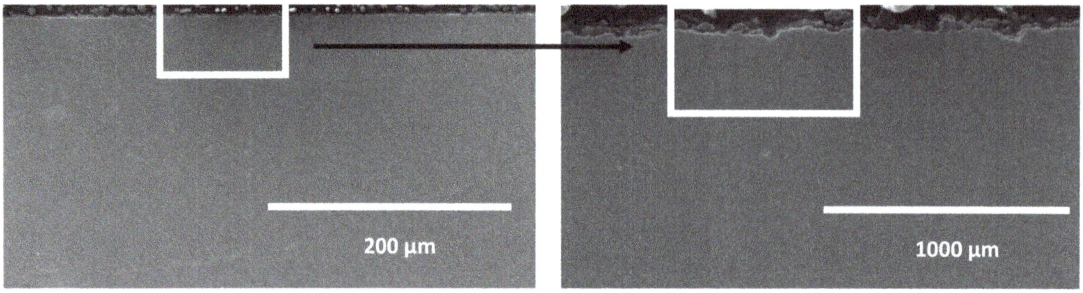

Figure 12. Cross section of L80-1 sample exposed to 65 °C (a long sample).

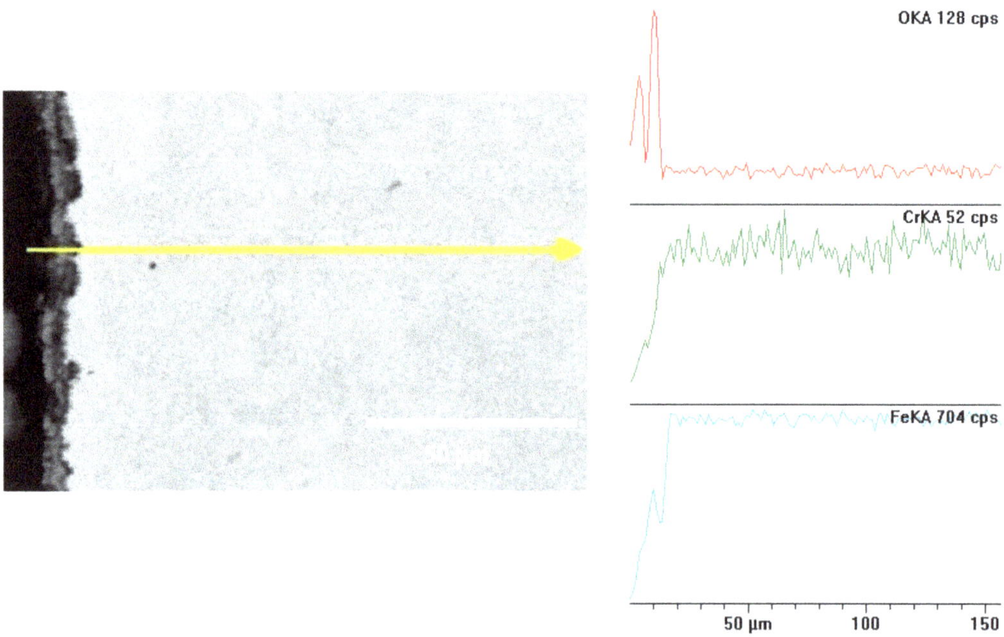

Figure 13. Cross section and linear cross scan of the elements in L80-1 steel at 65 °C.

Based on the obtained results, the occurrence of electrochemical corrosion was found in the tested samples, along with a propensity for pitting, resulting from the existence of chlorine ions from the reservoir water. The thickness of the layer of corrosion products, as well as the surface roughness and the pit depths, decrease with the reduction in temperature. At a temperature of 65 °C, a hard, rather thin adhesive layer of corrosion products forms. At the boundary with the steel substrate, practically no pits are observed. At 80 °C, and, in particular, at 95 °C, the corrosion layer is poorly adhesive, much thicker, and rough, with fractures and deep pits in the steel substrate. The existence of oxygen and chlorine was found in the corrosion layer. The chemical composition of the corroded layer, both on the surface and at the cross section, is uniform. The chlorine content is higher in the samples that corroded at a higher temperature. The tested steel L80-1 corrodes in the acid environment and at higher temperatures. The temperature plays a significant role in this case. The corrosion rate is affected by the type of product that forms under various temperature conditions. At low temperatures, iron and steel corrode, with the formation of a semi-protective layer of magnetite Fe_3O_4 on the surface, retarding the corrosion rate. The hitherto work shows that the corrosion rate of steel in CO_2-saturated brine is initially high, but it reduces over time, due to the formation of a siderite layer (iron II carbonate) of $FeCO_3$ on the surface (Zhang et al. 2013) [29,30]; studies were carried out that measured the loss of mass over time. In addition, H_2S has an adverse impact on the corrosion of steel, as was described by El Alami et al. (El Alami et al. 2011) [31]. At a temperature of 80 °C on the surface of the steel, a microscopic, thin layer of oxides practically stops the process of iron release. Increasing the temperature to 95 °C did not change the nature of the interactions and compounds, such as $FeOH^+$, which are subject to adsorption to drive the dissolution at a rate that changes with time, and—as H_2CO_3 and H^+ are reduced—they proceed simultaneously at a slower rate, but extensive deep pits are observed; this is a particular type of corrosion, featuring a point loss of steel mass [32]. The course of pitting is related to the action of a local cell, which forms between a large, passivated surface of steel, being a cathode, and a local depassivated zone, being an anode. The rate of metal dissolution on the anode is very high, resulting in perforation in a very short amount of time, without a bigger mass loss outside the attacked place [33]. The pitting corrosion of steel occurs most frequently in water environments containing halide ions, i.e., chlorine, bromine, or iodine ions, where its intensity mainly depends on temperature and the concentration of ions [34]. It usually occurs on any internal inhomogeneities of the metal (non-metallic inclusions, precipitations, deformations), as well as on external inhomogeneities (edges, scratches, indentations, remnants of scale, deposits, etc.). Instead, smooth and uniform surfaces are much more resistant to such corrosion [4]. If iron is the damaged metal (the main component of steel), then the role of the corrosion-causing metal is fulfilled by carbon, contained in steel in the form of graphite grains, or iron carbide [35].

4. Conclusions

The studied steel L80-1 corrodes in an acid environment, and the corrosion is proportional to the increase in temperature, which plays a significant role. By monitoring the degradation processes of L80-1 steel, using modern tests, it is possible to detect the early stages of this degradation, such as the initiation and development of cracks near non-metallic inclusions, and the formation of CO_{2-} and H_2S-induced bubbles on the surface of the material. You will be able to observe the course of such processes in the future, and determine the period of safe operation of structures operating at different temperatures and in the environment. The corrosion rate is affected by the type of corrosion product, which forms at various temperatures. At lower temperatures, the tested steel corrodes, with the formation of a semi-protective layer of magnetite Fe_3O_4 and siderite $FeCO_3$ on the surface, retarding the corrosion rate. At a temperature of 80 °C, a microscopic, thin layer of oxides on the steel surface practically stops the process of iron release. When the temperature is 95 °C, the content of chlorides in the corrosive medium exceeds a critical value, the hitherto 'natural' anti-corrosion coat is damaged, and pitting corrosion occurs. Corrosion at lower temperatures is uniform (general) and consists of a uniform attack on the entire steel surface. As a result, the pipe thickness decreases uniformly, with a simultaneous reduction in general strength. The prevention of uniform corrosion effects (by reducing its rate) consists of exchanging the steel for a material with better corrosion resistance, periodical passivation of the steel surface, and the application of protection, e.g., cathodic, or appropriate protective coats and layers.

Funding: This research received no external funding.

Institutional Review Board Statement: Not applicable.

Informed Consent Statement: Not applicable.

Data Availability Statement: Data is contained within the article.

Conflicts of Interest: The author declare no conflict of interest.

References

1. Ribeiro, L.P.; Paulo, C.A.S.; Neto, E.A. Compos Basin-Subsea Equipment: Evolution and Next Steps. In Proceedings of the Offshore Technology Conference, Richardson, TX, USA, 5–8 May 2003; p. 15223.
2. Estrella, G. The Importance of Brazilian Deepwater Activities to the Oil Industry Technological Development. In Proceedings of the Offshore Technology Conference, Richardson, TX, USA, 5–8 May 2003; p. 15049.
3. Yougui, Z. Electrochemical Mechanism and Model of H_2S Corrosion of Carbon Steel. Ph.D. Thesis, Ohio University, Athens, GA, USA, 2015.
4. Mahmoodian, M.; Qingi, C. Failure assessment and safe life prediction of corroded oil and gas pipelines. *J. Pet. Sci. Eng.* **2017**, *151*, 434–438. [CrossRef]
5. Wang, W.; Natelson, R.; Stikeleather, L.; Roberts, W. CFD simulation of transient stage of continuous countercurrent hydrolysis of canola oil. *Comput. Chem. Eng.* **2012**, *43*, 108–119. [CrossRef]
6. Nesic, S. Key issues related to modelling of internal corrosion of oil and gaspipelines—A review. *Corros. Sci.* **2007**, *49*, 4308–4338. [CrossRef]
7. Revie, R. Oil, Gas Pipelines. In *Integrity and Safety Handbook;* John Wiley & Sons, Inc.: Hoboken, NJ, USA, 2015.
8. Papavinasm, S. *Corrosion Control in the Oil and Gas Industry;* Elsevier: Houston, TX, USA, 2013.
9. Kermani, B.; Cochrane, R.; Dougan, M.; Linne, C.; Gonzalez, J. Development of low carbon Cr-Mo steels with exceptional corrosion resistance for oilfield applications. In Proceedings of the 56th NACE Annual Conference, Houston, TX, USA, 11–16 March 2001. Available online: https://www.onepetro.org/conferencepaper/NACE-03117 (accessed on 13 October 2021).
10. Stachowicz, A. Korozja rur wydobywczych odwiertów gazowych zawierających CO_2. *Nafta-Gaz* **2011**, *11*, 395–400. (In Polish)
11. Zhang, G.A.; Liu, D.; Li, Y.Z.; Guo, X.P. Corrosion behavior of N80 carbon steel in formation water under dynamic supercritical CO_2 condition. *Corros. Sci.* **2017**, *120*, 107–120. [CrossRef]
12. Stachowicz, A. Korozja rur w odwiertach oraz dobór ochrony inhibitowanej w płynach nadpakerowych. *Nafta-Gaz* **2013**, *7*, 525–531. (In Polish)
13. Atiwat, K. Lifetime Estimation of Steel Tubing by Corrosion Analysis for Petroleum. Master's Thesis, Geotechnology Suranaree University of Technology, Nakhon Ratchasima, Thailand, 2016.
14. Choi, Y.S.; Young, D.; Nesic, S. Effect of Impurities on the Corrosion Behavior of CO_2 Transmission Pipeline Steel in Supercritical CO_2-Water Environment. *Environ. Sci. Technol.* **2010**, *44*, 9233–9238. [CrossRef] [PubMed]
15. Choi, Y.S.; Nesic, S. Determining the Corrosive Potential of CO_2 Transport Pipeline in High pCO_2–Water Environments. *Int. J. Greenh. Gas Control* **2011**, *5*, 788. [CrossRef]
16. Zhang, Y.; Pang, X.; Qu, S.; Gao, X.; Li, K. The relationship between fracture toughness of CO_2 corrosion scale and corrosion rate of X65 pipeline steel under supercritical CO_2 condition. *Int. J. Greenh. Gas Control* **2011**, *5*, 1643–1650. [CrossRef]
17. Zhang, Y.; Pang, X.; Qu, S.; Gao, X.; Li, K. Discussion of the CO_2 corrosion mechanism between low partial pressure and supercritical condition. *Corros. Sci.* **2012**, *59*, 186–197. [CrossRef]
18. Suhor, M.F.; Mohamed, M.F.; Nor, A.M.; Singer, M.; Nesic, S. Corrosion of mild steel in high CO_2 environment: Effect of the $FeCO_3$ layer. In Proceedings of the Corrosion 2012, Salt Lake City, UT, USA, 11–15 March 2012; p. 0001434.
19. Cui, Z.D.; Wu, S.L.; Zhu, S.L.; Yang, X.J. Study on Corrosion Properties of Pipelines in Simulated Produced Water Saturated with Supercritical CO_2. *Appl. Surf. Sci.* **2006**, *252*, 2368. [CrossRef]
20. Osokogwu, U.; Oghenekaro, E. Evaluation of corrosion inhibitors effectiveness in oilfield production operations. *Int. J. Sci. Technol. Res.* **2012**, *1*, 19–23.
21. Banaś, J.; Mazurkiewicz, B.; Solarski, W. Problem korozji mikrobiologicznej w instalacjach geotermalnych. *Ochr. Przed Korozją* **2011**, *3*, 76–81. (In Polish)
22. Bęben, D. Ochrona chemiczna metali przed korozją na przykładzie wybranych kopalń gazu ziemnego. *Ochr. Przed Korozją* **2014**, *12*, 478–481. (In Polish)
23. Bęben, D. Badania skuteczności działania wybranych inhibitorów korozji stosowanych okresowo w przemyśle wydobywczym. *Ochr. Przed Korozją* **2019**, *62*, 376–381. (In Polish)
24. PN-70/H-04600 Korozja metali. Badanie odporności korozyjnej metali i stopów. Ogólne wytyczne. (In Polish)
25. PN-76/H-04601 Korozja metali. Badania laboratoryjne w cieczach i roztworach w temperaturze otoczenia. (In Polish)
26. PN-78/H-04608 Korozja metali. Skala odporności metali na korozji. (In Polish)
27. PN-78/H-04610 Korozja metali. Metody oceny badań korozyjnych. (In Polish)
28. Młynarski, A.; Piasecki, A.; Jakubowski, J. Badania mikroskopowe i strukturalne skorodowanych próbek stali L80-1. *Politech. Poznańska. Poznań* **2012**, 699–702. (In Polish)
29. Zhang, X.; Zevenbergen, J.; Benedictus, T. Corrosion Studies on Casing Steel in CO_2 Storage Environments. *Energy Procedia* **2013**, *37*, 5816–5822. [CrossRef]
30. Eliyan, F.F.; Alfantazi, A. Influence of temperature on the corrosion behavior of API-X100 pipeline steel in 1-bar CO_2-HCO_3^- solutions: An electrochemical study. *Mater. Chem. Phys.* **2013**, *140*, 508–515. [CrossRef]
31. El Alami, H.; Augustin, C.; Orlans, B.; Servier, J.J. Carbon Capture and Storage projects: Material integrity for CO_2 injection and storage. *Eurocorr* **2011**, *2011*, 4741.
32. Choi, Y.S.; Farelas, F.; Neši, S.; Magalhães, A.A.O.; Andrade, C.D.A. Corrosion behavior of deep water oil production tubing material under supercritical CO_2 environment: Part 1-effect of pressure and temperature. *Corrosion* **2014**, *70*, 38–47. [CrossRef]

33. Jakubowski, M. Influence of pitting corrosion on fatigue and corrosion fatigue of ship structures Part I Pitting corrosion of ship structures. *Pol. Marit. Res.* **2014**, *81*, 62–69. (In Polish) [CrossRef]
34. Li, D.; Liu, Q.; Wang, W.; Jin, L.; Xiao, H. Corrosion Behavior of AISI 316L Stainless Steel Used as Inner Lining of Bimetallic Pipe in a Seawater Environment. *Materials* **2021**, *14*, 1539. [CrossRef] [PubMed]
35. Zhang, D.; Gao, X.; Li, W.; Li, B.; Guo, J.; Zhang, J.; Pang, Q.; Xu, Z. CO_2 corrosion behavior of high-strength martensitic steel for marine riser exposed to CO_2-saturated salt solution. *Mater. Res. Express* **2021**, *8*, 076517. [CrossRef]

energies

MDPI

Article

A New Strategy for Pre-Selecting Gas Wells for the Water Shut-Off Treatment Based on Geological Integrated Data

Sławomir Falkowicz, Andrzej Urbaniec, Marek Stadtműller * and Marcin Majkrzak

Oil and Gas Institute—National Research Institute, 25A Lubicz Str., 31-503 Krakow, Poland; falkowicz@inig.pl (S.F.); urbaniec@inig.pl (A.U.); majkrzak@inig.pl (M.M.)
* Correspondence: stadtmuller@inig.pl

Abstract: This article presents a new analytical procedure for pre-selecting gas wells for water shut-off treatments based on available at hand results of an analysis of integrated geological, reservoir and exploitation data. Attention was paid to assess the possibility of the appearance of cross flows between layers in the near wellbore zone. Their appearance always eliminated the well as a candidate for the treatment. The basis for assessing the possibility of the emergence of cross flows was based on the assessment of the presence or absence of impermeable barriers (e.g., shales or anhydrite) in intervals, completed by perforation. For this assessment, well logging data were used, which were carried out in different years with the use of various types of probes. Based on modified quantitative and qualitative interpretation techniques, permeable and impermeable layers were separated in the analysed borehole sections. In some cases, in the absence of other data, well logs are the only source of information from which a vertical profile of horizontal permeability can be made. The article describes the verification process carried out for the eight wells situated in the area of the Carpathian Foredeep, which have been preselected by the operator. As part of this procedure, the available geological, reservoir and exploitation data were used, and a preliminary assessment of the possibility of implementing the obtained results into other wells and gas fields in the area of the Carpathian Foredeep was carried out. On the basis of the well ranking, two out of the eight analysed wells were recommended as candidates for water shut-off treatment.

Keywords: water shut-off treatment; gas production; water-gas ratio (WGR); formation permeability; well logging data quality and quantity interpretation; gas well ranking

Citation: Falkowicz, S.; Urbaniec, A.; Stadtműller, M.; Majkrzak, M. A New Strategy for Pre-Selecting Gas Wells for the Water Shut-Off Treatment Based on Geological Integrated Data. *Energies* **2021**, *14*, 7316. https://doi.org/10.3390/en14217316

Academic Editor: Marcin Kremieniewski

Received: 21 September 2021
Accepted: 29 October 2021
Published: 4 November 2021

Publisher's Note: MDPI stays neutral with regard to jurisdictional claims in published maps and institutional affiliations.

1. Introduction

One of the most serious problems arising during the exploitation of hydrocarbons is the high production of water in oil and gas wells. This has a negative impact on the economics of the exploitation of hydrocarbon formations. It is estimated that the world production of water in oil wells reaches a level of 75 billion barrels, and its disposal costs exceed 40 billion dollars [1]. In the United States alone, oil wells produced 21 billion barrels of reservoir water in 2007 [2], and in 2017 over 24.4 billion barrels [3].

Approximately 4.5 billion m^3 of natural gas is produced in Poland annually. Assuming a water-gas ratio (WGR) at a level of 0.1 kg/m^3, it can be assumed that gas production is accompanied by annual production of water in the amount of 450 thousand tonnes. Assuming the cost of managing/disposing of water at USD 40/tonne, this gives an amount equal to USD 18 million. In addition, especially in the late period of exploitation of the formation, the produced water significantly reduces gas production, often to an economically unacceptable level, which leads to the well being shut off [4]. While there may be many reasons for the high water production, for effective treatment, the source of the problem must first be correctly identified.

The next step involves selecting the most suitable technology for this and adapting it to local conditions. It should be remembered that no two cases are the same, not only for the same formation, but even for adjacent wells. Therefore, the selection of the target production/injection well is of key importance and should be based on the available formation and exploitation data [4].

Again, correctly determining the cause of a high WGR is the basis of success in water shut-off treatments. Unfortunately, operators often underestimate this important initial step. There are many reasons for this. For example, correct assessment requires time and effort, and it is not always known which of the solutions will turn out to be profitable. In addition, many service companies wrongly believe that there is only one way to fix a problem, or that all problems come from the same source. Some service companies argue that they have one universal solution that works for all cases. As previously emphasised, in water production limitation there is no one universal solution of the problem, and due to the complexity of the problem, no more than 80% of field operations can be classified as technically successful and no more than 60% as profitable [5].

In many cases of an excessive, uncontrolled inflow of water to a production well, the reason of the problem is incorrectly assumed. Often, specialists incorrectly believe that only one type of water production problem (e.g., three-dimensional coning) exists, while an appropriate analysis of the available data and a correct diagnosis may indicate the true cause, e.g., flow behind pipe or a two-dimensional coning. This knowledge enables the implementation of an appropriate "repair" program, ensuring a greater probability of success of the performed procedure.

After identifying the causes of water breakthrough to the production well, countermeasures should be introduced that will use the simplest possible solutions to resolve this problem. Based on an extensive analysis of formation and engineering data, 13 causes and water breakthrough mechanisms were identified. They were collected in four categories, from A to D, depending on the countermeasures to be taken to address the problem of excessive water production by the well, precisely described in literature [4,6]. There are two methods of reducing excess water production in hydrocarbon producing wells: mechanical and chemical. In chemical methods, special treatment fluids are injected into the formation in order to limit the permeability of rocks near the well in the water-producing layers. The most accepted by operators are the treatments defined as selective. In this case, without the mechanical isolation of the separated layers in the perforation interval, the treatment fluid is injected into the formation. As a result of contact with the formation water in the water bearing zone, it takes the form of a gel, completely or at least partially blocking the flow of water to the well.

Since the beginning of the 1980s, a popular method of reducing water production is the use of RPM (relative permeability modification) techniques. In these techniques, treatment fluids, most often based on polymers, are injected to a depth of 1 to 2 m into the formation, and adsorb on the formation rocks to reduce water production without reducing oil or gas production. [7]. Apart from the others, the effectiveness of RPM technique is related to the physical properties of pore walls of the formation rocks, which sometimes can be difficult to assess. In the case of fractured and pore-fractured formations, the injection of gels or gelants is applied, which results in the appearance of insulating barriers in specific parts of the formation. The factor initiating in-situ reactions in this type of treatment is the mixing of solutions, change of pH, precipitation of insoluble particles and others.

It is difficult to accept that the same technology is often used for the exploitation of both oil and gas. The justification for this assessment is that:

- In oil/water systems, the movement of a fluid with higher mobility (water or gas) should be controlled by restricting or blocking the flow;
- In gas/water systems, the flow of a fluid with lower mobility (water) should be controlled by restricting or blocking the flow.

Consequently, selective control of water production in oil/water and gas/water systems requires a different theoretical approach and the use of other technologies and surface facilities [8].

In recent years, interesting concepts of using working fluids based on microemulsions in WSO treatments have appeared [9]. Hungarian specialists developed an innovative method, classified as reservoir conformance control (RCC) technology, which was successfully implemented in 16 gas wells [8,10]. RCC technology differs significantly from RPM methods developed so far. The RCC method provides a more radical intervention in the formation, consisting of the creation of insulating gel barriers deeper in the formation. Barriers of this type are assumed to:

- Effectively limit the flow of water to the well;
- Be more durable due to how they emerge;
- Not cause losses in the production of hydrocarbons.

Regardless of what liquid is used in WSO treatments, the condition for success is always the correct selection of the well where the treatment will be performed. The basis for the assessment is the correct determination of the water breakthrough mechanism to the well, which is presented in detail in some publications e.g., [6,11]. In the case of finding water breakthrough, which was described by Chan [12] as a channel, it should be absolutely specified whether after the treatment there will be flows of reservoir fluids between the layers, so-called "*cross flows*". The occurrence of cross flows or lack thereof depends mainly on the geological structure of the formation. Cross flows will occur when the water bearing zone and the pay zone are not isolated with impermeable barriers, or the ratio of the horizontal to vertical permeability k_h/k_v of the perforated zone rocks is less than 10. This can be determined by several methods, but the results of pressure testing between zones are the most reliable. If it is not possible to perform such tests, flow profiles and well logs can also be helpful. In cases where it is not possible to determine whether there are impermeable barriers, it should be assumed that they do not exist. According to the authors of the article, the above assessment should be the first step in selecting a well as a candidate for WSO treatments. The further part of the article presents the procedure for confirming or excluding the presence of insulating impermeable barriers between the water and gas layers.

2. Geological Background and Specification of the Selected Wells

For the purpose of assessing the possibility of carrying out the procedure, the reduction of water production was initially designated by the operator of eight wells located in south-eastern Poland (Figure 1a). In these wells, various gas-bearing horizons (in some cases more than one horizon) were located at different depths and were associated with both clastic and carbonate collectors. Taking into account the geological units, seven of the selected wells (C-2, J-5K, J-8K, P-6, T-3, Z-1K, Z-74) were located in the Carpathian Foredeep, and one (J-7) in the marginal part of the Outer Carpathians (Figure 1b).

Intervals saturated with natural gas selected for the analysis in six of the above-mentioned wells (C-2, J-5K, J-8K, J-7, P-6, Z-74) were associated with multi-horizon gas fields, located within the profile of clastic formations of the autochthonous Miocene of the Carpathian Foredeep [16–23]. The gas-bearing horizons in the two remaining wells (T-3, Z-1K) represent massive gas formations, associated with a complex of Devonian carbonate rocks sealed with clay formations of the autochthonous Miocene or (locally) Carboniferous [24–26].

Generally, in the area of research, five main structural levels can be distinguished: (1) the series of Neoproterozoic anchimetamorphic rocks of the Ediacaran age, (2) the Meso-Paleozoic rock complex, characterised by strongly diversified thicknesses and lithology, (3) the autochthonous Miocene formations which were initially deposited in the Carpathian Foredeep Basin, (4) the Miocene-age deformed sediments located in front of the Carpathian orogen, (5) the allochthonous formations traditionally included within several tectonic units of the Outer Carpathians.

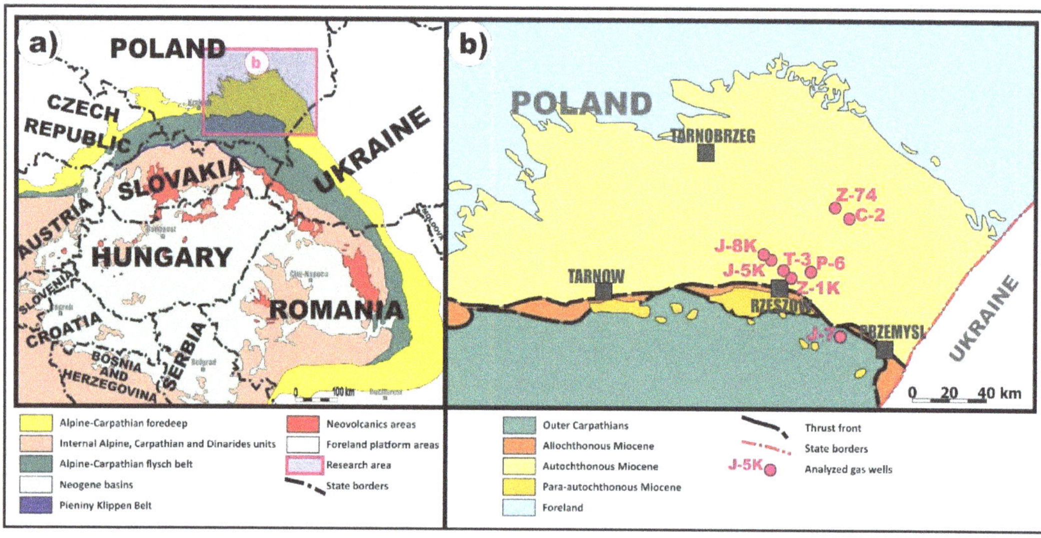

Figure 1. (**a**) Location of the study area in relation to the Carpathians and the Pannonian Basin System (PBS) (after Kováč et al. [13]; Golonka et al. [14]); (**b**) location of the analysed gas wells against the range of the eastern part of the Carpathian Foredeep in Poland; ranges of geological units according to Porębski and Warchoł [15].

Characteristics of the Selected Wells

All data used in the project were provided by the owner, Polish Oil and Gas Company (PGNiG SA), the operator of the studied wells. The collected data set included: exploitation data, parameters calculated for reservoir horizons, standard well logging data, geological documentation of fields. The quantity and quality of the data were sufficient to complete the study.

The C-2 well

The C-2 well is located in the NE part of the Carpathian Foredeep (Figure 1b). A total of three gas horizons were made available for exploitation in the upper part of the Miocene profile at a depth of about 250–300 m. Based on the interpretation of the well logs, it can be stated that the horizons provided by the perforations are quite homogeneous in terms of lithology and reservoir properties. Short sections of perforation (up to 4 m) provide only fragments of profiles of the identified gas-bearing horizons, characterised by better reservoir parameters.

The J-5K well

In the J-5K well, located in the central part of the research area (Figure 1b), one gas horizon, located in the middle part of the autochthonous Miocene profile, was included through a 17 m long perforation. Perforation provides the full profile of the gas horizon. The reservoir rock consists of sandy or mudstone rocks, genetically related to the delta environment [21]. Based on the interpretation of the well logs, it can be concluded that the profile of the analysed formation horizon is characterised by large lithological diversity and is composed of alternating layers of sandstone, mudstones and probably thin layers of claystone.

J-8K well

The J-8K well is located within the same gas field as the J-5K well. In this well, two perforation intervals (12 m and 6 m long) were selected for detailed analysis, which together complete the production of two gas horizons from the middle part of the autochthonous Miocene profile (including the same horizon as the J-5K, as well described above). The gas horizons clearly differ in terms of lithology, as the higher horizon is characterised by much greater lithological diversity and the associated differentiation of reservoir properties.

The J-7 well

The J-7 well is located in the SE part of the study area and as mentioned above, it is the only well located in the marginal part of the Outer Carpathians (Figure 1b). In this well, a wide (50 m) perforation interval covered two gas horizons identified in the autochthonous Miocene formations at a depth of less than 3000 m. Miocene clastic formations occur in this zone in the substrate of allochthonous Carpathian formations. The trap is interpreted as a compact anticline, developed above the elevation of the Neoproterozoic basement [17]. The reservoir rock consists of layers of sandstone, silt and heteroliths, and the profile of the interval completed for production is characterised by large lithological diversity.

The P-6 well

The P-6 well is located in the central part of the research area. The selected interval includes two perforations (18 m long and 5 m long) providing one gas-bearing horizon located in the middle part of the Miocene profile at a depth of approx. 1450–1500 m. Generally, the lithological profile of the interval is similar to the above-mentioned profiles of the J-5K and J-8K wells, and is characterised by significant lithological diversity. The dominant lithotypes are heteroliths dominated by mudstone and heteroliths with equal proportions of sandstone and mudstone. The reservoir rock consists of sandstone layers with good reservoir parameters and sequences of heteroliths.

The T-3 well

The T-3 well is located within a massive gas deposit, developed in the upper part of the Late and Middle Devonian carbonate rock complex, lying at a depth of slightly less than 2000 m. The reservoir rock consists of strongly fractured and sometimes also brecciated carbonate rocks (dolomites and limestones). The genesis of this reservoir formation is related to the intensive development of paleokarst processes, therefore, the gas-bearing horizon is very nonhomogeneous and is characterised by significant diversification of reservoir properties. The formation is sealed with Miocene claystone and mudstone formations. A 20 m long top part of the horizon has been completed for production.

The Z-1K well

The Z-1K well is located in close proximity to the T-3 well, on an adjacent gas field, which is very similar to the one described above. The reservoir rock consists of strongly brecciated and fractured dolomites and limestones, and the top seal is a series of impermeable fine-grained clastic sediments of the Lower Carboniferous. The reservoir properties are very diverse. An 18 m interval, located in the upper part of the horizon, at a depth below 2100 m, was completed for the production in this well.

The Z-74 well

The Z-74 well is located in the NE part of the Carpathian Foredeep, not far from the C-2 well described above (Figure 1b). This well produced gas from a deeper part of the Miocene profile, which, according to the literature data, is generally associated with the complex of submarine fans [19,27,28]. The perforation is 8 m long and the gas-bearing horizon is composed of thin-layer rocks (most likely the dominant lithotype is heterolith dominated by mudstone).

3. Initial Assessment of the Suitability of Wells for WSO Treatments Based on the Analysis of Diagnostic Plots

One of the commonly used methods for identifying the water breakthrough mechanism to the well is the analysis of the history of changes in the WGR and its derivative as a function of time. The correct diagnosis of the "water problem" is a key element in indicating an effective method (mechanical or using treatment fluids) for limiting the production of formation water [6,8,29]. The method proposed by Chan [12] consists of comparing the changes in the WOR/WGR and its derivative obtained as a result of computer simulations in idealised operating conditions, with the history of changes in the indicated parameters in the analysed well. According to the cited methodology, there are three basic mechanisms of water inflow to the well, which have a significant negative impact on the production of hydrocarbons [1,12,30]:

- Flow behind pipe or by fracture/fault, characterised by a sharp increase in the water ratio curve and its rapid stabilisation at a high level;
- Channel flow, also distinguished by a rapid increase in the water ratio (as a result of the flow of the formation water through the full cross section of the layer with high permeability) while the period of reaching the maximum values (curve stabilisation level) is longer, is interpreted as a gradual saturation of the pore space of the productive layer with formation water. The course of the derivative curve of the water exponent is similar, being parallel to the course of the ratio curve;
- Cone flow, characterised by a slow increase in the value of the water ratio with a simultaneous decrease in the value of its derivative.

This article presents an assessment of the formation water breakthrough mechanism for four out of eight analysed wells: J-5K, J-8K, P-6 and Z-1K. Apart from using the graphs of the history of the WGR and its derivative (WGR') change in individual wells, the daily history of gas and formation water production was also analysed. Many professionals agree that diagnostic plots cannot be the only basis for determining the water breakthrough mechanism [29]. The method of computer simulations has shown that in the case of multilayer formations, there may be situations in which cone water inflow can be confused with multilayer channel inflow [29,31]. In the analysed cases, it was assumed that the inflow of formation water to the well is not in the form of a 3D cone, which results directly from the lithology of the reservoir intervals in the layered autochthonous Miocene deposits in completed wells (with the exception of the Z-1K well). They are mostly thin-layer deposits, in which the thickness of individual layers with different lithology rarely exceeds 30 cm.

As already mentioned in the assessment of the water breakthrough mechanism, in addition to the diagnostic plots, the history of gas and water extraction was also used, assuming both graphs with identical periods of operation in a logarithmic system. This made it possible to easily link the changes in the diagnostic plots with the changes in the production history of the well. The analysis of diagnostic plots in order to determine the water breakthrough mechanism is not always an easy matter and does not always give unambiguous, indisputable results. First of all, the required situation is such that no well/reservoir interventions take place in the analysed period of time. Only data included in the range of the so-called "diagnostic window" is included, i.e., the stage of production in which according to the operator's declaration, no changes were made to the method of exploitation of the interpreted formation horizon (Figure 2). The diagnostic plots shown in Figures 3–6 were plotted for the entire extraction period, ignoring the first hundred days of operation. Data from that period have practically no diagnostic value. This made the graphs more readable for points of significant change in the water ratio and its derivative of between the 800th and 1000th day of production, due to the specific nature of the distribution of values on the axes of the graph provided in a logarithmic scale.

The J-5K well

In the J-5K well, a 17 m long profile fragment was perforated. The analysis of the WGR chart (Figure 3a) shows that in the initial period of the well's operation, the value of the water exponent remained constant (on average about 40 g/m^3), which is characteristic of the layered flow of formation water. A significant increase in the ratio was recorded at around the 800th day of operation, which should be interpreted as an increase in phase permeability for brine. As production continued, no change in the trend of the WGR curve was observed—the last analysed point shows a value of approximately 490 g/m^3. The course of the curve of the water ratio derivative is parallel to the WGR curve almost throughout the entire analysed period of production. A slight decrease in the presented values takes place from the moment of the increase in the value of the WGR (800th day of the well's operation). The chronological summary of the production curves indicates that after the accompanying rapid increase in the production of formation water, the increase in the amount of exploited gas was only temporary (Figure 3b). Despite the subsequent decline in gas production, the trend in the amount of produced brine did not change. On the basis of the presented data, it should be concluded that the nature of the inflow of formation water to the J-5K well changed from layered to channel.

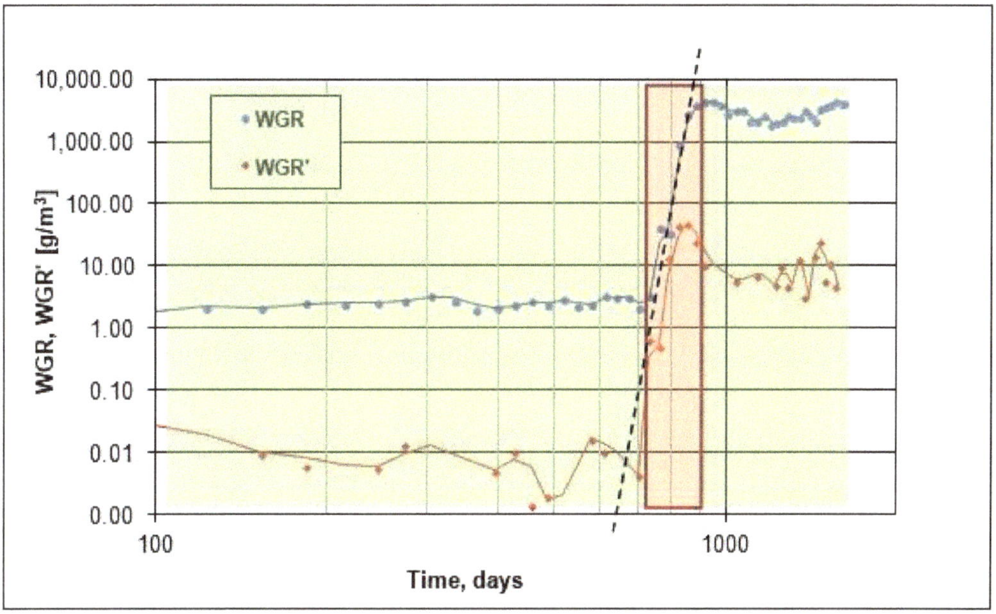

Figure 2. The diagnostic window technique for the assessment of the water breakthrough mechanism to the well.

The J-8K well

The J-8K well was perforated with two intervals (12 m + 6 m). In the initial years of operation, the trajectory of the WGR history plot (Figure 4a) remained at a constant and low level, i.e., not exceeding a few g/m^3. After approximately 800 days of the well's production, the beginning of a slow but systematic increase in the WGR was recorded. With the ratio equal to about 10 g/m^3, after about 2000 days of extraction, a sharp 10-fold increase in the WGR to about 100 g/m^3 was recorded. The next period was characterised by a continued, but at a much slower pace, increase in the WGR until it temporarily stabilised at a level of about 320 g/m^3. After this stage, another increase in the WGR was noted—the available data indicate the maximum recorded value of the exponent ratio to 676 g/m^3. The curve of the WGR derivative is characterised by a parallel, with a slight downward tendency, course up to about the 800th day of production. After this period, the WGR′ curve increased—The course of both curves is parallel. A comparison of the production curve with the WGR and WGR′ graphs indicates the beginning of the decrease in gas production when the WGR increase begins (800th day of operation). The temporary increase in gas production was preceded by a sharp increase in the share of formation water in the total volume of the exploited media (Figure 4b). Despite another decline in production, no reduction in water production was recorded, but on the contrary, its quantity increased dramatically. On the basis of the presented data, mainly the curves of the water ratio and its derivative, it is possible to indicate a change in the nature of the water inflow to channel inflow. The presented and described increases in WGR correspond to the commencement of inflow of formation water from subsequent layers and are characteristic of multilayer channelling. The course of changes in the water ratio and its derivative from the moment when the ratio value exceeded 10 g/m^3 is very similar to the results of numerical simulations obtained by Seright [32], when successive layers separated by impermeable barriers were flooded. This is a very favourable situation for WSO treatments because the probability of cross flows is minimal [32].

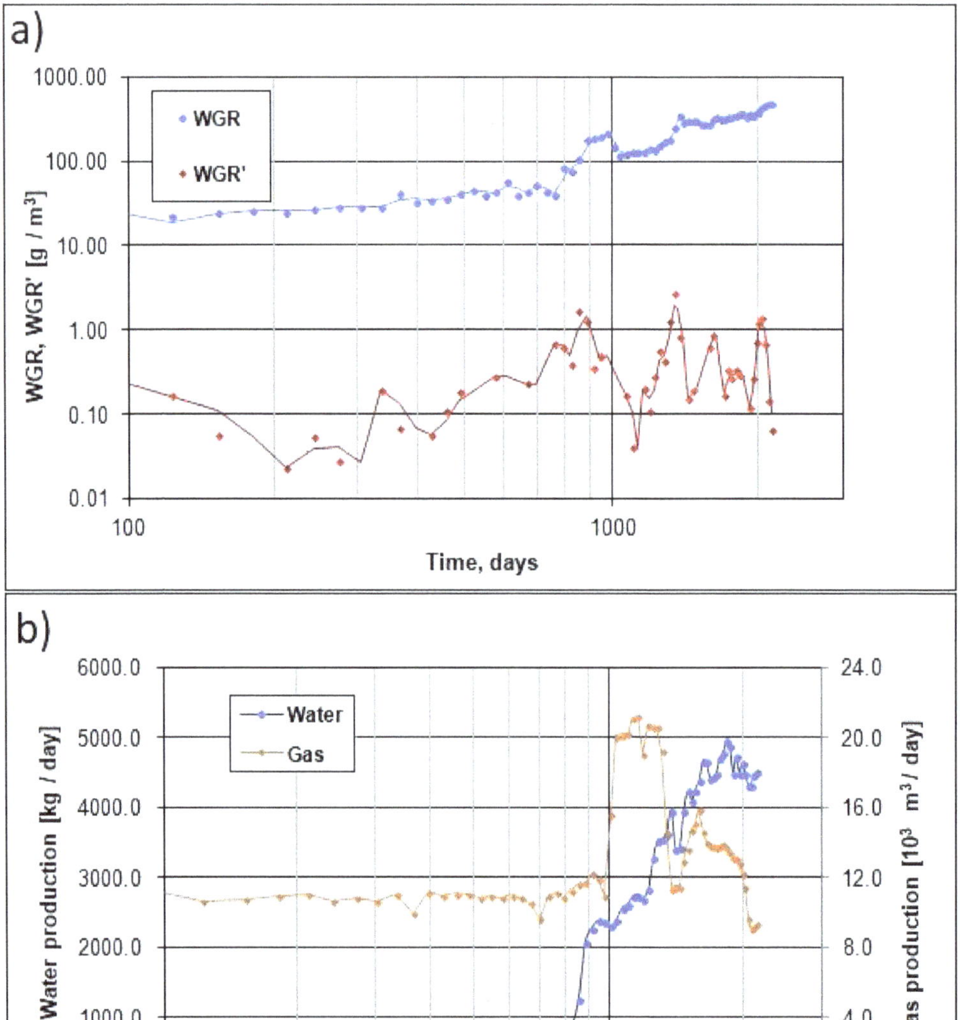

Figure 3. Changes in the WGR and its derivative over time (**a**) and the history of gas and formation water extraction (**b**) for the J-5K well.

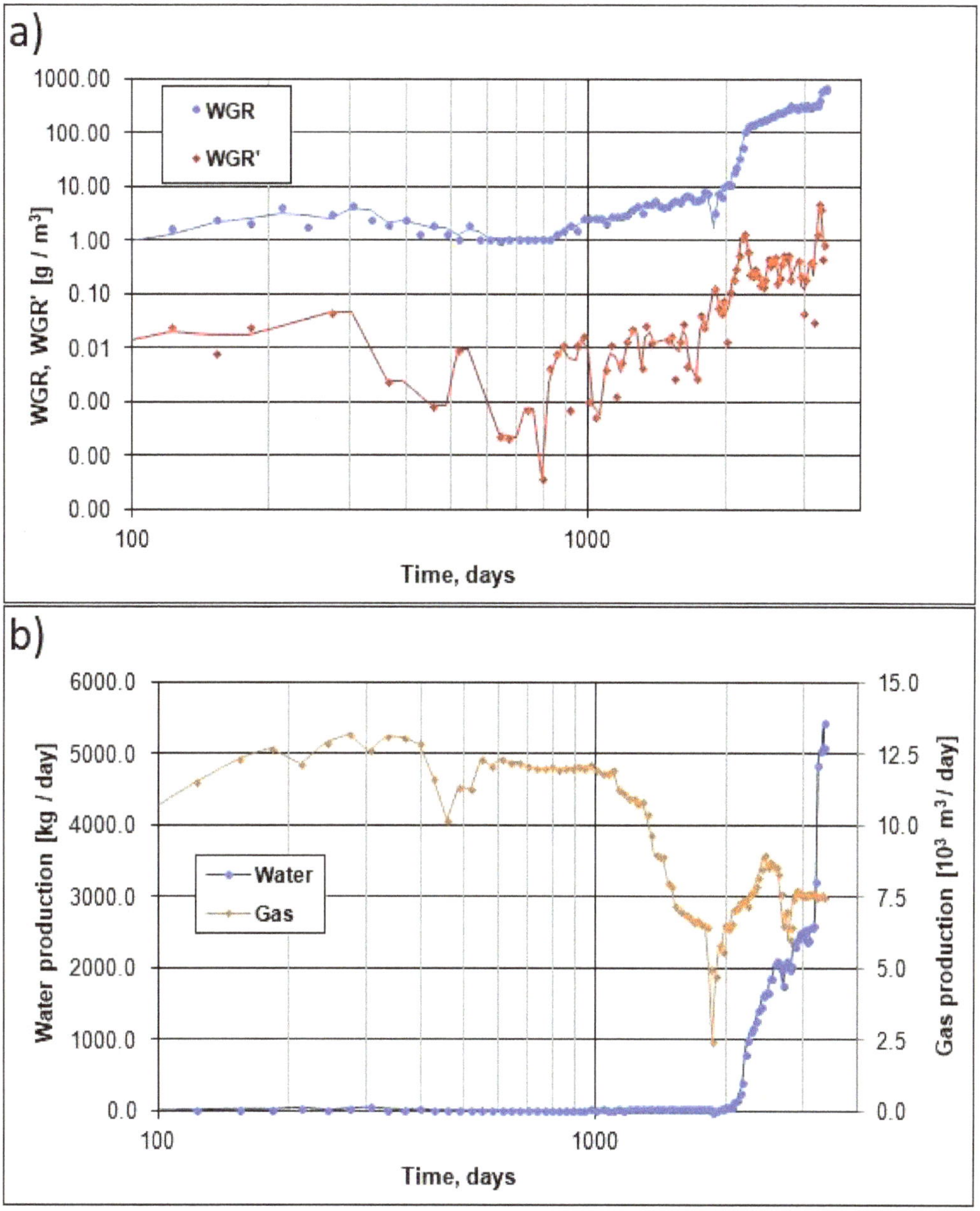

Figure 4. Changes in the WGR and its derivative over time (**a**) and the history of gas and formation water extraction (**b**) for the J-8K well.

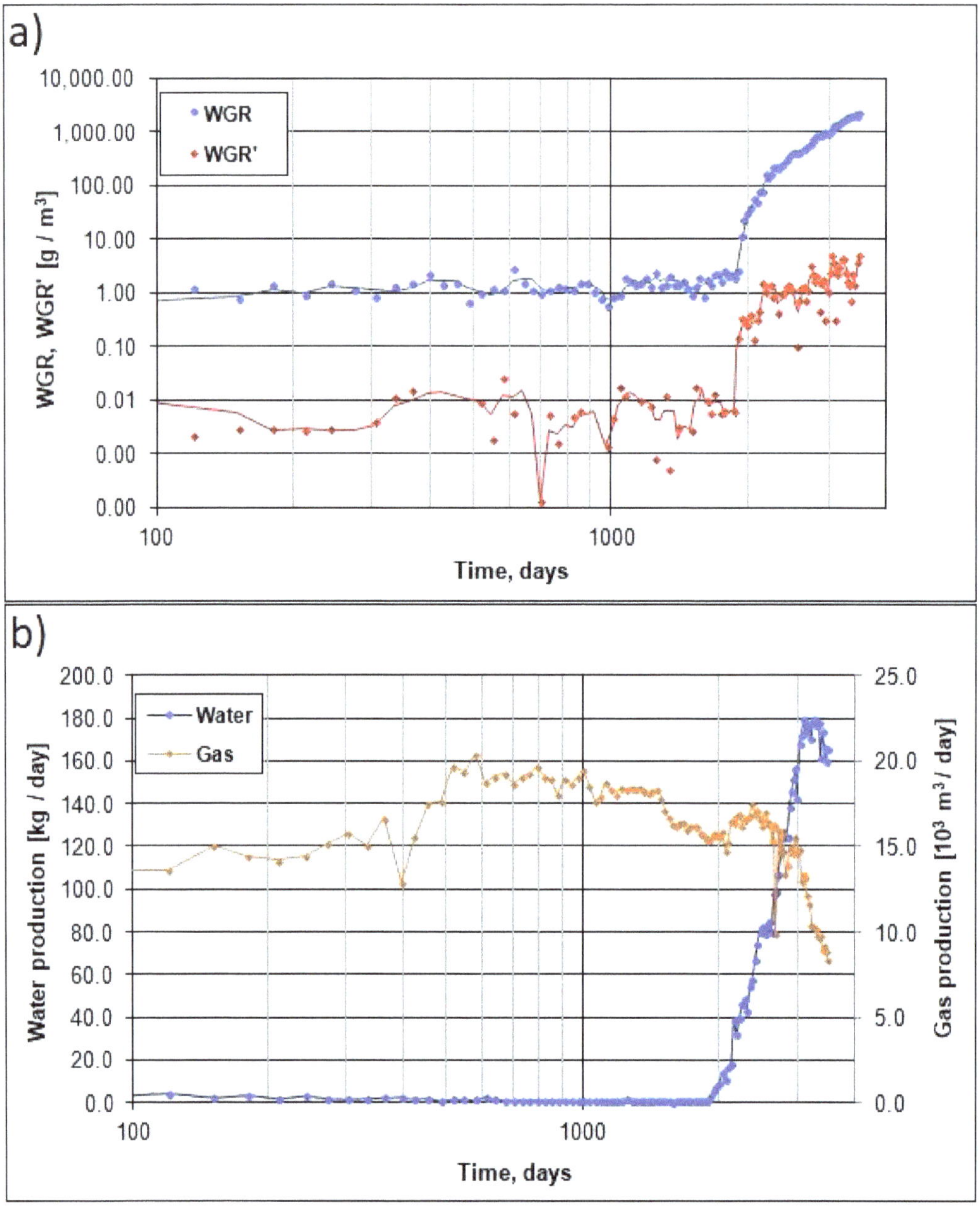

Figure 5. Changes in the WGR and its derivative over time (**a**) and the history of gas and formation water extraction (**b**) for the P-6 well.

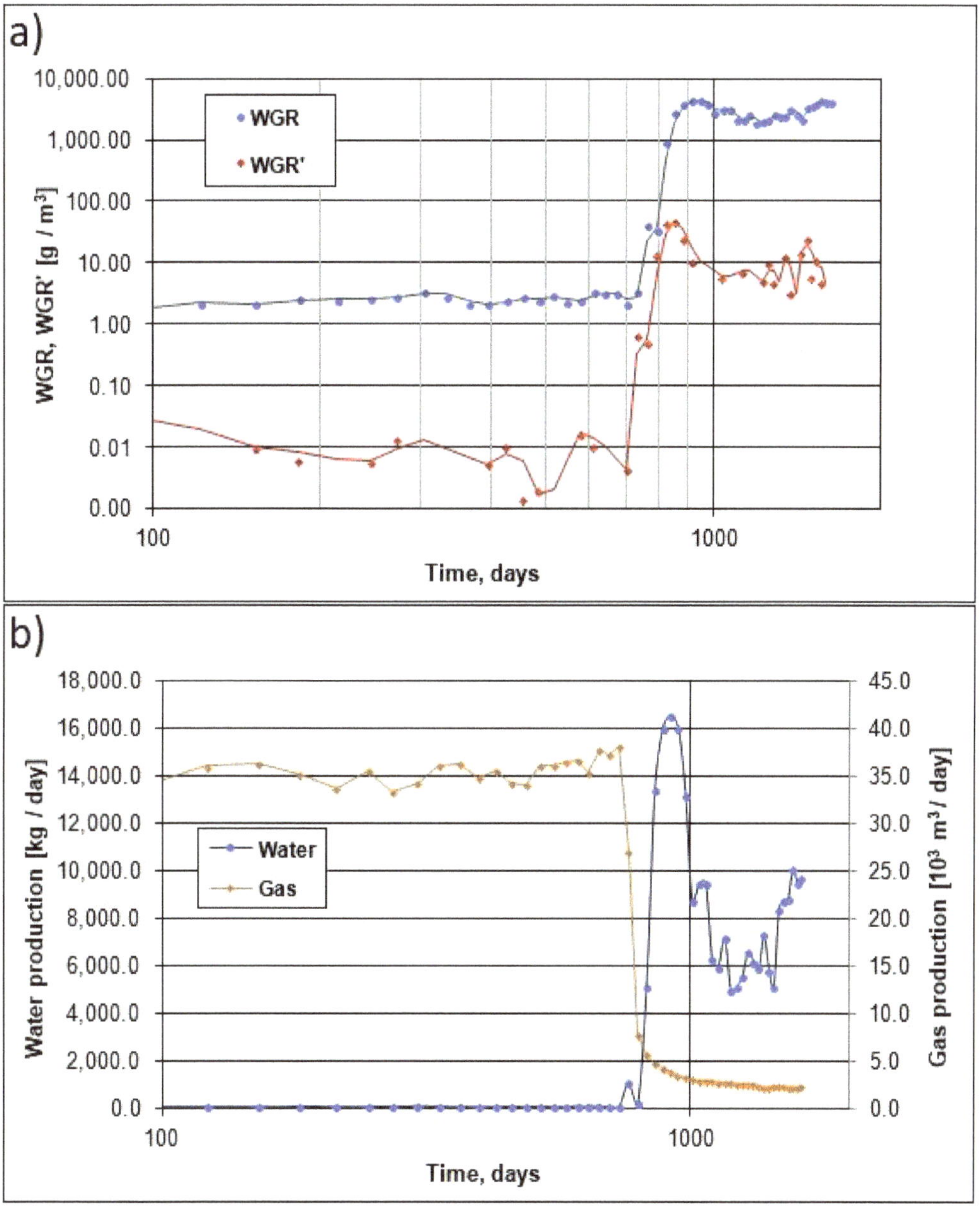

Figure 6. Changes in the WGR and its derivative over time (**a**) and the history of gas and formation water extraction (**b**) for the Z-1K well.

The P-6 well

Two intervals of 18 m and 5 m were completed for production through the P-6 well. Based on the trajectory of the determined curves (Figure 5a,b), it is clear that in the first years of the well's production (up to about the 2000th day), the exploitation was carried out with a very low water ratio, with the level of condensation water not exceeding 3 g/m^3. The analysed period in the case of the derivative of the water ratio was also characterised by a constant and parallel course of the WGR curve. The increase in the phase permeability of reservoir rocks to the formation water is characterised by a sudden and rapid increase in the water ratio. During about 1500 days of production, the WGR increased to a value of about 2300 g/m^3, and the WGR′ curve is characterised by an analogous and almost parallel course (Figure 5a). Changes in the gas and formation water production curves fully correlate with the changes in the ratio and its derivative. The increase in the production of formation water very quickly manifests itself in the decreasing volume of the exploited gas. The P-6 well is a typical example of a channel mechanism for inflow of formation water to a production well.

The Z-1K well

The Z-1K well provides the production horizon with an 18 m long perforation. The analysis of data from the determined curves (Figure 6a) indicates a constant and low level of the water ratio, not exceeding 4 g/m^3 up to about the 730th day of operation. During this period, based on the analysis of the course of the water ratio derivative, it should be concluded that there is a conical mechanism of the inflow of formation water. The water cone increases vertically and radially, thereby increasing the water saturation in its zone. In the next stage, there is a sudden and rapid increase in the WGR value to about 4000 g/m^3, followed by an almost immediate stabilisation of the ratio at a similar level. The course of the WGR′ curve for the growth period is parallel to the course of the water ratio curve. The information contained in the curves of the production of formation water and gas confirm a significant change in the amount of gas produced when the volume of exploited formation water increases (Figure 6b). This nature of the course of changes in the value of the water ratio clearly indicates the inflow of water to the well through a fracture or behind the pipe. Based only on the above data, it is not possible to clearly determine which of these mechanisms occurred in the Z-1K well.

4. The Use of Well Logging Data to Estimate the Permeability Profile of the Reservoir Horizons in Production Wells

The problem of estimating the permeability of porous materials has a long research history, dating back to the beginning of the 20th century. Of fundamental importance are the theoretical works by Kozeny [33] and Carman [34,35], which allowed for the development of a very general Kozeny–Carman (KC) formula for the calculation of permeability, linking many structural and textual parameters of the rock medium with the parameters of the migrating fluid. In practical terms, this approach is difficult to apply due to the general lack of information about the values of specific parameters, and especially their effective values in the case of polymineral media with a complex system of double porosity. At the same time, empirical models developed for a specific research area were developed based on macroscopic parameters determined from drill core samples, including: porosity (φ), irreducible water content (S_{wi}), capillary water content (S_{wirr}) and electrical parameters. Models: Tixier [36], Timur [37], Coates and Dumanoir [38] and Coates [39] are useful tools for the continuous estimation of changes in rock medium permeability along the well profile. However, the cited solutions require each time to take into account the specificity of the studied region and the application of the procedure of calibrating the results on representative data for the current analysed rock formations.

As a part of the project to select the optimal well for the WSO treatment, the most useful approach was the approach using the existing quantitative solutions of clay mineral content (V_{il}), mineral composition (V_{mi}), porosity (*PHI*) and saturation (S_w), and available geophysical profiling of natural potentials (*SP*), diameter (*CALI*) and nominal diameter (*BSM*). The developed methodology integrates the qualitative approach, based on the analysis of anomalies registered by individual logs and their mutual relations, with a purely quantitative approach based on the Zawisza model developed for Polish conditions [40]. For obvious reasons, such an approach is needed only in causes when the Production Log (PL) measurements are impossible to perform due to technical or financial reasons, because the interpretation of PL would significantly help in the selection of a well for the planned treatment.

4.1. Assessment of Relative Changes in Permeability—Methodological Aspects

The methodology for determining the permeable and impermeable zones was based on the qualitative analysis of profiling the diameter in relation to the nominal diameter. The presence of permeable zones is manifested by the presence of a narrowing of the diameter, i.e., the appearance of the so-called mud cake, which manifests its presence in the zones of intensive filtration of the drilling mud deep into the rock mass. Thus, a criterion can be constructed:

$CALI \leq BSM$ − potential permeable zones;
$CALI > BSM$ − potential sealing zones.

The above criterion is rather of a qualitative nature and cannot be used alone except where no other option is available.

The second possibility is a strict quantitative interpretative approach based on the absolute permeability estimation according to the Zawisza model. This model allows the calculation of the irreducible water content (Swi) according to Formula (1) and the absolute permeability according to Formula (2):

$$Swi = V_{il}^{a}(1-\varphi)^{b} \tag{1}$$

$$Kzaw = A\varphi^{c}(1-Swi)^{d} \tag{2}$$

where:

V_{il}—volumetric content of clay minerals;
φ—porosity;
A, a, b, c, d—calibration constants.

The calibration constants in the above equations are determined each time for the region and formation in which the estimation is carried out and modified to best reflect the trends of changes in permeability in the available laboratory tests of drill cores, or the average values of this parameter determined on the basis of data from performed well tests.

The results of the permeability estimation allow the definition of another criterion of division into permeable and sealing layers, assuming the limit value Kgr = 1 mD for water inflow to the well. This condition will be as follows:

$Kzaw \geq Kgr$ − potential permeable zones;
$Kzaw < Kgr$ − potential sealing zones.

The sum of the above criteria is a collective criterion and is the basis for the final division of the profile into:

$CALI \leq BSM$ and $Kzaw \geq Kgr$ − permeable zones;
$CALI > BSM$ and $Kzaw < Kgr$ − sealing zones.

The characteristics of the remaining parts of the profile are ambiguous in relation to the above methodological assumptions.

4.2. Analysis Results

The overall results of the analyses are presented in the figures below (Figures 7–10). The following columns show:

Track 1—profiling of gamma GR (API) and natural potentials SP (mV);
Track 2—MD depth (m);
Track 3—perforation intervals;
Track 4—borehole diameter (caliper);
Track 5—results of lithological and porosity analysis;
Track 6—results of the analysis of primary saturation with formation media;

Track 7—the results of the absolute permeability estimation according to the Zawisza model (*Kzaw*) (mD);
Track 8—analysis of permeable and sealing zones according to the collective criterion;
Track 9—MD depth (m).

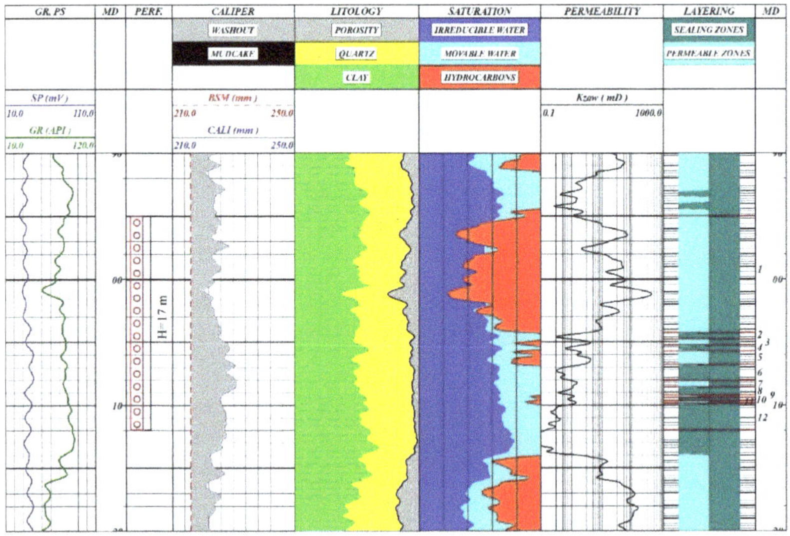

Figure 7. Results of the interpretation of the profile of permeability changes within the analysed perforation interval in the J-5K well.

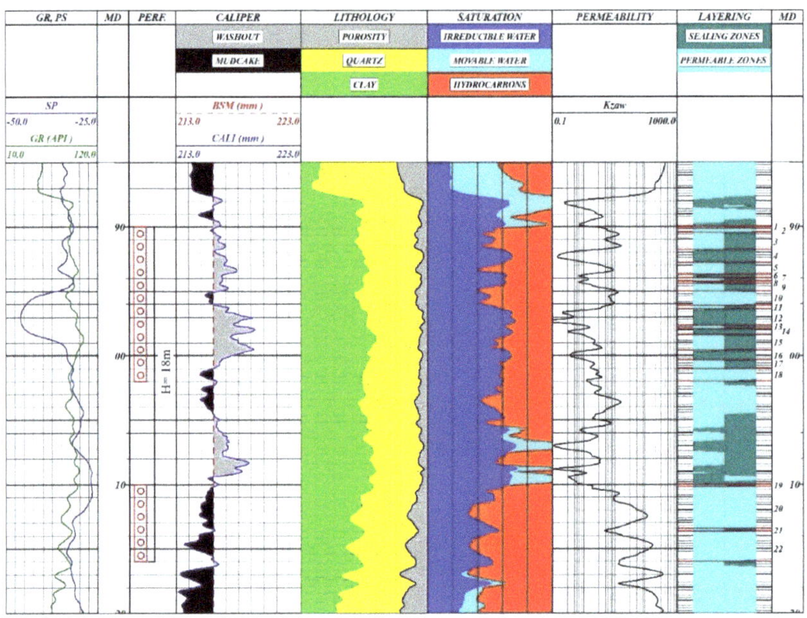

Figure 8. The results of the interpretation of the permeability change profile within the analysed perforation intervals in the J-8K well.

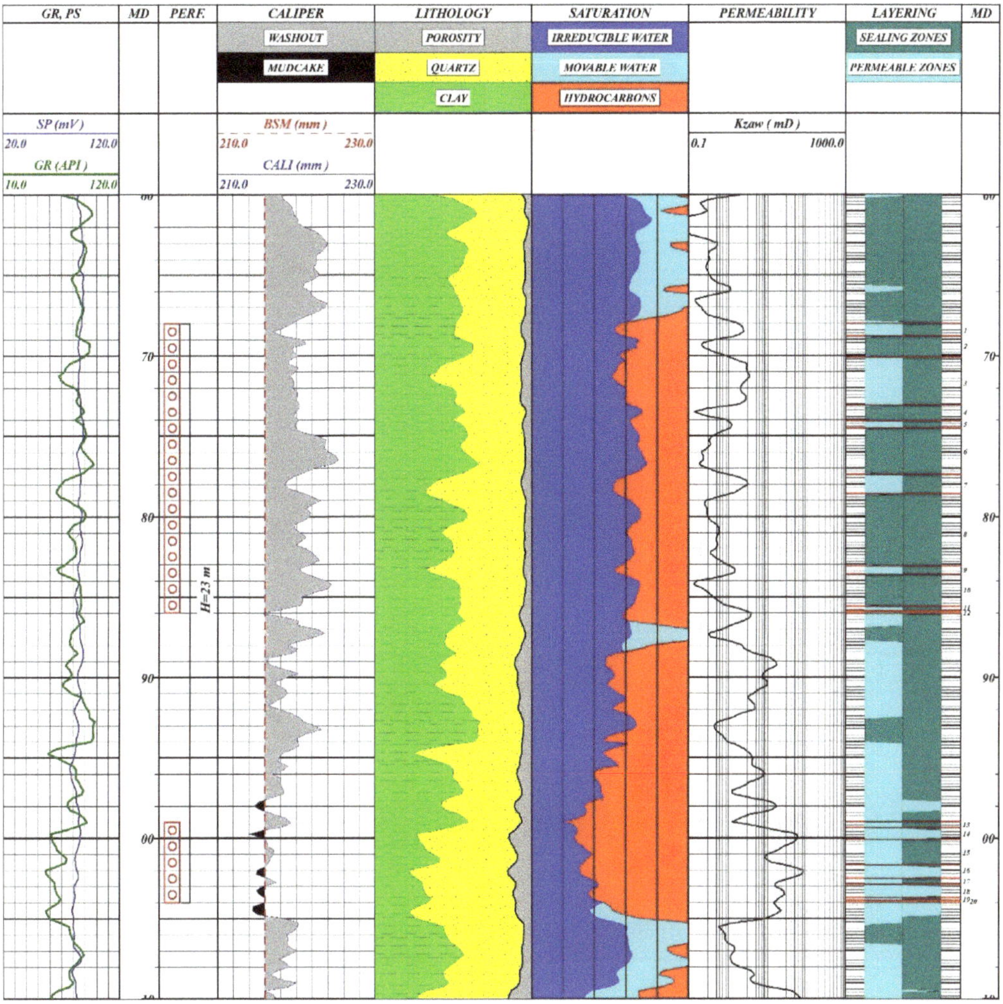

Figure 9. Results of the interpretation of the permeability changes profile within the analysed perforation intervals in the P-6 well.

The description of the results obtained for four of the analysed wells is presented below, with the emphasis on the number of interpreted layers (permeable and impermeable).

The J-5K well (Figure 7)

In the interval of the perforation providing access to the 17 m section of the gas horizon profile, an analysis was performed which confirmed the presence of 12 zones with variable filtration characteristics, most of them with unambiguous sealing properties, especially in the bottom part of the horizon.

The J-8K well (Figure 8)

Two perforation intervals were analysed in the J-8K well: the 12 m long upper interval profile is characterised by much greater diversity in terms of filtration properties. The lower perforation interval (6 m) almost entirely covers the unambiguously permeable zone. In total, 22 layers with different reservoir parameters were distinguished in this well.

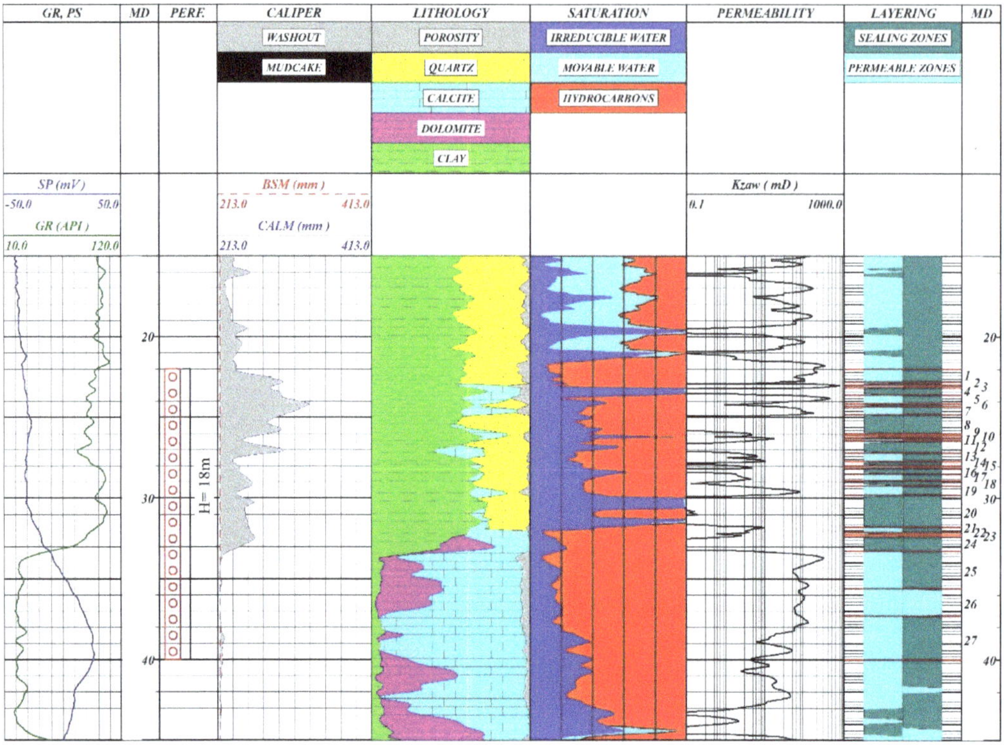

Figure 10. Results of the interpretation of the permeability changes profile within the analysed perforation interval in the Z-1K well.

The P-6 well (Figure 9)

The subject of the analysis in the P-6 well were two perforation intervals (18 m + 5 m) completed in one gas horizon. These two fragments of the gas horizon profile differ significantly in terms of reservoir characteristics. The lower perforation covers a fairly homogeneous sandstone complex which is practically one permeable formation zone. On the other hand, the part of the profile completed in the upper perforation is characterised by much more variety in terms of reservoir properties, with a large number of unambiguously sealing zones. In this well, it is possible to distinguish a total of 20 zones with different filtration properties.

The Z-1K well (Figure 10)

The 18 m long perforation, consisting of the top part of the gas horizon in the Z-1K well, includes the top part of an Upper Devonian carbonate complex and the bottom part of Lower Carboniferous clastic formations, which directly translates into the two-fold nature of the interval provided. The lower part of the perforation interval (carbonate rocks) is a fairly homogeneous zone in terms of filtration properties, while the clastic formations lying above are composed of a large number of layers which are unambiguously and ambiguously sealing. The performed interpretation allowed for distinguishing 27 zones with different filtration properties.

4.3. Ranking of Wells for the Performance of the WSO Treatment Based on the Interpretation of Well Logging Data

For the purpose of performing a ranking of wells for comparative purposes, a quantitative parameter the called "perforation overlayering index" (U) (m^{-1}) was defined, which specifies the average number of distinguished layers in relation to a running metre of perforation made in the well. The calculations were carried out according to the following Equation (3)

$$U = \frac{w}{H}\left[m^{-1}\right] \tag{3}$$

where:

w—total number of recognised layers (both unambiguous and ambiguous);
H—total number of metres of perforation made in the well (m).

The introduced U parameter determines the diversification of the formation in terms of the number of layers with different filtration properties. It was found that the highest values of this parameter would be the most beneficial in terms of performing WSO procedures. Table 1 summarises the systematised results of the U parameter estimation for the above-mentioned wells. When analysing the compiled values of the U parameter for clastic rocks, the first place in the ranking was achieved by the J-8K well (U = 1.22), the second by P-6 (U = 0.87) and then J-5K (U = 0.7). For the Z-1K well, representing carbonate rocks, the U parameter value was 1.5.

Table 1. Ranking of the wells for the performance of the WSO treatment based on the interpretation of permeability.

Well	w	H (m)	$U = w/H$ (m^{-1})
J-5k	12	17	0.7
J-8k	22	18	1.22
P-6	20	23	0.87
Z-1k	27	18	1.5

5. Well Ranking Based on the Well Evaluation Card

As part of the project, in the first stage of the ranking, eight wells from the Carpathian Foredeep and the marginal part of the Carpathians were evaluated (Figure 1). The initial selection of these wells was made by the operator. In six wells (C-2, J-5K, J-8K, J-7, P-6, Z-74), gas-bearing horizons related to clastic reservoir levels in Miocene formations were selected for a possible treatment, while in the other two cases (T-3, Z-1K) gas accumulations were located in Palaeozoic carbonate rocks. The ranking of selected wells and reservoir horizons was based on the collected geological, reservoir and exploitation data, as well as a preliminary interpretation of the well logs. As part of the project, a special tool was created, which was called the Well Evaluation Card (WEC). It is a special spreadsheet created in MS Excel, which in a relatively simple technical way allows one to assess the suitability of a given gas well for water shut-off treatment on a 100-point scale. Thus, when evaluating several wells, it is possible to rank them in terms of their suitability for the treatment based on a simple criterion, i.e., the more points the well achieves, the more suited it is for the treatment. It should be emphasised that in the event of failure to meet (or lack of data) one of the so-called "critical conditions" (included in the first part of the card), a well is disqualified as a candidate for water shut off. The interface of this tool is relatively simple. It has been constructed in such a way as to be as easy as possible to use for the user, and at the same time to minimise the possibility of making mistakes by the user, enabling multi-stage evaluation of individual criteria through the use of advanced MS Excel functions.

The category of "critical conditions" necessary to be met includes: water production mechanism, the length of the perforation interval (not more than 50 m), water salinity and the formation temperature. The second group of scored criteria is "other conditions", which include: geometry of the water inflow, presence or absence of impermeable barriers, lithological profile differentiation, type of reservoir, water saturation coefficient, thickness of the gas-bearing horizon and effective permeability. The third group of criteria (not scored) is "supplementary information", which is intended to facilitate the possible decision regarding the validity of the procedure in the well, if the aforementioned scoring criteria are met. The third group of criteria includes parameters, such as field pressure, water-gas ratio, recovery efficiency, porosity, possibility of acid treatment, possibility of increasing the pressure drawdown and additional tests performed.

From among the eight analysed wells, the four wells with the highest scores were selected in the first stage based on the criteria and scores resulting from the WEC: C-2 (81.3), J-5K (76.6), J-8K (76.6) and P-6 (73.0). The T-3 (53.4) and Z-1K (50.0) wells had the lowest scores, mainly due to the unfavourable mechanism of water inflow through a fracture or behind pipe.

In the next stage of the ranking, only the four above-mentioned wells, which obtained the highest number of points in the previous stage, were considered. The detailed data contained in the geological documentation of the gas fields and the detailed results of the interpretation of the well logs were taken into account. The analysis of these data allowed for obtaining more detailed information on the lithological diversity and the spread of reservoir properties (especially permeability) in the vertical profiles of the selected formation horizons. In three of the four analysed wells (J-5K, J-8K and P-6), the profiles of the interpreted reservoir horizons are characterised by a clear lithological differentiation and the presence of impermeable inserts of fine-clastic rocks within them. In the fourth case (C-2), the reservoir horizons are characterised by lower lithological diversity, and above all by good reservoir properties in the entire intervals completed by pipe perforations. Such horizons should not be taken into account for the test procedure of limiting water production, due to the possibility of cross flows resulting from the homogeneity of the layers completed by perforations. In the final ranking, carried out on the basis of updated information obtained from research and analyses carried out during the project implementation, two wells were considered the most optimal in terms of the planned treatment to reduce water production: P-6 (73.0) and J-8K (68.2). In both of these wells, the reservoir horizons are completed by two perforation intervals with different properties: the upper intervals are characterised by a much greater lithological variability and differentiation of reservoir parameters (within them, a number of permeable and impermeable zones can be distinguished), while the intervals of lower perforations constitute zones with good permeability (Figures 8 and 9).

6. Discussion

The volume of natural gas consumption in Poland is growing at a much faster pace than production from gas fields that are newly connected for exploitation. An important issue is therefore the idea of implementing secondary methods aimed at increasing the depletion rate over the largest possible area of the field. However, it should be remembered that in the case of small, highly depleted formations, expenses and investment expenditures (OPEX and CAPEX) may exceed the expected returns. In such cases, a more logical decision may be to implement one of the possible methods of improving exploitation efficiency, i.e., the implementation of technologies aimed at the immediate vicinity of production or injection wells, as WSO technologies are. Technologies of this type, including the innovative RCC technology, have been successfully used in Hungary [8,10]. The first treatments performed will provide an answer to the effectiveness of this type of treatment in the Polish part of the Carpathian Foredeep. It should be emphasised that the injection of the treatment fluid into the formation is a relatively simple operation from a technical point of view. However, this type of treatment cannot be performed in every gas well, therefore an extremely important factor for determining the success of the operation is the selection of an appropriate well that meets certain criteria.

This article only discusses the issues related to the selection of wells based on broadly understood geological, reservoir and exploitation data. However, economic factors were not mentioned. Therefore, it should be emphasised that while the assessment of the technological effects of the treatment is relatively simple (by comparing the production of water and gas before and after the treatment), the assessment of the economic effects is not easy or unambiguous. When assessing economic aspects, many factors should be taken into account, such as the costs of sanding, the method of transporting and managing water in specific technical and legal conditions, benefits related to the possibility of continuing operation under unchanged conditions of pressure, etc.

Many authors [1,5,12,41–44] agree that one of the most important criteria for assessing wells is the interpretation of the history of changes in the water ratio and its derivative, leading to the identification of the well water breakthrough mechanism. Correct identification of the water breakthrough mechanism allows the rejection of those wells which do not qualify for any chemical water shut-off treatment. Another important criterion, which significantly determines the possibility of performing the treatment is the vertical lithological differentiation of the profile of perforated interval. Other important criteria include:

- Formation water salinity;
- The reservoir temperature;
- Length of the perforation.

However, it is worth noting that as research and technological development progresses, perhaps new solutions will be developed and in the future some of these limitations will no longer be an obstacle to the implementation of water production reduction measures.

Based on the ranking, two wells were finally selected from the eight wells assessed for the water production limitation treatment (P-6 and J-8K). Bearing in mind the procedure of selecting and recommending wells for the possible treatment of limiting water production, it is worth noting that it prefers specific features or specific ranges of values of particular reservoir and exploitation parameters. It can be assumed that the currently proposed technology, based on working fluids prepared on the basis of microemulsions, will be suitable for gas horizons characterised by significant lithological differentiation and located mainly in the middle part of the autochthonous Miocene profile in the Carpathian Foredeep. In practice, this differentiation translates into the presence of a number of permeable and impermeable alternate layers in the profile of the gas interval completed by perforation. The lack of such impermeable barriers disqualifies the well as a candidate for the treatment. The discussed technology will certainly not be used in collectors developed in thick-bedded formations (e.g., amalgamated sandstone beds), massive formations or reservoirs located in stratigraphic traps.

In each case of an attempt to reduce water production in a gas well, it is necessary to assess such a well in terms of water breakthrough mechanisms and its verification based on the Well Evaluation Card. This will help to exclude cases where the procedure cannot be performed, or it will not bring the expected results.

In the case of horizons located in the deeper part of the Miocene profile of the Carpathian Foredeep, as well as natural gas fields located in other petroleum basins, such features as lithological variability, the degree of compaction as well as formation water salinity may be so different that the proposed technology will require significant modification.

7. Conclusions

Well pre-selection procedure is the first step in the selection of wells for WSO treatment. This was developed to quickly and definitively reject a well as a candidate for WSO treatment. Wells that were not rejected in the pre-selection procedure are not candidates for treatment, yet further verification is required.

In the porous formations, the lack of natural impermeable barriers in the completed interval strongly reduces the probability of WSO treatment success in a given well.

It is important to base the diagnosis on time intervals (so-called diagnostic windows) when assessing the water inflow mechanism, when the well was producing without operator intervention.

Finally, to recommend a well as a candidate for WSO treatment, it is not possible to rely only on one of the criteria, but it is necessary to analyse all available geological, formation and exploitation data for successful outcome.

Author Contributions: Conceptualisation: S.F., A.U., M.S. and M.M.; formal analysis: S.F.; investigation and methodology: S.F., A.U., M.S. and M.M.; writing—original draft preparation: S.F., A.U., M.S. and M.M.; writing—review and editing: S.F. and A.U.; visualisation: M.S. and M.M.; supervision: S.F. and A.U. All authors have read and agreed to the published version of the manuscript.

Funding: The research leading to these results received funding from the Project titled "Procedure: well selection, water shut-off treatment scenario in gas production wells and preparing microemulsion based treatment fluid", under grant agreement No. DK-4100-190/18.

Institutional Review Board Statement: Not applicable.

Informed Consent Statement: Not applicable.

Data Availability Statement: Not applicable.

Acknowledgments: The authors express their thanks to the Polish Oil and Gas Company, PGNiG SA, Warsaw, Poland, for granting permission for the use of geological and exploitation data and consent to publish the results.

Conflicts of Interest: The authors declare no conflict of interest.

References

1. Bailey, B.; Crabtree, M.; Tyrie, J.; Elphick, J.; Kuchuk, F.; Romano, C.; Roodhart, L. The Challenge of Water Control. *Oilfield Rev.* **2000**, *12*, 30–51.
2. Veil, J.A.; Clark, C.E. Produced Water Volumes and Management Practices. In Proceedings of the SPE International Conference on Health, Safety and Environmental in Oil and Gas Exploration and Production, Rio de Janeiro, Brazil, 12–14 April 2010. [CrossRef]
3. Veil, J.A. Produced water volumes and management in 2017—An update. In Proceedings of the GWPC Annual Forum, Oklahoma City, OK, USA, 15–17 September 2019; pp. 15–17.
4. Seright, R.S.; Lane, R.H.; Sydansk, R.D. A strategy for attacking excess water production. *SPE Prod. Facil.* **2003**, *18*, 158–169. [CrossRef]
5. Lakatos, I.J.; Lakatos-Szabo, J.; Szentes, G.; Jobbik, A.; Vago, A. Application of Self-Conforming well Stimulation technology in Oil and Gas Fields—Fundamentals and Case Histories. In Proceedings of the IOR 2017—19th European Symposium on Improved Oil Recovery, Stavanger, Norway, 24–27 April 2017. [CrossRef]
6. Sydansk, R.; Seright, R.S. When and Where Relative Permeability Modification Water-Shutoff Treatments Can Be Successfully Applied. *SPE Prod. Oper.* **2007**, *22*, 236–247. [CrossRef]
7. Zaitoun, A.; Kohler, N.; Guerrini, Y. Improved Polyacrylamide Treatments for Water Control in Producing Wells. *J. Pet. Technol.* **1991**, *43*, 862–867. [CrossRef]
8. Lakatos, I.J.; Lakatos-Szabo, G.; Szentes, G.; Bodi, T.; Vago, T.; Karaffa, Z. Multifunctional Stimulation of Gas Wells Operating in Gas Cap over a Depleted oil Reservoir. In Proceedings of the SPE International Conference & Exhibition on Formation Damage Control, Lafayette, LA, USA, 24–26 February 2016. [CrossRef]
9. Sobanova, O.B.; Fridman, G.B.; Arefyev, Y.N. Laboratory and Oil Field Testing for Application of Compositions Including Hydrocarbonsand Surfactants for Restricting Water Influx into Producing Wells. In Proceedings of the Paper Presented at the 7th European Symposiumon IOR, Moscow, Russia, 5–7 July 1993.
10. Lakatos, I.J.; Lakatos-Szabo, J.; Szentes, G.; Vago, A. Restriction of Water Production in Gas Wells by Induced Phase Inversion: Field Case Studies. In Proceedings of the SPE International Symposium and Exhibition on Formation Damage Control, Lafayette, LA, USA, 26–28 February 2014. [CrossRef]
11. Alfarge, D.K.; Wei, M.; Bai, B. Numerical simulation study of factors affecting relative permeability modification for water-shutoff treatments. *Fuel* **2017**, *207*, 226–239. [CrossRef]
12. Chan, K.S. Water Control Diagnostic Plots. In Proceedings of the SPE Annual technical Conference and Exhibition, Dallas, TX, USA, 22–25 October 1995. [CrossRef]
13. Kováč, M.; Nagymarosy, A.; Oszczypko, N.; Ślączka, A.; Csontos, L.; Marunteanu, M.; Matenco, L.; Márton, E. Palinspastic reconstruction of the Carpathian–Pannonian region during the Miocene. In *Geodynamic Development of the Western Carpathians*; Rakúš, M., Ed.; Slovak Geological Survey (GSSR): Bratislava, Slovakia, 1998; pp. 189–217.
14. Golonka, J.; Pietsch, K.; Marzec, P. Structure and plate tectonic evolution of the northern Outer Carpathians. In *Tectonics*; Closson, D., Ed.; INTECH: Rijeka, Croatia, 2011; pp. 65–92.
15. Porębski, S.J.; Warchoł, M. Hyperpycnal flows and deltaic clinoforms—Implications for sedimentological interpretations of late Middle Miocene fill in the Carpathian Foredeep Basin. *Prz. Geol.* **2006**, *54*, 421–429.
16. Borys, Z. Aktualne problemy poszukiwań węglowodorów we wschodniej części przedgórza Karpat. *Prz. Geol.* **1996**, *44*, 1019–1023.
17. Myśliwiec, M. Poszukiwania złóż gazu ziemnego w osadach miocenu zapadliska przedkarpackiego na podstawie interpretacji anomalii sejsmicznych—podstawy fizyczne i dotychczasowe wyniki. *Prz. Geol.* **2004**, *52*, 299–306.

18. Myśliwiec, M. Typy pułapek gazu ziemnego i strefowość występowania ich złóż w osadach miocenu wschodniej części zapadliska przedkarpackiego. *Prz. Geol.* **2004**, *52*, 657–664.
19. Myśliwiec, M. Żołynia—Leżajsk—stare złoże, nowe zasoby. *Nafta-Gaz* **2006**, *3*, 97–105.
20. Myśliwiec, M.; Madej, K.; Byś, I. Złoża gazu ziemnego w osadach miocenu rejonu Rzeszowa (zapadlisko przedkarpackie) odkryte na podstawie wyników nowoczesnej kompleksowej interpretacji danych sejsmicznych. *Prz. Geol.* **2004**, *52*, 501–506.
21. Myśliwiec, M.; Borys, Z.; Bosak, B.; Liszka, B.; Madej, K.; Maksym, A.; Oleszkiewicz, K.; Pietrusiak, M.; Plezia, B.; Staryszak, G.; et al. Hydrocarbon resources of the Polish Carpathian Foredeep: Reservoirs, traps, and selected hydrocarbon fields. In *The Carpathians and Their Foreland: Geology and Hydrocarbon Resources*; Golonka, J., Picha, F.J., Eds.; AAPG: Tulsa, OK, USA, 2006; Volume 84, pp. 351–393.
22. Rzemieniarz, A. Analiza zmienności wybranych parametrów petrofizycznych utworów struktury Palikówka. *Nafta-Gaz* **2004**, *11*, 561–569.
23. Chmielowska, A.; Sowiżdżał, A.; Tomaszewska, B. Prospects of Using Hydrocarbon Deposits from the Autochthonous Miocene Formation (Eastern Carpathian Foredeep, Poland) for Geothermal Purposes. *Energies* **2021**, *14*, 3102. [CrossRef]
24. Górka, A.; Gliniak, P.; Madej, K.; Maksym, A. Oil and gas fields in the Carpathians and the Carpathian Foredeep. *Prz. Geol.* **2007**, *55*, 993–998.
25. Sowiżdżał, K.; Słoczyński, T.; Matyasik, I.; Stadtmüller, M. Analiza paleozoicznego systemu naftowego w strefie brzeżnej nasunięcia Karpat fliszowych. *Nafta-Gaz* **2015**, *9*, 632–646.
26. Wójcik, K.; Peryt, T. *Pakiet Danych Geologicznych do Postępowania Przetargowego na Poszukiwanie Złóż Węglowodorów, Obszar Przetargowy "Błażowa"*; Państwowy Instytut Geologiczny—Państwowy Instytut Badawczy: Warszawa, Poland, 2016.
27. Maksym, A.; Dziadzio, P.; Liszka, B.; Staryszak, G. Środowisko sedymentacji utworów miocenu autochtonicznego w brzeżnej strefie Karpat, a interpretacja geologiczno-złożowa w obszarze Husów-Albigowa-Krasne. *Naft. Gaz* **1997**, *9*, 407–414.
28. Dziadzio, P. Sekwencje depozycyjne w utworach badenu i sarmatu w SE części zapadliska przedkarpackiego. *Prz. Geol.* **2000**, *48*, 1124–1138.
29. Seright, R.S.; Lane, R.H.; Sydansk, R.D. A Strategy for Attacking Excess Water Production. In Proceedings of the SPE Permian Basin Oil and Recovery Conference, Midland, TX, USA, 15–17 May 2001. [CrossRef]
30. Chan, K.S.; Bond, A.J.; Keese, R.F.; Lai, Q.J. Diagnostic Plots Evaluate Gas Shut-Off Gel treatments at Prudhoe Bay, Alaska. In Proceedings of the Technical Conference and Exhibition, Denver, CO, USA, 6–9 October 1996. [CrossRef]
31. Seldal, M. Using Water/Oil Ratios to Diagnose Excessive Water Production Mechanisms. Master's Thesis, New Mexico Institute of Mining and Technology, Socorro, NM, USA, 1997.
32. Seright, R.S. Improved Methods for Water Shutoff. In *Annual Report DOE/PC/91008-14*; New Mexico Institute of Mining and Technology: Socorro, NM, USA, 1998.
33. Kozeny, J. Ueber kapillare Leitung des Wassers im Boden. *Sitz. Akad. Wiss. Wien* **1927**, *136*, 271–306.
34. Carman, P.C. Fluid flow through granular beds. *Inst. Chem. Eng.* **1937**, 32–48. [CrossRef]
35. Carman, P.C. *Flow of Gases through Porous Media*; Butterworths: London, UK, 1956.
36. Tixier, M.P. Evaluation of Permeability from Electric-Log Resistivity Gradients. *Oil Gas J.* **1949**, *48*, 113–122.
37. Timur, A. An Investigation of Permeability, Porosity, and Residual Water Saturation Relationship for Sandstone Reservoirs. *Log Anal.* **1968**, *9*, 8.
38. Coates, G.R.; Dumanoir, J.L. A New Approach to Improved Log-Derived Permeability. *Log Anal.* **1974**, *1*, 17.
39. Schlumberger Ltd. *Log Interpretation Charts*; Schlumberger Ltd.: Houston, TX, USA, 1987.
40. Zawisza, L.; Nowak, J. *Metodyka Określania Parametrów Filtracyjnych Skał na Podstawie Kompleksowej Analizy Geofizyki Otworowej*; AGH University of Science and Technology Press: Kraków, Poland, 2012.
41. Botermans, C.W.; Calrymple, E.W.; Dahl, J.; Smith, D. Chemical system for water and gas control: Terminology, evaluation methods, candidate selection and expectations. In Proceedings of the SPE International Symposium on Oilfield Chemistry, Houston, TX, USA, 13–16 February 2001. [CrossRef]
42. Lakatos, I.J.; Lakatos-Szabo, J.; Szentes, G.; Vago, A. New Alternatives in Conformance Control: Nanosilica and Liquid Polymer Aided Silicate Technology. In Proceedings of the SPE European Formation Damage Conference and Exhibition, Budapest, Hungary, 3–5 June 2015. [CrossRef]
43. Lakatos, I.J.; Szentes, G.; Toro, M.; Karaffa, Z.; Vago, A. Mitigation of Formation Damage Caused by Chemical Overdosing in Water Shut-Off Treatments. In Proceedings of the SPE International Conference and Exhibition on Formation Damage Control, Lafayette, LA, USA, 19–21 February 2020. [CrossRef]
44. Seright, R.S.; Brattekas, B. Water shutoff and conformance improvement: An introduction. *Pet. Sci.* **2021**, *18*, 450–478. [CrossRef]

energies

Article

Influence of Hblock Fine-Grained Material on Selected Parameters of Cement Slurry

Marcin Kremieniewski

Oil and Gas Institute—National Research Institute, 25A Lubicz Str., 31-503 Krakow, Poland; kremieniewski@inig.pl

Abstract: Fine-grained materials are used to seal the microstructure and improve the mechanical parameters of the formed cement sheath. They are used in cement slurries designed to seal casing columns at great depths and in geothermal boreholes to improve thermal conductivity. Such additions shorten the setting time and the transition time from the value of the initial time of the setting to the final of the setting. This allows the shortening of the time needed to bind the slurry and move on to further work. Additionally, it helps to eliminate the possibility of gaseous medium intrusion into the structure of the setting cement slurry. The goal of the work is to determine the influence of the Hblock fine-grained material on selected parameters of the cement slurry. Hblock is used in the research ranging from 1% to 20% (bwoc (bwoc—by weight of cement)), and the two types of cement most commonly used in the petroleum sector. Then the technological parameters of fresh and hardened cement slurries were tested. Research on rheological parameters, determined rheological models and flow curves are performed. On the basis of the obtained test results, the influence of the additive on the technological parameters of the cement slurry is discussed. The research results allow for further work and then the application of the fine-grained material in the petroleum sector.

Keywords: cement slurry; fine-grained material; borehole sealing efficiency; technological parameters

Citation: Kremieniewski, M. Influence of Hblock Fine-Grained Material on Selected Parameters of Cement Slurry. *Energies* **2022**, *15*, 2768. https://doi.org/10.3390/en15082768

Academic Editors: Izabela Hager and F. Pacheco Torgal

Received: 18 January 2022
Accepted: 8 April 2022
Published: 9 April 2022

1. Introduction

The column of casing pipes is sealing a cement slurry, which is a mixture of water, cement and additives or admixtures modifying the technological parameters of the cement slurry. Its properties are adjusted to the geological and technical conditions prevailing in the borehole where the hydration process takes place and the formation of a cement sheath in the annular space [1–3]. In this case, the decisive factor is the type of drilled geological layers, the final drilling depth, dynamic and static temperature, as well as formation pressure and fracturing pressure [4–7]. Therefore, it is necessary to adjust the parameters of the cement slurry properly. For this purpose, various types of chemicals, additives and fine-grained material are used, the use of which allows to obtain the parameters required for the given conditions and to obtain the highest possible tightness and durability of the cement sheath formed from the cement slurry [2,8,9]. However, the application of a new type of agent must be preceded by a test cycle, after which it is possible to determine the benefits or the impact of using a given additive [10,11]. Based on the analysis of the obtained results, it is possible to determine the suitability of the tested agent and then modify the slurry recipe for specific borehole conditions. The use of such an advantageously modified cement slurry contributes to the elimination of bad cementing. Such actions are necessary because poor cementation may result in complications and difficulties in liquidating possible outflows of the formation medium from the annular space [12,13]. Due to the high costs of additional works related to sealing, research on the impact and the possibility of using a new type of agents is a necessary element in the continuous improvement of the properties of drilling fluids.

The Role of Fine-Grained Material in the Technology of Cement Slurries

The use of hydraulic binders and modification of the properties of the developed cement slurries with the use of additives and mineral admixtures is an issue subject to a wide scope of research [14,15]. It is related to the pursuit of continuous improvement of technological parameters of cement slurries as a result of the use of a new type of agent. Thanks to this, it is possible to seal the matrix of the cement sheath formed in the annular space. Moreover, the modified cement slurry recipe has a more compact structure with a lower porosity value and a much lower gas permeability [2,16,17]. Such conditions make it possible to limit, and most often eliminate, gas migration through the cement sheath. Gas migration is a phenomenon that can occur both during the setting and hardening of the cement slurry, but also during the exploitation of the hole in which the slurry is in the solid state. After binding the cement slurry, the greatest risk of gas migration is in the capillary pore spaces, microcracks in the structure of the cement sheath or in the crevices formed [2,18–20]. Therefore, it is necessary to design a cement slurry recipe that, after setting, will be characterized by a lack of gas permeability. For this purpose, fine-grained material and admixtures in the form of mineral dust are used. By increasing the volume of the slurry and hydraulic activity, fine-grained material increases the tightness of the hardened cement slurry [21–23]. It is important because high permeability values of the cement sheath cause corrosion and decrease the mechanical parameters [2,24,25]. The mechanical strength, which is related to the microstructure of the cement sheath and the value of the total porosity, should also be taken into account. When designing the cement slurry recipe, its density is selected depending on the geological and technical conditions, and the mechanical strength decreases in proportion to the reduction in the cement slurry density. Therefore, in order to develop a cement slurry with a well-dense and gas-impermeable microstructure, macropores are filled with a solid substance of finer graining than that of the cement used [26–28]. As a result, it is possible to increase the value of the mechanical strength of the hardened cement slurry and shorten the transition time from the liquid phase to the solid phase. A very important parameter is also the setting time of the cement slurry and the transition time from the point of the beginning of the binding to the point of the end of the binding. It is one of the basic parameters allowing to pre-define whether the liquid cement slurry binds as quickly as possible before gas intrusion into the structure of the liquid cement slurry after being pushed into the cemented annular space,. The shortest possible time between the value of the beginning of the setting and the end of the setting is important [29,30]. Fine-grained materials are commonly used in the building materials industry to improve selected technological parameters of fresh and hardened cement slurries. However, in the technology of cement slurries, these additives are treated with caution. This is due to the much more restrictive parameter values for slurries. The beneficial effect of fine-grained material used to improve the parameters of the hardened cement slurry may adversely affect the rheological parameters of the liquid slurry, which is related to the increase in the water needed for the designed slurry [31–34]. It is important in terms of the pumpability of the cement slurry during the cementing procedure. Therefore, the possibility of using fine-grained material is being investigated in terms of improving the parameters of both the liquid and hardened cement slurry. This is an innovative issue as fine-grain materials from the construction industry are rarely used in the oil industry. Such preliminary work allows to broaden the scientific horizons in this direction and open a new path in the technology of cement slurries.

2. Materials and Methods

2.1. Materials

CEM I 42.5R Portland cement was used to prepare slurries. The cement contained 3.22% SO_3 and 0.069% Cl^-. The cement specific surface was 4200 cm^2/g, the specific density was 3.09 g/cm^3, and the alkalis content was (eq Na_2O) 0.61%. The tap water contained 0.064 mg/L NH_4 (ammonia), 2.95 mg/L NO_3 (nitrates), 0.048 mg/L NO_2 (nitrites), 29.2 μg/L Fe (iron), 5.81 μg/L Mn (manganese), and 344.0 mg/L $CaCO_3$ (calcium carbonate). The second cement used for the tests was the German G class cement (Dyckerhoff). The specific surface area (measured by the Blaine method) is 2800 cm^2/g. The alkali content is kept at 0.89%, chloride at 0.061%, and sulphate at 3.23%. The loss on ignition is 3.03%, and the insoluble residue is 0.54%. The volume stability was found to be 0.44 mm.

Hblock is a fine-grained material with the name and chemical composition coded by the manufacturer, in which the dominant oxides are: CaO (35.81–37.53%) and Fe_2O_3 (26.52–32.45%). The share of MgO is 7.65%, Al_2O_3 3.62%, MnO 2.87%, and P_2O_5 1.07%.

2.2. Methods

To determine the effect of Hblock on selected technological parameters of the fresh cement slurry, the density test, rheological parameters of the cement slurry, and the slurry setting time test along with the transition time from the beginning of setting to the end of the setting are performed. Compressive strength tests and gas permeability tests are carried out for samples of hardened cement slurry.

2.2.1. Slurry Preparation

In order to determine the effect of Hblock on the parameters of the cement slurry, 10 recipes (5 slurries per one type of cement) were tested. The points of reference are two control recipes consisting of cement and water with a constant water-cement ratio. These are samples No. 1 and No. 6. Two types of cement materials are selected for research work: CEM I 42.5R Portland cement with a specific surface area of 4200 cm^2/g, and the C_3A content in clinker of approx. 7% and class G drilling cement. The specific surface area is 2800 cm^2/g, and the C_3A content in the clinker is approx. 3%. Modified slurries are 4 recipes containing the tested Hblock fine-grained material, the specific surface of which is 3200 cm^2/g. Starters 2 to 5 are made of CEM I 42.5R cement. Starters 7 to 10 were made on the basis of drill cement G. During the preparation of the cement slurry, a certain amount of water is measured with a measuring cylinder. The right amount of Hblock is dosed into the water. Water is poured into the mixer. The speed is then set to 1600 rpm, and mixing takes 5 min. After this time, the cement is poured into the mixing water mixture and mixed for another 25 min. Mixing at low speed is the same as mixing the slurry under the conditions of the well. The basic cement slurry, marked as No. 1 in Table 1, was prepared as a control sample. In order to eliminate the measurement error, the same water-cement coefficient of 0.46 was adopted for all recipes. In subsequent recipes of the cement slurry, increasing concentrations of Hblock in the amount of 1%, 3%, 6%, and 20% (bwoc) were used. Such action is aimed at checking the changes under the influence of the amount of the agent dosed to the starter. The slurry recipes are listed in Table 1 below, while the selected technological parameters are presented in Table 2. Table 3 summarizes the shear rate at 12 speed ranges from 600 rpm to 1 rpm, which corresponds to the shear rate from 1022.040 to 1.703 [1/s]. Table 4 summarizes the apparent viscosity values. On the other hand, Table 5 presents the rheological parameters of the tested slurries, described according to the presented rheological models.

Table 1. Recipes of cement slurries with the addition of Hblock agent (water-cement coefficient 0.46).

	Composition No.									
Type of Cement	**1**	**2**	**3**	**4**	**5**	**6**	**7**	**8**	**9**	**10**
Portland cement CEM I 42.5R, %			100.0					-		
Drilling cement G, %			-					100.0		
Hblok small-particle agent, %	-	1.0	3.0	6.0	20.0	-	1.0	3.0	6.0	20.0

The share of all measures is given in relation to the weight of cement (bwoc).

Table 2. Parameters of the tested cement slurrie.

Composition No.	Type of Cement	Temp, °C Hydration Pressure, MPa	Density	Cement Slurry Setting Time, min Initial Time of the Setting PW	Final Time of the Setting KW	Transition Time PW to KW	Compressive Strength after 24 h of Hydration	Gas Permeability
			[kg/m^3]	[min]	[min]	[min]	[MPa]	[mD]
1	Portland cement CEM I 42.5R	30 °C 3 MPa	1890	338	532	194	15.5	0.245 ± 0.002
2			1900	269	343	74	20.0	0.043 ± 0.002
3			1910	278	340	62	24.5	0.032 ± 0.002
4			1920	272	344	72	17.5	0.021 ± 0.002
5			1960	339	382	43	25.5	0.013 ± 0.002
6	Drilling cement G		1870	322	538	216	14.5	0.337 ± 0.002
7			1880	401	546	145	17.5	0.126 ± 0.002
8			1890	400	513	113	18.0	0.042 ± 0.002
9			1910	387	532	145	15.0	0.033 ± 0.002
10			1970	451	555	104	16.0	0.021 ± 0.002

Table 3. Values of the compressive strength measurements for samples P1 to P10.

	Mesurement 1	2	3	4	5	Average Value
P1	16	15.8	14.7	15.5	15.5	15.5
P2	19.9	20	20.1	20	20	20
P3	24	24.5	25	24.5	24.5	24.5
P4	18.5	17	17	17.5	17.5	17.5
P5	24.6	25	26.9	25.5	25.5	25.5
P6	13	15	15.5	14.5	14.5	14.5
P7	17.8	17.3	17.4	17.5	17.5	17.5
P8	17	18	19	18	18	18
P9	15.5	15.6	13.9	15	15	15
P10	14.9	15.4	17.7	16	16	16

Table 4. Within-subject comparison (comparison of samples to baseline P1) Two-sample *t*-test for the mean.

	P1	**P2 − P1**	**P3 − P1**	**P4 − P1**	**P5 − P1**
Average	15.5	20	24.5	17.5	25.5
Variance	0.245	0.005	0.125	0.375	0.755
Observations	5	5	5	5	5
Pearson's correlation		0.92857	0.92857	0.61859	0.9998
Difference of means according to the hypothesis		0	0	0	0
df		4	4	4	4
t Stat		**17.92843**	**24.13988**	**9.035079**	**16.39565**
P(T ≤ *t*) unilateral		2.84×10^{-5}	8.73×10^{-6}	0.000416	4.05×10^{-5}
Test T unilateral		2.131847	2.131847	2.131847	2.131847
P(T ≤ *t*) bilateral		5.69×10^{-5}	1.75×10^{-5}	0.000831	8.1×10^{-5}
Test *t* bilateral		2.776445	2.776445	2.776445	2.776445

2.2.2. Density of the Cement Slurry

Density is measured using a Baroid pressure balance (Figure 1) in accordance with EN-PN ISO10426-2 [35]. The balance consists of an arm with a slurry vessel on one side and a calibrated counterweight on the other. The weighing arm is equipped with a sliding weight and has a g/cm^3 scale in the range of 0.8–2.75. The weight is read from the position of the sliding weight when the scale is leveled by the appropriate positioning of the bubble on the balance arm.

Table 5. Within-subject comparison (comparison of samples to baseline P6) *t*-test: pairwise pairing for the mean.

	P6	P7 − P6	P8 − P6	P9 − P6	P10 − P6
Average	14.5	17.5	18	15	16
Variance	0.875	0.035	0.5	0.455	1.115
Observations	5	5	5	5	5
Pearson's correlation		0.92857	0.944911	0.61413	0.771966
Difference of means according to the hypothesis		0	0	0	0
df		4	4	4	4
t Stat		**6.036327**	**22.13594**	**0.7706**	**4.918694**
P(T ≤ *t*) unilateral		0.001899	1.23×10^{-5}	0.241967	0.003968
Test T unilateral		2.131847	2.131847	2.131847	2.131847
P(T ≤ *t*) bilateral		0.003798	2.47×10^{-5}	0.483935	0.007937
Test *t* bilateral		2.776445	2.776445	2.776445	2.776445

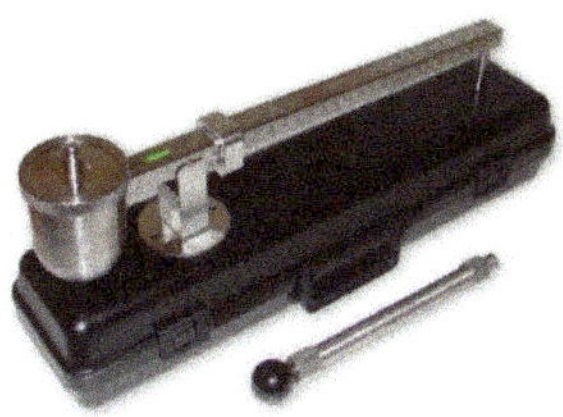

Figure 1. Baroid weight.

2.2.3. Rheological Parameters

The study of rheological properties is based on the determination of shear curves. The tests are performed for the following rotational speeds: 600, 300, 200, 100, 60, 30, 20, 10, 6, 3, 2, and 1 rpm. These values correspond to the shear rates (Ỹ): 1022,04; 511,02; 340.68; 170.34; 102.20; 51.10; 34.07; 17.03; 10.22; 5.11; 3.41; and 1.70 s^{-1}. The tests are carried out at a temperature of 20 ± 2 °C. The dosing liquid is tap water, free from contamination. The viscometer-Model 900 FANN (Figure 2) with coaxial cylinders is used to test rheological properties. To facilitate the calculations (Table 5), the numerical software "Rheo Solution" is used. The software is owned by the Faculty of Drilling, Oil and Gas of the AGH University of Science and Technology and is used in research and development [33,34].

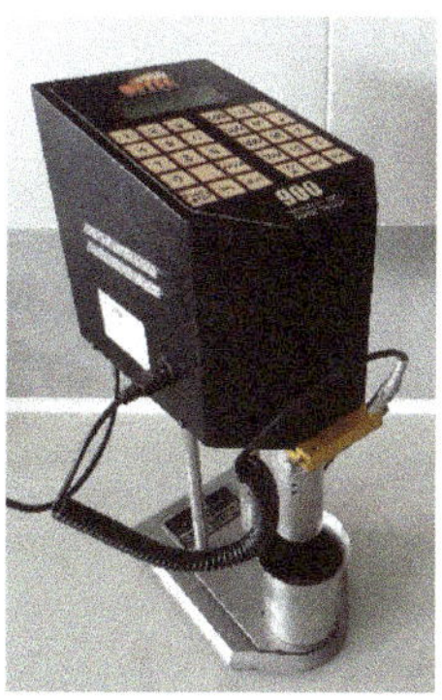

Figure 2. Fann model 900 rotary viscometer.

2.2.4. Setting Time

The examination is performed with the use of an automatic six-chamber Vicat apparatus (Figure 3). The test is performed according to IS:4031-PART 4-1988 VICAT APPARATUS [36]. The measurement consists of determining the depth of immersion of the needle in the binding cement slurry. The points of the initial time of the setting and the final time of the setting are measured. The test is carried out continuously at the prescribed time. The camera is electronically controlled from a computer using software where the measurement conditions are set. The delay time after which the setting time test is to start is determined, and then the measurement is performed at specified time intervals ranging from 1 min to 20 min. Each subsequent measurement is carried out in a different place on the sample in order to avoid the needle sinking into the same place resulting in an incorrect value.

Figure 3. Automated Vicat apparatus.

On the basis of the obtained measurement results, a graph is drawn of the dependence of the depth of the needle plunging into the start of the needle to the binding time of the cement slurry (Figure 4). After each measurement, the needle returns to its base position, where it is cleaned and waits for the next measurement of a given sample or for the next sample. Up to 6 samples of different cement slurries can be measured at the same time. The samples are placed in a water bath that is temperature-controlled from 25 °C to 50 °C. When the initial time of the setting end is reached, further measurements are stopped, and the results are saved graphically.

2.2.5. Compressive Strength

Measurements are made on beams in accordance with the EN-PN ISO10426-2 [35] standard. The tests are performed using a Chandler Engineering model 4207 testing machine (Figure 5). After the cement slurry is prepared, it is placed in 4 cm × 4 cm × 16 cm molds. The molds are stored in water for 24 h in an autoclave at a temperature of 22 ± 2 °C and at atmospheric pressure.

2.2.6. Gas Permeability

The test is performed using a gas permeability meter. The measurement is the basic measure of the ability of a porous medium to transport the liquid and gaseous media it contains. The gas permeability meter (Figure 6) used in the test is used to measure the permeability of core or cement samples one inch in diameter and length (Figure 7). The sample is placed in a "modified Hassler" test cylinder [27]. Nitrogen is then supplied, and the differential pressure is measured at a constant flow velocity obtained. The flow velocity is measured with a calibrated flowmeter. The variables, when substituted for Darcy's law, allow the permeability to be calculated in hardened cement slurry samples.

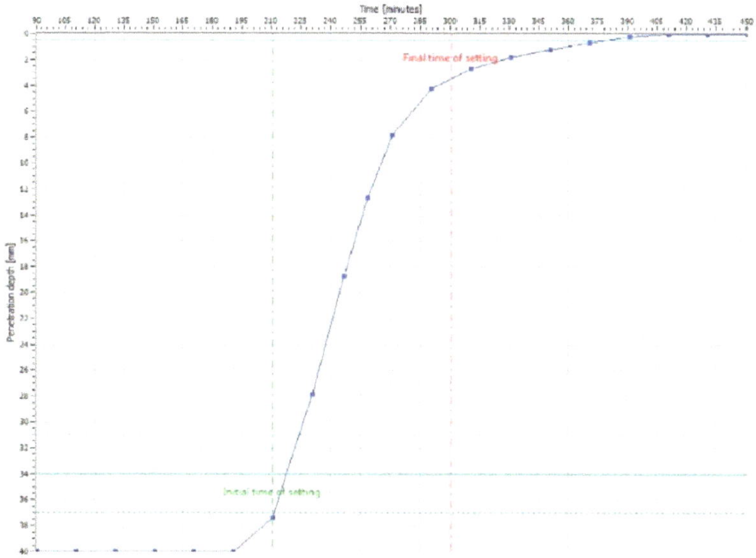

Figure 4. Graph of the dependence of the needle immersion depth in initial time of the setting as a function of time.

Figure 5. Chandler Engineering testing machine.

Figure 6. OFITE gas permeability meter.

Figure 7. A sample of hardened cement slurry for gas permeability testing.

3. Results and Discussion

The presence of the fine Hblock agent causes the density of the slurry to increase. When analyzing the P1 sample prepared on the basis of CEM I 42.5R cement, the density increased by 10 kg/m^3 after adding 1% Hblock compared to the sample without Hblock addition. The addition of 3% Hblock increases the density by 10 kg/m^3 compared to the previous sample. The next amount of 6% Hblock in sample P4 is a further increase in density by 10 kg/m^3 compared to sample P3, which has 3% Hblock. On the other hand, the use of 20% Hblock in sample P5 causes an increase in density by 40 kg/m^3 compared to sample P4.

However, when analyzing the P6 sample prepared on the basis of class G cement, the same trend is visible, but for some time. So 1% Hblock in sample P7 causes an increase in density by 10 kg/m^3 compared to the sample without the addition of Hblock, i.e., P6. The addition of 3% Hblock in sample P8 increases the density by 10 kg/m^3 compared to sample P7. However, the amount of 6% Hblock in sample P9 increases the density by 20 kg/m^3 compared to sample P8, which has 3% Hblock. On the other hand, the use of 20% Hblock in sample P10 causes an increase in density by 60 kg/m^3 compared to sample P9.

On the basis of the above, the increase in density is more evident in the formula of the slurry prepared on the basis of G drilling cement than in the case of the CEM I 42.5R Portland cement grout (Figure 8). At the maximum 20% concentration of the Hblock agent in the CEM I 42.5R cement-based slurry, an increase in density by 70 kg/m^3 compared to the control recipe can be seen, while in the case of the cement-based slurry, the increase in density is 100 kg/m^3 at only a concentration of 20% Hblock (Table 2). This is due to the larger grain diameter of class G drilling cement than CEM I 42.5R class cement, which is related to a better filling of the microstructure of the G cement-based cement slurry.

It should be noted here that the P1 sample has a density of 1890 kg/m^3, and the base sample, made on the basis of G cement, has a density of 1870 kg/m^3. The difference of 20 kg/m^3 more in sample P1 compared to sample P2 results from the greater specific surface area, which for cement in sample P1 is 4200 cm^2/g, while cement G used in sample P2 has a specific surface area of 2800 cm^2/g.

The addition of 1% Hblock fine-grained material to the composition of the CEM I 42.5R Portland cement-based cement (sample No. 2) accelerates the setting, as shown in Figure 9. The setting time is shortened by 69 min, while the setting time is shortened by 189 min (Table 2). A further increase in Hblock concentration (values 3% and 6%) does not cause a significant difference in the setting time of the slurry; only at a 20% concentration of the agent (sample P5) is an extension of the thickening time visible compared to the previous sample (P4). However, this value is lower than the base sample (P1). When analyzing the slurries based on drilling cement G, it can be seen that the presence of the Hblock agent in the slurry results in an upward trend for the analyzed values of the slurry setting time (Figure 9). This effect is a result of the larger grain diameter of class G cement than CEM I 42.5R cement.

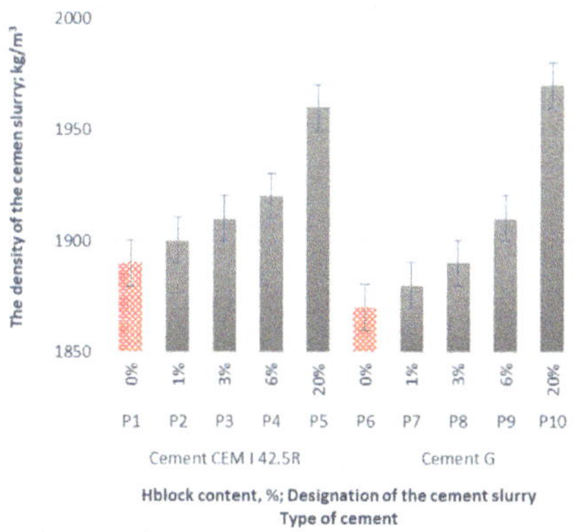

Figure 8. Summary of the density of the tested cement slurries.

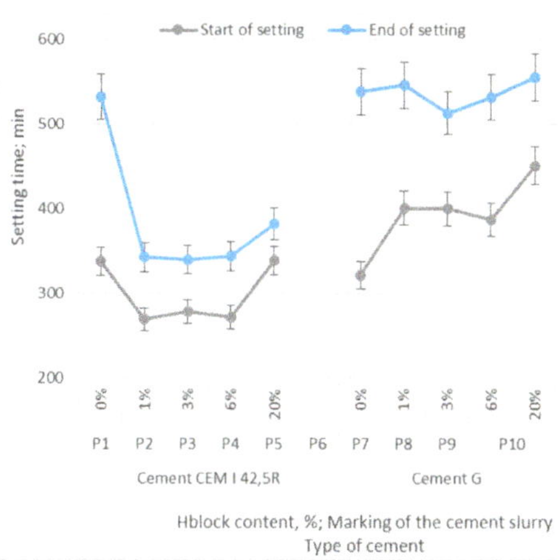

Figure 9. Summary of the setting time values of the tested cement slurries.

The transition time of the tested recipes was significantly shorter in the case of CEM I 42.5R Portland cement slurries than in the case of G cement-based slurries (Figure 10).

The transition from the value of the beginning of the setting to the end of the setting for the base slurry (sample P1) is strongly stretched in both time and amount to 194 min, as shown in Figure 11.

On the other hand, after using the Hblock agent, much shorter values were obtained in the range from 43 min for the P5 sample to 74 min for the P2 sample. The short transition time from the value of the beginning of the setting to the value of the end of the setting for the P5 cement slurry is shown in Figure 12 and Table 2.

In the case of G cement-based slurry, a reduction in the value of the transition time between the initial time of the setting and the final time of the setting is also obtained. However, this change takes place to a lesser extent than in the previous group of slurries. The cement G-based control slurry has a transit time of 217 min, while the Hblock-containing slurries have transition times ranging from 104 min (sample P10) to 145 min (samples P7, P9). When analyzing the changes in the value of the transition time of the slurries containing the Hblock agent (Figure 10), the most favorable change is visible when using the 1% Hblock concentration in relation to the cement mass.

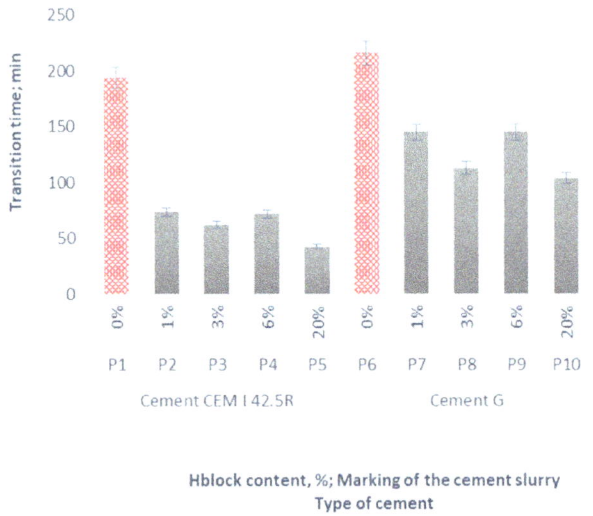

Figure 10. Summary of the values of the transition time from the initial time of the setting to final time of the setting of the tested cement slurries.

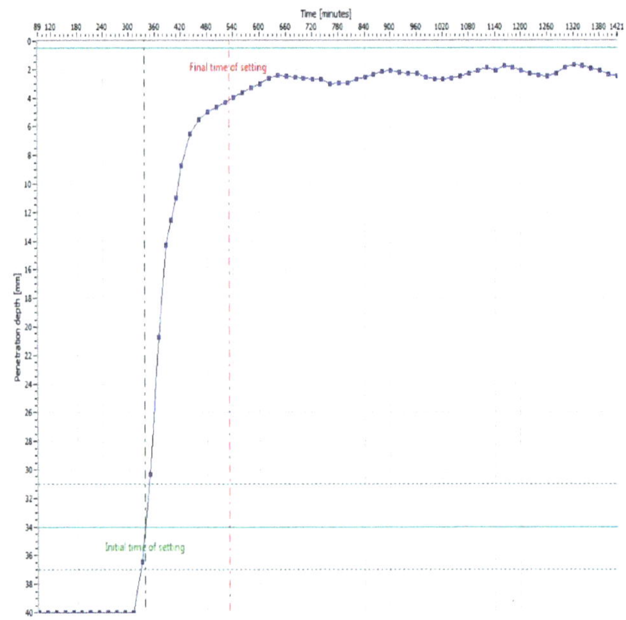

Figure 11. The setting process of the cement slurry based on CEM I 42.5R Portland cement (long transition time); Composition No. 1.

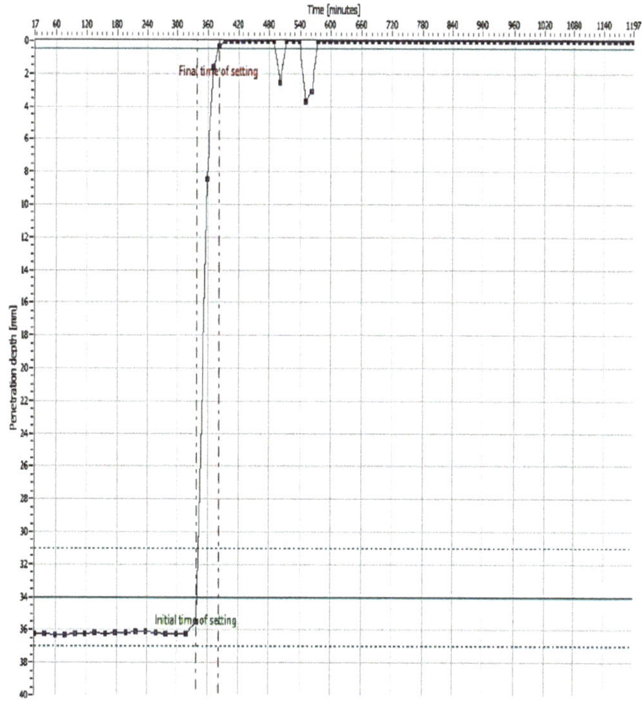

Figure 12. The setting process of the cement slurry with the addition of 20% Hblock (short transition time) Composition No. 5.

The addition of Hblock in the tested cement slurries has a beneficial effect on the improvement of the mechanical parameters of hardened cement slurries. When analyzing the results of compressive strength tests after 24 h of hydration for the cement-based CEM I 42.5R slurry, an increase in the range from 17.5 MPa (sample P4) to 25.5 MPa (sample P5) compared to the base value of 15, 5 MPa for the P1 hardened cement slurry (Figure 13, Table 2). For slurries based on G drilling cement, an increase in the value of compressive strength is also visible, but to a lesser extent. From the value of 14.5 MPa for the base cement slurry (sample P6), the strength value increased to 18 MPa for the P8 slurry containing 3% Hblock. A further increase in the concentration of the tested agent does not improve the value of the compressive strength (Figure 13). The most favorable concentration range is visible, ranging from 1% to 3% of Hblock (marked in Figure 13). It significantly improves the compressive strength of the hardened cement slurry based on the analyzed cement.

Statistical Analysis of the Value of Compressive Strength

Based on the obtained results, a statistical analysis is performed, taking into account the variability of the data. Table 3 presents the measurement values. However, in Tables 4–6 the uncertainties of measurement and calculated values are estimated. They come from direct measurements based on the accuracy of the measuring devices. Value uncertainties are estimated from the calculated standard deviations from the mean, and a pairwise *t*-test for the mean is estimated from the appropriate formulas.

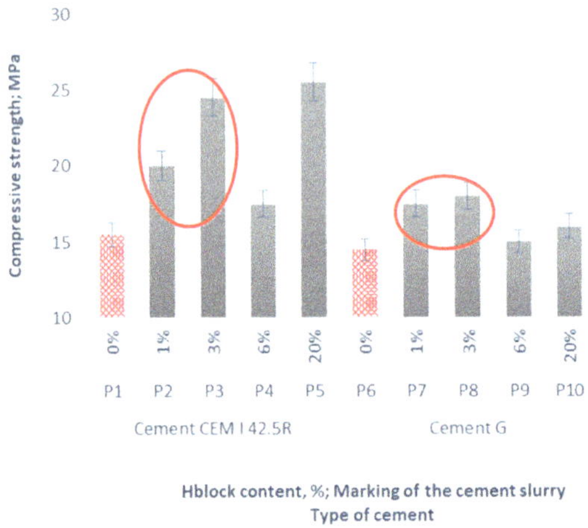

Figure 13. Compressive strength values of the samples of hardened cement slurries.

Table 6. Among-subject comparison (comparison of samples with the same amounts of Hblock) *t*-test: pairwise pairing with two samples for the mean.

	P1 − P6	P2 − P7	P3 − P8	P4 − P9	P5 − P9
Average	14.5	20	17.5	17.5	25.5
Variance	0.875	0.005	0.375	0.375	0.755
Observations	5	5	5	5	5
Pearson's correlation	0.75593	0.75593	0.86603	0.453921	0.999989
Difference of means according to the hypothesis	0	0	0	0	0
df	4	4	4	4	4
t Stat	**−1.65748**	**22.82177**	**−0.87706**	**8.287419**	**113.5467**
$P(T \leq t)$ unilateral	0.086381	1.09×10^{-5}	0.214987	0.000579	1.8×10^{-8}
Test T unilateral	2.131847	2.131847	2.131847	2.131847	2.131847
$P(T \leq t)$ bilateral	0.172762	2.18×10^{-5}	0.429973	0.001157	3.61×10^{-8}
Test *t* bilateral	2.776445	2.776445	2.776445	2.776445	2.776445

On the basis of the mean equality test for $\alpha = 0.05$, the T-Student's statistic is calculated and compared with the data from the distribution. When analyzing the within-subject comparison for samples P1 − P6 made on the basis of CEM I 42.5R cement (Table 4), all Tstat values are in the critical area and range from 9.03 to 24.14, assuming a two-sided T-test of 2.78. Therefore, the samples are statistically different.

The second comparison in Table 5 for the samples (P7 − P10) based on drill cement G shows that only sample P9 is statistically similar to the base samples (P9 − P6) null hypothesis. However, the remaining samples are statistically different.

However, in the among-subject analysis (Table 6), where samples prepared on different types of cement are compared, it can be seen that the P1 − P6 base samples and samples with 3% Hblock additive (P3 − P8) show statistical similarity. The 3% Hblock amount may be the optimal dosing value to achieve the required mechanical performance.

The gas permeability of the analyzed cement slurries decreased from 0.245 mD (sample P1) to 0.043 mD (sample P2) after using 1% Hblock in the case of CEM I 42.5R Portland cement slurries and from 0.337 mD (sample P6) to the value of 0.042 mD (sample P8). The obtained values are summarized in Table 2.

Analyzing the influence of the Hblock agent on the rheological parameters of the cement slurry prepared on the basis of CEM I 42.5R Portland cement, an increase in the yield point after the application of the Hblock additive can be seen. The base slurry P1 had the value of the yield point equal to 6.1 Pa, described by the Herschel–Bulkley model; the introduction of 1% Hblock causes the increase in the yield limit HB to 10.08 Pa, while the 3% Hblock reduces the yield limit to 5.8 Pa. Increasingly larger amounts of the tested agent cause an increase in the yield point value, which can be seen in Figure 14,. The addition of Hblock also increases the consistency coefficient according to the Herschel–Bulkley model. The control slurry (P1) has a consistency coefficient HB of 9.59 Pa s^n. After introducing Hblock into the cement slurry, the HB consistency coefficient increases in the range from 12.29 Pa s^n to 31.21 Pa·s^n. Significantly lower values of the discussed rheological parameters were obtained for the cement slurry prepared on the basis of drilling cement G (Figure 15, Tables 7–9).

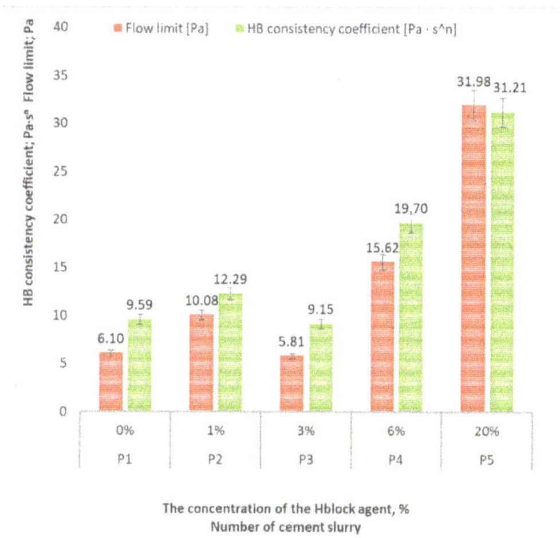

Figure 14. Summary of selected test results for rheological parameters of cement slurries (cement class CEM I 42.5R).

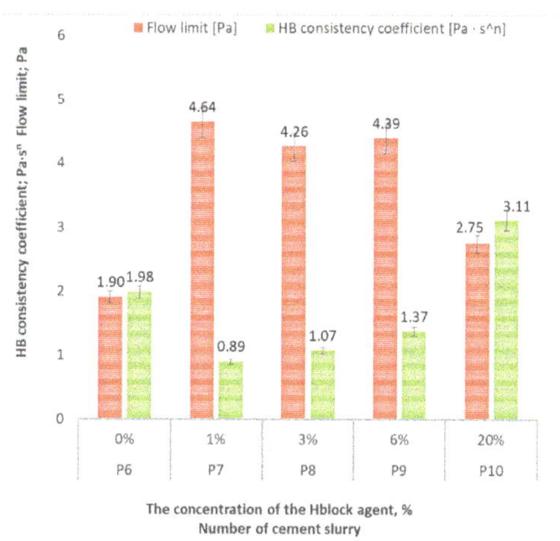

Figure 15. Summary of selected test results for rheological parameters of cement slurries (cement class CEM G).

Table 7. Results of measurements and calculated values of shear stress for slurries.

Shear Rate, s^{-1}	Shear Stress, Pa									
	Composition No.									
	1	2	3	4	5	6	7	8	9	10
1.703	8.176	6.643	6.643	9.709	10.220	3.577	3.577	3.066	3.577	5.110
3.407	10.220	8.687	8.687	12.775	14.308	4.599	4.599	4.599	5.621	7.665
5.110	12.775	10.731	10.731	15.330	17.374	5.621	6.132	6.643	7.154	8.687
10.220	17.885	15.841	15.841	21.462	24.017	9.198	9.198	9.709	10.220	12.775
17.034	23.506	21.462	21.462	30.149	32.704	12.264	12.775	12.775	13.286	16.863
34.068	33.215	32.193	32.193	42.924	44.968	16.863	15.330	17.374	19.418	26.061
51.102	42.924	43.946	45.479	58.254	55.699	18.907	16.863	18.396	21.462	30.660
102.204	65.408	55.699	55.188	71.029	85.337	25.039	24.528	24.528	26.572	38.325
170.340	79.716	65.408	63.875	81.249	102.200	31.682	27.083	30.660	33.215	46.501
340.680	104.244	84.315	81.249	100.156	131.327	44.457	38.325	43.946	45.990	64.386
511.020	121.618	100.156	92.491	116.508	155.344	55.188	51.100	56.210	57.743	80.227
1022.040	170.674	134.393	130.816	157.388	174.762	81.760	80.738	89.425	87.892	120.085

Table 8. Results of measurements and calculated values of apparent viscosity for slurries.

Shear Rate, s^{-1}	Apparent Viscosity, Pa·s									
	Composition No.									
	1	2	3	4	5	6	7	8	9	10
1.703	4.7998	3.8998	3.8998	5.6998	5.9998	2.0999	2.0999	1.7999	2.0999	2.9999
3.407	2.9999	2.5499	2.5499	3.7499	4.1998	1.3499	1.3499	1.3499	1.6499	2.2499
5.110	2.4999	2.0999	2.0999	2.9999	3.3999	1.1000	1.2000	1.2999	1.3999	1.6999
10.220	1.7499	1.5499	1.5499	2.0999	2.3499	0.9000	0.9000	0.9500	1.0000	1.2500
17.034	1.3799	1.2600	1.2600	1.7699	1.9199	0.7200	0.7500	0.7500	0.7800	0.9900
34.068	0.9750	0.9450	0.9450	1.2600	1.3199	0.4950	0.4500	0.5100	0.5700	0.7650
51.102	0.8400	0.8600	0.8900	1.1400	1.0900	0.3700	0.3300	0.3600	0.4200	0.6000
102.204	0.6400	0.5450	0.5400	0.6950	0.8350	0.2450	0.2400	0.2400	0.2600	0.3750
170.340	0.4680	0.3840	0.3750	0.4770	0.6000	0.1860	0.1590	0.1800	0.1950	0.2730
340.680	0.3060	0.2475	0.2385	0.2940	0.3855	0.1305	0.1125	0.1290	0.1350	0.1890
511.020	0.2380	0.1960	0.1810	0.2280	0.3040	0.1080	0.1000	0.1100	0.1130	0.1570
1022.040	0.1670	0.1315	0.1280	0.1540	0.1710	0.0800	0.0790	0.0875	0.0860	0.1175

Table 9. Rheological parameters of cement slurries.

Rheological Model	Rheological Parameters	Composition No.									
		1	2	3	4	5	6	7	8	9	10
Newton's model	Newton's dynamic viscosity, Pa·s	0.2022	0.1625	0.1559	0.1917	0.2281	0.0933	0.0891	0.0988	0988	0.1368
	Correlation coefficient. -	0.8231	0.7599	0.7812	0.6736	0.6622	0.8717	0.8953	0.9047	0.8765	0.8624
Bingham's model	Plastic viscosity, Pa·s	0.1600	0.1240	0.1202	0.1403	0.1677	0.0756	0.0730	0.0817	0.0799	0.1100
	Flow limit, Pa	27.2736	24.8480	23.0535	33.2245	38.9787	11.4776	10.3816	10.9878	12.1882	17.3186
	Correlation coefficient, -	0.9428	0.9276	0.9346	0.9162	0.8868	0.9679	0.9801	0.9801	0.9742	0.9661

Table 9. *Cont.*

Rheological Model	Rheological Parameters	Composition No.									
		1	2	3	4	5	6	7	8	9	10
Ostwald–de Waele's model	Coefficient of consistency, Pa·s^{s}	5.2967	5.2854	5.2419	8.0588	8.3847	2.7714	2.8778	2.7498	3.2356	4.2045
	Exponent, -	0.4929	0.4841	0.4740	0.4442	0.4685	0.4840	0.4628	0.4895	0.4649	0.4797
	Correlation coefficient, -	0.9962	0.9865	0.9920	0.9844	0.9740	0.9983	0.9882	0.9929	0.9920	0.9973
Casson's model	Casson's viscosity, Pa·s	0.1146	0.0866	0.0830	0.0915	0.1184	0.0519	0.0477	0.0554	0.0523	0.0748
	Flow limit, Pa	11.7086	11.2039	10.5368	16.5485	17.6915	5.2151	5.0616	5.0643	5.9163	7.9848
	Correlation coefficient, -	0.9652	0.9539	0.9600	0.9483	0.9217	0.9852	0.9928	0.9926	0.9893	0.9839
Herschel–Bulkley's model	Flow limit, Pa	6.1029	10.0769	5.8080	15.6180	31.9790	1.8997	4.6380	4.2584	4.3869	2.7504
	Coefficient of consistency, Pa·s^{n}	9.5936	12.2924	9.1489	19.6989	31.2099	1.9809	0.8940	1.0721	1.3721	3.1114
	Exponent. -	0.4196	0.3541	0.3880	0.3104	0.2791	0.5313	0.4381	0.6284	0.5903	0.5207
	Correlation coefficient, -	0.9989	0.9986	0.9967	0.9971	0.9950	0.9988	0.9958	0.9971	0.9955	0.9972

The P6 base slurry has a 1.9 Pa Herschel–Bulkley flow limit. The presence of Hblock results in an HB flow limit in the range of 2.75 Pa (sample P10) to 4.64 Pa (sample P7). The HB consistency coefficient ranges from 0.89 Pa·s^{n} (sample P7 with 1% Hblock content) to 3.11 Pa·s^{n} (sample P10 with 20% Hblock content). Casson viscosity of slurries prepared on the basis of CEM I 42.5R Portland cement ranges from 0.08 Pa·s (sample P3 with the addition of 6% Hblock) to 0.12 Pa·s (sample P5 with the addition of 20% Hblock) (Figure 16). In turn, the exponent n decreases with the increase in the Hblock addition (Figure 16). In the case of slurries prepared on the basis of drilling cement G, the Casson viscosity values are lower than in the previous group of slurries and range from 0.05 Pa·s to 0.07 Pa·s (sample P10 with 20% Hblock) (Figure 17). The exponent n has an irregular course and does not show a satisfactory correlation with the concentration of the Hblock measure (Figure 17).

For a thorough analysis of the rheological parameters and determination of the effect of the addition of fine-grained Hblock on the rheology of the cement slurry, diagrams of the flow curves of the tested slurries were prepared. The decrease in the dynamic viscosity of the slurry with the increase in the Hblock agent in the range from 1% to 6% (samples P2, P3, P4) in the slurries prepared on the basis of CEM I 42.5R Portland cement is visible from the slope of the curves (Figure 18), i.e., the decreasing tangent of the angle that forms with the line of abscissa. This is confirmed in previous studies of rheological parameters, where the Casson viscosity values and the plastic viscosity described by the Bingham model decrease due to the increasing amount of Hblock in the slurry. Only in the case of the slurry with the highest amount of Hblock (20% bwoc) (sample P5) is the flow curve above the base slurry curve (P1—red line) visible. This proves the increase in the dynamic viscosity of the cement slurry with a significant (20%) concentration of the Hblock agent.

Figure 16. Summary of selected test results for rheological parameters of cement slurries (cement class CEM I 42.5R).

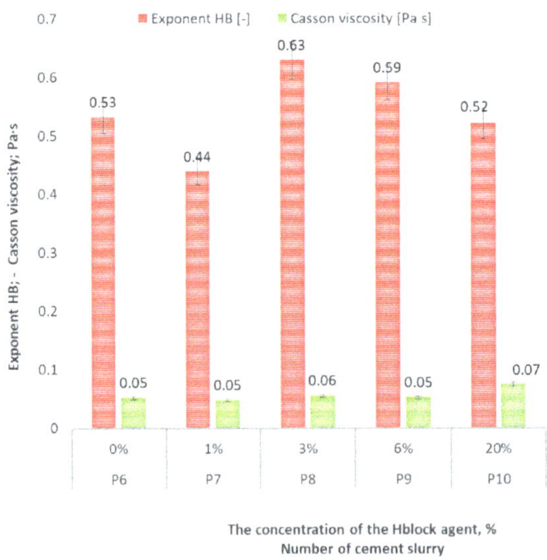

Figure 17. Summary of selected test results for rheological parameters of cement slurries (cement class CEM G).

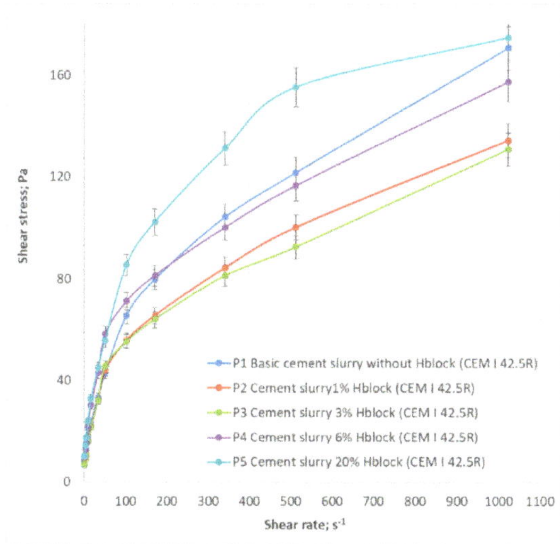

Figure 18. Flow curves for cement slurries with the addition of Hblock (cement class CEM I 42.5R).

A different course of the flow curves takes place during the interpretation of slurries with the addition of the Hblock agent prepared on the basis of drilling cement G (Figure 19). The curve for the base slurry (sample P6—red color) runs almost the lowest, and the recipe of this slurry has the lowest value of dynamic viscosity. A slightly lower course of the flow curve occurred only in the sample with the addition of 1% Hblock agent (sample P7), while successive increasing amounts of Hblock in the range from 3% to 6% caused an increase in viscosity manifested by a smaller slope of the curves. The greatest increase in viscosity is seen at the highest 20% concentration of Hblock (sample 10).

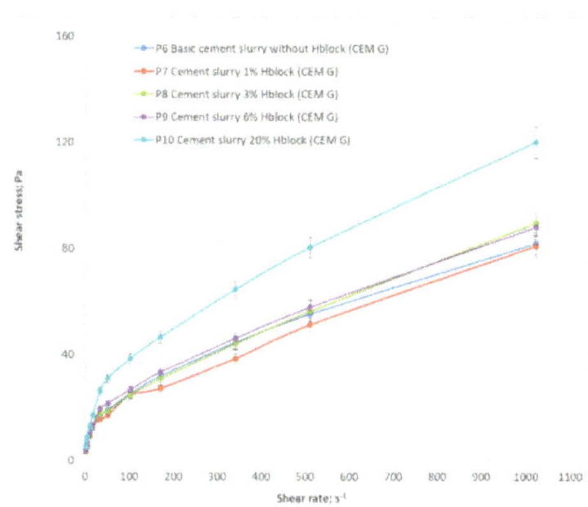

Figure 19. Flow curves for cement slurries with the addition of Hblock (cement class CEM G).

When analyzing the flow curves, a significant increase in shear stresses is visible at low shear rates in the group of slurries made on the basis of CEM I 42.5R Portland cement. This result confirms the thixotropic nature of cement slurries, which have high values of the yield point and structural strength (Table 9; Figure 18). In the case of slurries prepared on the basis of G drilling cement, comparable values of shear stresses in the low ranges of shear rates can be seen, which confirms the low values of rheological parameters in this group of slurries (Table 9; Figure 19).

The fine-grained material of Hblock significantly improves the parameters of both fresh and hardened cement slurry. The use of Hblock reduces the time of transition from the PW value to the KW value in all tested cement slurries (Table 2; Figures 9, 11 and 12). Compressive strength increases after 24 h of sample hydration with small (1% to 3%) amounts of Hblock (Table 2; Figure 13). The use of the tested Hblock agent results in the sealing of the cement matrix, which will result in a reduction in the gas permeability value (Table 2). The use of the Hblock fine-grained material is advantageous due to the fact that a compact and impermeable microstructure of the cement sheath is obtained, characterized by a higher value of compressive strength.

4. Conclusions

Based on the results of the work aimed at determining the effect of the fine-grained Hblock agent on selected parameters of the cement slurry, it is concluded that:

- The fine-grained material of Hblock causes an increase in the density of the slurry compared to the corresponding control samples without this additive.
- The use of Hblock in the formulation of the cement slurry based on CEM I 42.5R Portland cement shortens the setting time. On the other hand, when using G drilling cement, the presence of Hblock causes a slight increase in setting time.
- In all formulations, the presence of Hblock reduces the transit time from the initial time of the setting to the final time of the setting. This is very advantageous when designing cement slurries to seal columns of pipes in openings with an increased risk of gas migration.
- When analyzing the mechanical parameters of cement slurries with the addition of Hblock, a significant improvement in compressive strength is observed with the use of small amounts of Hblock (1% to 3% bwoc).
- The use of the Hblock agent in cement slurries tightens the structure of the hardened cement slurry, which is manifested by a reduction in the gas permeability of the tested samples.
- The Hblock fine-grained material causes a slight increase in the yield point according to the Herschel–Bulkley model, which may contribute to the improvement of the sedimentation stability of the cement slurry.

Funding: The work was financially supported by Ministry of Science and Higher Education Warsaw (Internal order Oil and Gas Institute—National Research Institute Project No. 0021/KW/21).

Institutional Review Board Statement: Not applicable.

Informed Consent Statement: Not applicable.

Data Availability Statement: Not applicable.

Acknowledgments: The author thanks the anonymous reviewers for their constructive comments and the editor for handling the paper.

Conflicts of Interest: The author declares no conflict of interest.

Nomenclature

Symbol	Explanation
mD	Milidarcy (Gas premability)
cm^2/g	Specific surface area
MPa	Megapascal (Compressive strength)
s^{-1}	Shear rate
Pa	Shear stress
Pa·s	Apparent viscosity

References

1. Li, L.; Yuan, X.; Sun, J.; Xu, X.; Li, S.; Wang, L. Vital Role of Nanotechnology and Nanomaterials in the Field of Oilfield Chemistry. In Proceedings of the International Petroleum Technology Conference, Beijing, China, 26 March 2013. [CrossRef]
2. Kremieniewski, M. Drobnoziarniste materiały wypełniające w technologii zaczynów cementowych. *Wiadomości Naft. i Gazow.* **2021**, *5*, 4–7.
3. Stryczek, S.; Wiśniowski, R.; Gonet, A.; Złotkowski, A. Wpływ rodzaju cementu na właściwości reologiczne zaczynów uszczelniających stosowanych w technologiach wiertniczych. *Wiert. Naft. Gaz.* **2010**, *27*, 721–739.
4. Rzepka, M.; Stryczek, S.; Kremieniewski, M.; Wiśniowski, R.; Dębińska, E. Recipes of cement slurries for sealing casing in deep wellbores. *AGH Drill. Oil Gas* **2016**, *33*, 455. [CrossRef]
5. Stryczek, S.; Małolepszy, J.; Gonet, A.; Wiśniowski, R.; Kotwica, Ł. Wpływ dodatków mineralnych na kształtowanie się właściwości technologicznych zaczynów uszczelniających stosowanych w wiertnictwie i geoinżynierii Wydawnictwo. *S.C.M.R. Kraków* **2011**, 1–164. Available online: http://katalog.nukat.edu.pl/lib/item?id=chamo:4133282&fromLocationLink=false&theme=nukat (accessed on 21 May 2015).
6. Stryczek, S.; Kotwica, Ł.; Wiśniowski, R.; Złotkowski, A.; Skrzypaszek, K.; Kremieniewski, M.; Rzepka, M. Analysis of technological parameters of cementing slurries for horizontal casing works in Pomeranian Basin. *AGH Drill. Oil Gas* **2015**, *32*, 431–442. [CrossRef]
7. Stryczek, S.; Wiśniowski, R.; Kremieniewski, M.; Rzepka, M.; Kotwica, Ł.; Złotkowski, A. Analysis of technological parameters of cement slurries used for sealing casing in the Lublin Basin area. *AGH Drill. Oil Gas* **2016**, *33*, 145. [CrossRef]
8. Kremieniewski, M.; Badawczy, I.N. Wpływ drobnoziarnistej krzemionki na parametr czasu oczekiwania na cement—WOC. *Nafta-Gaz* **2019**, *75*, 683–690. [CrossRef]
9. Mahmoud, A.A.; Elkatatny, S.; Mahmoud, M. Improving Class G Cement Carbonation Resistance Using Nanoclay Particles for Geologic Carbon Sequestration Applications. In Proceedings of the Dhabi International Petroleum Exhibition & Conference, Abu Dhabi, United Arab Emirates, 12 November 2018. [CrossRef]
10. Kremieniewski, M. O konieczności prowadzenia serwisowych badań parametrów technologicznych zaczynów uszczelniających. *Nafta-Gaz* **2019**, *75*, 48–55. [CrossRef]
11. Kremieniewski, M. Analiza parametrów technologicznych wybranych zaczynów lekkich stosowanych do uszczelniania płytkich otworów o temperaturze do 45 °C. *Nafta-Gaz* **2020**, *76*, 710–718. [CrossRef]
12. Murtaza, M.; Rahman, M.K.; Al-Majed, A.A.; Samad, A. Mechanical, Rheological and Microstructural Properties of Saudi Type G Cement Slurry with Silica Flour Used in Saudi Oil Field under HTHP Conditions. In Proceedings of the SPE Saudi Arabia Section Technical Symposium and Exhibition, Al-Khobar, Saudi Arabia, 19 May 2013. [CrossRef]
13. Kremieniewski, M.; Stryczek, S. Zastosowanie cementu wysokoglinowego do sporządzania zaczynów uszczelniających w technologiach wiertniczych. *Cem. Wapno Beton* **2019**, *22*, 215–226. [CrossRef]
14. Bensted, J.; Smith, J.R. Oilwell Cements Part 5: Applications of fly ash in well cementing. *Cem. Wapno Beton* **2008**, *1*, 17–30.
15. Stryczek, S.; Wiśniowski, R.; Gonet, A.; Złotkowski, A. Influence of specific surface of lignite fluidal ashes on rheological properties of sealing slurries. *Arch. Min. Sci.* **2012**, *57*, 313–322.
16. Stryczek, S.; Brylicki, W.; Małolepszy, J.; Gonet, A.; Wiśniowski, R.; Kotwica, Ł. Potential use of fly ash from fluidal combustion of brown coal in cementing slurries for drilling and geotechnical works. *Arch. Min. Sci.* **2009**, *54*, 775–786.
17. Brylicki, W.; Małolepszy, J.; Stryczek, S.; Wiśniowski, R.; Kotwica, Ł. Effects of modification of alkali activated slag cementing slurries with natural clinoptilolite. *Miner. Resour. Manag.* **2009**, *25*, 61–76.
18. Kotwica, Ł.; Chorembała, M.; Kapeluszna, E.; Stępień, P.; Deja, J.; Illikainen, M.; Gołek, Ł. Effect of metakaolinite on properties of alkali activated slag materials. *Key Eng. Mater.* **2018**, *761*, 69–72. [CrossRef]
19. Kremieniewski, M. Influence of Graphene Oxide on Rheological Parameters of Cement Slurries. *Energies* **2020**, *13*, 5441. [CrossRef]
20. Sun, Y.F.; Zhou, T.S.; Gao, P.W.; Chen, M.; Liu, H.W.; Xun, Y. Microstructure and Microwave Absorption Properties of Cement-Based Material Reinforced with Reduced Graphene Oxide and Nanoparticles. *Strength Mater.* **2019**, *51*, 601–608. [CrossRef]
21. Kremieniewski, M. Modyfikacja przestrzeni porowej kamieni cementowych. *Nafta-Gaz* **2012**, *3*, 165–170.
22. Kremieniewski, M. Modyfikacje receptur zaczynów uszczelniających w celu zminimalizowania przepuszczalności powstałych kamieni cementowych. *Nafta-Gaz* **2014**, *3*, 170–175.
23. Tarabani, S.; Hareland, G. New cement additives that eliminate cement body permeability. *SPE* **1995**, *29269*, 20–22.
24. Oskarsen, R.T.; Wright, J.W.; Walzel, D. Analysis of gas flow yields recommendations for best cementing practices. *World Oil* **2010**, *231*, 33–39.
25. Kremieniewski, M. Ocena przepuszczalności kamieni cementowych pod kątem ograniczenia migracji gazu. *Pr. Nauk. INiG-PIB* **2014**, *196*, 1–155.
26. Radecki, S.; Witek, W. Dobór technik i technologii cementowania w aspekcie występowania migracji gazu. *Nafta-Gaz* **2000**, *56*, 487–497.
27. Kremieniewski, M. Korelacja wynik & oacute;w badań wytrzymałości na ściskanie i przyczepności do rur stalowych płaszcza cementowego z zaczynu o obniżonej gęstości. *Nafta-Gaz* **2019**, *75*, 613–624. [CrossRef]
28. Szaj, P. Wpływ wybranych dodatków mineralnych na właściwości reologiczne zaczynów cementowych. *Pr. Nauk. Inst. Górnictwa Politech. Wrocławskiej* **2012**, *134*, 285–294.

29. Stryczek, S.; Wiśniowski, R.; Gonet, A.; Ferens, W. Parametry reologiczne świeżych zaczynów uszczelniających w zależności od czasu ich sporządzania. *Wiert. Naft. Gaz* **2009**, *26*, 369–382.
30. Kremieniewski, M.; Rzepka, M. Oil and Gas Institute-National Research Institute Poprawa szczelności płaszcza cementowego za pomocą innowacyjnych dodatków antymigracyjnych. *Nafta-Gaz* **2018**, *6*, 8–15. [CrossRef]
31. Bentz, D.; Garboczi, E.J.; Haecker, C.J.; Jensen, O.M. Effects of cement particle size distribution on performance properties of Portland cement-based materials. *Cem. Concr. Res.* **1999**, *29*, 1663–1671. [CrossRef]
32. Stryczek, S.; Wiśniowski, R.; Gonet, A.; Złotkowski, A. The influence of time of rheological parameters of fresh cement slurries. *AGH Drill. Oil Gas* **2014**, *31*, 123–133. [CrossRef]
33. Stryczek, S.; Gonet, A.; Wiśniowski, R. Wpływ wybranego dodatku mineralnego na własności technologiczne zaczynów cementowych. *Wiert. Naft. Gaz* **2005**, *22*, 333–341.
34. Stryczek, S.; Wiśniowski, R.; Kumala, B. Wpływ superplastyfikatora na parametry technologiczne zaczynów uszczelniających sporządzonych na osnowie cementów portlandzko popiołowych. *Wiert. Naft. Gaz* **2008**, *25*, 717–731.
35. *PN-EN ISO 10426-2:2006*; Przemysł Naftowy i Gazowniczy-Cementy i Materiały do Cementowania Otworów Wiertniczych-Część 2: Badania Cementów Wiertniczych. Polish Committee for Standardization: Warsaw, Poland, 2006.
36. *IS: 4031-PART 4-1988*; VICAT APPARATUS Metchods of Physical Tests for Hydraulic Cement, Part 4 Determination of Consistency of Standard Cement Paste. Available online: https://www.scribd.com/doc/41585859/IS-4031-Part4 (accessed on 18 March 2022).

energies

MDPI

Article

Strategy of Compatible Use of Jet and Plunger Pump with Chrome Parts in Oil Well

Oleg Bazaluk [1], Olha Dubei [2], Liubomyr Ropyak [3], Maksym Shovkoplias [3], Tetiana Pryhorovska [4] and Vasyl Lozynskyi [5,*]

1 Belt and Road Initiative Institute for Chinese-European Studies (BRIICES), Guangdong University of Petrochemical Technology, Maoming 525000, China; bazaluk@ukr.net
2 Department of Petroleum Production, Ivano-Frankivsk National Technical University of Oil and Gas, 076019 Ivano-Frankivsk, Ukraine; olgadubej@gmail.com
3 Department of Computerized Engineering, Ivano-Frankivsk National Technical University of Oil and Gas, 076019 Ivano-Frankivsk, Ukraine; l_ropjak@ukr.net (L.R.); maks.shovkoplias@gmail.com (M.S.)
4 Department of Engineering and Computer Graphics, Ivano-Frankivsk National Technical University of Oil and Gas, 076019 Ivano-Frankivsk, Ukraine; t.pryhorovska@nung.edu.ua
5 Department of Mining Engineering and Education, Dnipro University of Technology, 49005 Dnipro, Ukraine
* Correspondence: lvg.nmu@gmail.com

Abstract: During oil fields operation, gas is extracted along with oil. In this article it is suggested to use jet pumps for utilization of the associated oil gas, burning of which causes environmental degradation and poses a potential threat to the human body. In order to determine the possibility of simultaneous application of a sucker-rod pump, which is driven by a rocking machine, and a jet pump (ejector) in the oil well, it is necessary to estimate the distribution of pressure along the borehole from the bottomhole to the mouth for two cases: when the well is operated only be the sucker-rod pump and while additional installation of the oil-gas jet pump above its dynamic level. For this purpose, commonly known methods of Poettman-Carpenter and Baksendel were used. In addition, the equations of high-pressure and low-pressure oil-gas jet pumps were obtained for the case, when the working stream of the jet pump is a gas-oil production mixture and the injected stream is a gas from the annulus of the well. The values which are included in the resulting equations are interrelated and can only be found in a certain sequence. Therefore, a special methodology has been developed for the practical usage of these equations in order to calculate the working parameters of a jet pump based on the given independent working parameters of the oil well. Using this methodology, which was implemented in computer programs, many operating parameters were calculated both for the well and for the jet pump itself (pressures, densities of working, injected and mixed flows, flow velocities and other parameters in control sections). According to the results of calculations, graphs were built that indicate a number of regularities during the oil well operation with such a jet pump. The main result of the performed research is a recommendation list on the choice of the oil-gas jet pump location inside the selected oil well and generalization of the principles for choosing the perfect location of such ejectors for other wells. The novelty of the proposed study lays in a systematic approach to rod pump and our patented ejector pump operation in the oil and chrome plating of pump parts. The result of scientific research is a sound method of determining the rational location of the ejector in the oil well and the calculation of its geometry, which will provide a complete selection of petroleum gas released into the annulus of the oil well. To ensure reliable operation of jet and plunger pumps in oil wells, it is proposed to use reinforcement of parts (bushings, plungers, rods, etc.) by electrochemical chromium plating in a flowing electrolyte. This has significantly increased the wear resistance and corrosion resistance of the operational surfaces of these parts and, accordingly, the service life of the pumps. Such measures will contribute to oil production intensification from wells and improve the environmental condition of oil fields.

Keywords: jet pump; oil; well; sucker-rod pump; gas-water-oil mixture; chrome coating

Citation: Bazaluk, O.; Dubei, O.; Ropyak, L.; Shovkoplias, M.; Pryhorovska, T.; Lozynskyi, V. Strategy of Compatible Use of Jet and Plunger Pump with Chrome Parts in Oil Well. *Energies* **2022**, *15*, 83. https://doi.org/10.3390/en15010083

Academic Editor: Marcin Kremieniewski

Received: 13 November 2021
Accepted: 21 December 2021
Published: 23 December 2021

1. Introduction

It is almost impossible to avoid associated gases extraction during the operation of oil field. These associated gases consist of gas mixture, which contains methane and propane-butane fraction. It requires separation into fractions in special gas processing plants, which are not always present near oil producing areas. Therefore, in oil fields these gases are often burnt in flares and the products of their combustion are a potential threat to the human body.

The practice of associated oil gas utilization using flare units is very dangerous for the environment [1]. For each ton of extracted oil there is from 25 m^3 to 800 m^3 of associated gas [2]. During the combustion of oil gas, a large amount of mutagenic and toxic substances (carbon oxides, nitrogen oxides, sulfur dioxide, hydrocarbons, soot) are released into the environment, which, when they enter the living organism, lead to the development of irreversible changes caused by the bioaccumulation effect [3–5]. In particular, in the regions that are influenced by the combustion plants, the population is suffering from an increased disease incidence of circulatory, respiratory, digestive and endocrine systems. The effect from combination of several compounds is synergistically upset while the response from the influence of several pollutants is greater than the summation of separate effects [6–9].

Oil producing countries and companies have created a partnership association "Global Gas Flaring Reduction Partnership" GGFR. The purpose of this association is to reduce the environmental impact from gas combustion at flare units through the creation of an appropriate legislative framework and expansion of markets for the disposal of associated gases [10]. Additionally, the promising method of associated gas utilization is using jet pumps for its capturing. The potential energy of the associated oil gas, when properly used, can serve to raise the production fluid to the surface [11–14]. Due to the gas adding to the water–oil mixture of the well, it is possible to reduce the fluids density to such level that it will even reduce the minimum required outlet pressure of the sucker-rod pump, which is used for oil well operation [15,16]. The paper in [17] describes the fundamental possibility of joint operation of the ground jet pump in the conditions of sinusoidal change of the working flow and the rod pump.

A perspective solution for the creation of a gas–water–oil mixture inside the well is inserting of a jet pump in it. Such jet pump, due to the flow of production fluid that will pass through it, will offtake free oil gas from the annulus, mix it with fluid and such mixture of a significantly lower density will be pumped to the mouth of the well. Jet pump can be installed either inside the tubing or in the annulus. With a correctly selected location for such an oil–gas jet pump in the well it will reduce the tubing load and decrease the electric energy consumption of the plunger pump. In addition, to ensure efficient operation of wells, it is necessary to increase the service life of pumps, as their replacement requires significant resources and leads to the cessation of oil production.

Analysis of the method of calculating the pressure distribution in the well, as well as its improvement proposed to clarify the results by the methods of Poettman-Carpenter and Baksendel was given by the authors in the previously published article [18].

1.1. State of the Current Development of the Problem and Formulating a Research Task

Oil and gas equipment, plunger, centrifugal and jet pumps are operated in extreme conditions during oil production. They are subject to high alternating loads in aggressive environments with the content of solid abrasive particles of rock at elevated temperatures [19]. The action of these factors leads to intensive corrosion and mechanical wear and destruction of machine parts, so the design, materials for their manufacture and technologies for strengthening parts are subject to increased requirements. Jet pumps are widely spread in different areas of modern technology. In recent decades, they have been used directly in oil production. For many cases, during oil wells operation with electrical submersible pump (ESP), additional jet pump usage ensured the stabilization of the ESP work. At the same time, this allowed to use the energy of free petroleum gas, which is always present in the injected flow and, to a certain extent, this provides an increase in the well's flow rate [20]. In this case the jet pump was installed directly above the ESP.

To create reliable and efficient jet pumps for oil production, it is necessary to develop a theory of their operation. Theoretical issues related to the operation of jet pumps in oil wells were considered in the works in [21,22]. Practical issues related to the efficiency and reliability of pumping equipment are considered in [23–25]. A case of placing the jet pump above the dynamic level of the well is also used in order to extract gas from its annulus, which made it possible to stabilize the dynamic level and increase the reliability of the ESP system [26]. In this case, the injected flow of the jet pump is oil gas, which is accumulated in the annulus.

After this technology showed its effectiveness, such oil-gas jet pumps were proposed to be used together with sucker-rod pumps [27] or individually in complicated production conditions [28,29]. It should be noted here that a sucker rod string could be subjected to abnormal static and dynamic loads [30–32] especially for complex profile wells. Some authors [33–35] consider vibration reduction of oil equipment as a guarantee of dynamic stability of pump rod strings and ensuring the stability of submersible pumps. Actual problems of anti-vibration device designing for long structures (rod strings, pipe strings, etc.) were considered in [36–38], recommendations for over-loaded pump reliability are given in [39–41].

The high profitability of oil-gas jet pumps was also proven in case of their application for gas-lifting operation of wells [42–44]. In general, jet pumps can improve the operation of the equipment they work with, extend its overhaul period and increase oil recovery. Due to such a list of advantages, jet pumps are installed in different places inside of oil wells (near the bottomhole, above and below the dynamic level, on the surface and near the mouth) [45].

A technique for selection of the geometrical parameters for the downhole ejection system's body with an external placement of several jet pumps is presented [46]. To assess the strength of the jet pump's bodies, the theory of membrane-free shells with via openings or cutouts is used [47,48]. The presence of technological openings, designed to attach jet pumps, contributes to the uneven distribution of stresses and, as a result, reduce the fatigue life of such structures. The problems of stress concentration in shell systems under conditions of contact interaction while complex loading were considered in [49–51]. When designing such equipment, engineers pay special attention to the accuracy and tightness of threaded joints [52–54], increasing thread [55–57] and preventing self-unscrewing of threaded joints [58,59].

Concerning the features of the design of modern ejection systems that are operated under aggressive conditions, some authors suggest the use metallic ceramics or ceramic coatings to provide the required operational life of the nozzle [60–62]. Recommendations have been made on the technology of forming, optimal design and calculation of the operational properties of single layer [63,64], flexible [65,66] or multilayer ceramic coatings [67–69]. Thermoplastic composite materials are also widely used [70–73]. For surface hardening of parts of elite models of downhole ejection systems, chromium plating in a flowing electrolyte is used, as well as functionally gradient coatings of required parts of surfaces [74–76]. New technological schemes for hydrocarbon development from used wells to use additionally non-commercial and closed deposits and extend mining enterprise operation are presented in [77–80].

Therefore, one of the most important tasks before the implementation of jet pump technologies in production was the development of a methodology for calculating their working and geometric parameters, as well as technologies to strengthen their parts. Literature review, analyses of patents and regulations on electrochemical chromium plating of parts in quiet and flowing electrolytes showed that chromium plating in quiet electrolyte does not provide a uniform thickness of coatings, and the formation of coatings in flowing electrolyte occurs at higher current densities [62]. This ensures uniform application of the coating layer with low roughness, as well as achieving high coating performance and increased current yield of chromium. Increasing the service life of pumps can be achieved through the use of reinforcing technologies in the manufacture of their parts by chrome plating.

The issues of obtaining theoretical dependences and their usage for practical calculations of jet pumps are described in [81]. Since the theory of jet devices, installed above the dynamic level of oil wells, which are operated by sucker-rod pumps, has already been developed [82], it is possible to proceed directly to determining the rational location of such jet pumps in wells and calculating their geometry with strengthening of their details.

1.2. Purpose and Tasks of Research

Thus, the purpose of this article is to determine such location of the oil-gas jet pump inside of the oil well, which will provide the maximum possible reduction of outlet pressure for the sucker-rod pump and decrease the tubing load; as well as being used in development of technology of strengthening of details of jet and plunger pumps by chrome coverings. The formulated purpose can be achieved by solving the following tasks:

- calculating the distribution of operating parameters along the borehole of the oil well, operated by a sucker-rod pump;
- performing the calculation of geometric and thermobaric parameters of oil-gas jet pumps, installed at different depths in the oil well;
- analyzing which of the considered operation modes is the most profitable and determine which variables have the strongest affect on the efficiency of jet pumps' usage in the oil well;
- calculating the technologically possible reduction of the outlet pressure for the sucker-rod pump and tubing load for optimal operation mode;
- developing a technology of electrochemical chromium plating in the flow electrolyte of the working surfaces of the parts of jet and plunger pumps.

2. Materials and Methods

2.1. Features of Design, Manufacture and Strengthening by Chrome Plating of Details of Pumps

To strengthen the parts of jet and plunger pumps, an installation for electrochemical chromium plating in a flowing electrolyte and a technological process of coating the parts of these pumps were developed. The installation for electrochemical chrome plating of pump parts in a flowing electrolyte contains interchangeable electrochemical cells used for coating the outer surfaces of the plunger or rod, respectively, or the inner surfaces of the plunger pump sleeve and the jet pump ejector. The electrochemical cell, after mounting the part and the electrode in it, was sealed and connected by means of current leads to the power supply, and by pipes—to the system of electrolyte supply to the annular cavity between the electrodes of the cell. The unit is equipped with an automated control system for electrochemical chromium plating in the flow electrolyte, which allows to maintain independently at a given level the technological parameters of electrolysis: temperature, electrolyte flow rate, electrolysis operating current and the ratio of electrolyte components. Chrome coatings were applied to the working surfaces of pump parts made of steel 40KHN (GOST 4543-2016), which were surface hardened with high frequency currents and subjected to grinding. The anode was an alloy of lead with sulfur and tin. Chrome plating of pump parts (ejectors, bushings, plungers, rods) was performed in a standard electrolyte based on chromic anhydride and sulfuric acid with nano-additives.

To reduce the harmful effects of chromium electrolyte on the environment during the application of electrochemical chrome coating on the pump parts the developed design of the electrochemical sealed cell was used. The system for supplying electrolyte to the annular cavity formed between the electrodes of the electrochemical sealed cell is equipped with a unit for neutralization and purification of gaseous electrolysis products according to our invention [83]. The amount of harmful emissions released into the atmosphere does not exceed the maximum allowable concentration.

2.2. Determination of Rational Placement of the Jet Pump in the Well

Instead of directing the extracted gas from the well space directly to the flare for its combustion, the article proposes mixing oil gas with well products in the downhole ejector, which allows to use its potential energy of this gas. In the future, this gas is separated at oil treatment plants, and then it can be sent for further use or for the needs of production, or for the current needs of maintenance personnel. The innovation proposed in the article eliminates the combustion of emitted gas during the operation of oil wells, which reduces the amount of emissions of carbon dioxide and other combustion products into the atmosphere. Such measures improve the environmental situation in the oil and gas fields.

In order to determine the rational placement of the oil-gas jet pump in the oil well 753-D Dolyna Oil Field "Dolynanaftogaz" PJSC "Ukrnafta", first of all, it is necessary to calculate the distribution of thermobaric and other operating parameters along its borehole and also to find the daily amount of oil gas, entering its annulus [18].

The scheme of the general layout and the mutual placement of the sucker-rod pump and the oil-gas jet pump is shown in Figure 1 [84].

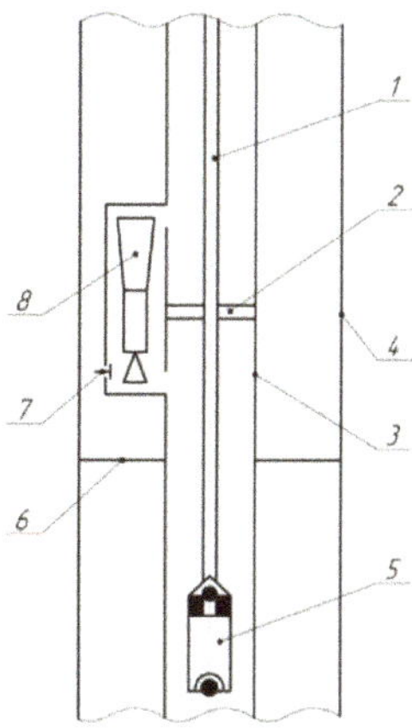

Figure 1. General layout of the tandem installation [84]: 1—rods; 2—separator; 3—tubing string; 4—casing; 5—sucker-rod pump; 6—dynamic level; 7—check valve; 8—oil-gas jet pump.

The wellbore fluid, which is pumped by the sucker-rod pump (5) into the tubing string (3), enters the oil-gas jet pump (8), which is placed in the annulus, where, due to its high velocity, the associated gas from the annular space enters the ejector through the return valve (7). Then this gas–liquid mixture returns back to the tubing and then moves to the mouth of the well. Inside the tubing at the jet pump's level, a separator (2) must be installed to ensure that the ejector's input and output are not connected. More detailed view of the suggested design for the oil-gas jet pump is shown in Figure 2.

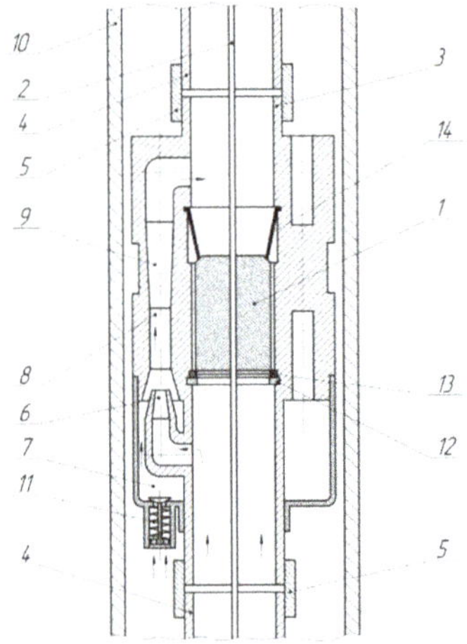

Figure 2. Construction scheme of the downhole oil-gas jet pump [85]: 1—separator; 2—rod; 3—jet pump's body; 4—tubing; 5—collar; 6—working nozzle; 7—receiving chamber; 8—mixing chamber; 9—diffuser; 10—casing; 11—check valve; 12—spring ring; 13—packing seal; 14—retainer.

In order to determine the location of the jet pump inside the oil well, first of all the range of possible depths of the jet pump placement was selected: from 610 m to 890 m. Low-pressure jet pumps with a ratio of the mixing chamber and the nozzle cross-section areas equal to 6.25. With the help of a created computer program, the outlet pressure for the jet pump was found, as well as a number of its variable parameters (like geometric dimensions), based on the improved Poettman-Carpenter method, which is used for calculation of thermobaric parameters' distribution in wells. In such case, the working flow for the jet pump will be the gas–oil mixture of the well and the injected flow—oil gas from the annulus.

The study [18] considered in details all the values included in the Poettman-Carpenter and Baxendel equation and explains the procedure for implementing this calculation. The scientific novelty of the proposed methods is to solve nonlinear differential equations on which the Poettman-Carpenter and Baxendel methods are based, the Adams-Krylov numerical method, which excludes a number of assumptions made in the original methods (including assumptions about the linear relationship between temperature and pressure). wells). Comparison of the results of the proposed and well-known methods is presented in [86].

The phenomenon of cavitation and ejector efficiency have not been studied by the authors. The work [87] is devoted to obtaining the characteristics of high-pressure and low-pressure ejectors operating in an oil well; this article also presents the curves of characteristics of these jet pumps to represent dependences of dimensionless relative difference of pressure on coefficient of injection are resulted.

As input parameters in this program we have taken: pressure of the working flow before entering the jet pump, its density, consumption gas content, the density of free gas. All of these parameters were equal to the correspondent parameters of considered oil well. In addition, the velocity of the working flow before the nozzle is also known as far as it was previously determined through the average integral velocity of the production liquid in the tubing. Finally, we have values of pressure in the receiving chamber of the jet pump, the finding of which was described in article [61].

The basic parameter of any jet pump, which influences all the rest geometric and working parameters of it, is the cross-section area of the working nozzle. In the course of mathematical transformations, a formula for determining cross-section area of the nozzle was obtained [84]:

$$f_1 = w_w f_w[(1-\beta_w)p_1 + \beta_w p_w]\sqrt{\frac{(1+\xi_n)\rho_w}{[2p_w(1-\beta_w)+\rho_w w_w^2]p_1^2 - 2(1-\beta_w)p_1^3}} \tag{1}$$

where w_w is velocity of the working liquid before entering the nozzle, m/s; f_w is cross-section area in the outlet of a nozzle, m^2; β_w is consumption gas content under working conditions of a well; p_1 is pressure of the working liquid at the outlet of a nozzle, Pa; p_w is pressure of the working liquid before entering the nozzle, Pa; ξ_n is resistance coefficient of the nozzle; ρ_w is density of the working liquid before entering the nozzle, kg/m^3.

Based on the found area of the nozzle, the cross-section area of the mixing chamber was calculated ($f_3 = 6.25 f_1$) and the diameters of the nozzle d_n and the mixing chamber d_{mc}. The cross-section area in the outlet of the diffuser and its diameter were obtained from the condition that the angle of the diffuser's cone is $\alpha = 8°$ and the length of the diffuser is equal to the twelve diameters of the mixing chamber.

First, the Bernoulli equations for the working flow within the nozzle, the receiving chamber and the mixed flow for the diffuser, as well as the equation of change of the amount of motion (pulse conservation) for the working and injected flows in the cylindrical mixing chamber are recorded. In the last equation, all flow rates are expressed in terms of volume costs. From Bernoulli's equations are expressed the pressures at the inlet and outlet of the mixing chamber and substitute in the equation of change in the amount of motion. We express the terms in the right part of the obtained equation, which contain the flow velocities in different sections of the ejector due to the flow of working and injected flows at the entrance to the mixing chamber. The law of conservation of mass is written down and substituted into the previously obtained equation (the product of the cross section at the entrance to the mixing chamber at the expense of the workflow). As a result, we obtain the equation of the oil and gas ejector. This derivation is given in more detail in [84,87].

The variable parameters of the jet pump were determined in a certain sequence and as a result of solving characteristic equations for jet pumps [58] and to find the outlet pressure the equation of a low-pressure oil-gas jet pump was used:

$$\begin{gathered}\frac{2f_1^2\rho_1^2}{(1+\xi_n)\rho_w f_3}\left(p_w - p_1\frac{\rho_w}{\rho_1} + \frac{\rho_w w_w^2}{2}\right) \times \left\{\frac{1}{f_1\rho_1} + \frac{\rho_{in2}}{\rho_1^2 f_{in2}}u_2^2\left[1 - \frac{k-1}{2k}(1+\xi_{in})\right] - \right. \\ \left. -(\xi_d + \xi_{mc} + 1)\frac{1}{2f_3\rho_3}\left(1 + \frac{\rho_{in2}}{\rho_1}u_2\right)^2\right\} - p_m\frac{\rho_3}{\rho_m} - \frac{\rho_3 w_m^2}{2} + p_1 \times \\ \times\left[1 - \left(1 - \frac{\rho_{in2}}{\rho_{in}}\right)\frac{f_{in2}}{f_3}\right] + \frac{k-1}{2k}\frac{f_{in2}}{f_3}\rho_{in2}w_{in}^2 = 0,\end{gathered} \tag{2}$$

where f_1 is cross-section area in the outlet of a nozzle, m^2; ρ_1 is density of the working liquid at the outlet of a nozzle, kg/m^3; f_3 is cross-section area at the outlet of a mixing chamber, m^2; ρ_{in2} is density of the injected gas before entering the mixing chamber, kg/m^3; f_{in2} is cross-section area of the injected gas before entering the mixing chamber, m^2; u^2 is injection ratio before entering the mixing chamber; k is adiabatic exponent; ξ_{in}, ξ_{mc}, ξ_d are resistance coefficients for the inlet of injected gas to the mixing chamber, in the mixing chamber and diffuser; ρ_3 is density of the mixed flow at the outlet of the mixing chamber, kg/m^3; ρ_m is density of the mixed flow at the outlet of the diffuser, kg/m^3; p_m is pressure of the mixed flow at the outlet of the diffuser, Pa; w_m is velocity of the mixed flow at the outlet of the diffuser, m/s; ρ_{in} is density of the injected gas while entering the mixing chamber, kg/m^3; w_{in} is velocity of the injected gas while entering the mixing chamber, m/s.

Equation (2) was obtained based on Bernoulli equation and mass conservation law. It can be used for jet pumps, where the working and mixed flows are compressible two-phase liquids [84].

In order to establish the pressure distribution between the jet pump and the mouth based on the known parameters of the mixed flow after the jet pump, a computer program was created, where the Baksendel method was applied [61].

After building up the pressure distribution curves along the wellbore, taking into account the presence of the jet pump, it is necessary to analyze the obtained value of the wellhead pressure. If the wellhead pressure is higher than technologically required, that means that it is allowed to reduce the pressure in every point of the wellbore starting from the outlet of the sucker-rod pump as long as the form of the curve remains the same. However, if the wellhead pressure turned to be lower than technologically required it is essential to consider other installation depth of even another jet pump (means with other geometrical parameters).

In addition to low-pressure jet pumps, which are able to prove high flow rates of liquid and gas but with a great pressure drop in the jet pump, there are high-pressure jet pumps with cross-section areas ratio $f_3/f_1 = 3$ and less. These jet pumps have smaller pressure losses but can inhaust less gas.

The methodology of choosing the rational place of such jet pumps in oil well is exactly the same as for the low-pressure jet pumps. The only difference is in the characteristic equation for finding outlet pressure of jet pump:

$$\begin{aligned} &\frac{2f_1^2\rho_1^2}{(1+\xi_n)\rho_w f_3}\left(p_w - p_1\frac{\rho_w}{\rho_1} + \frac{\rho_w w_w^2}{2}\right) \times \left[\frac{1}{f_{w2}\rho_{w2}} + \frac{\rho_{in2}}{\rho_{w2}^2 f_{in2}}u_2^2 - \right. \\ &\left. -\frac{1}{2}\left(1 + \frac{\rho_{in2}}{\rho_{w2}}u_2\right)^2\left(\frac{\xi_d+\xi_{mc}+1}{f_3\rho_3} + \frac{f_3\rho_3}{f_m^2\rho_m^2}\right)\right] - p_m\frac{\rho_3}{\rho_m} + p_2 = 0, \end{aligned} \tag{3}$$

where f_{w2} is cross-section area of the working liquid before entering the mixing chamber, m^2; ρ_{w2} is density of the working liquid before entering the mixing chamber, kg/m^3; p_2 is pressure of the working liquid before entering the mixing chamber, Pa.

3. Results

Table 1 shows some of the results obtained using mentioned programs for low-pressure oil-gas jet pumps.

As can be seen from the Table 1, if the inlet parameters of the jet pump (p_w, ρ_w, β_w, w_w) in the cross-section of the well where it is installed were the same as in the well without jet pump, then the pressure at the wellhead would be less than necessary (should be 0.5 MPa). The only exception is the jet pump installed at a depth of 890 m. This jet pump will provide a slightly higher wellhead pressure (p_{wh} = 0.52 MPa).

As soon as the pressure at the wellhead should be p_{wh} = 0.5 MPa, that requires installation depth of jet pump not less than 890 m.

What is then the cause of this phenomenon? After all, the relation between the flow densities at the inlet and at the outlet of the jet pump it is quite significant (ρ_w/ρ_m = 2.06–2.246). In addition, outlet pressure of the jet pump is significantly less than the inlet pressure, which leads to a substantial flash gas liberation. Yet, despite this, during the movement of production liquid to the mouth, the pressure is dropping faster in case of jet pump installation compared to the option without it.

Figure 3 shows the pressure distribution between jet pumps, placed in different depth, and the wellhead for the case when the inlet pressure of jet pumps would be equal to the pressure in the corresponding cross-sections of the well before jet pumps installation (curves 2, 3, 4 and 5).

Table 1. Input and calculated parameters of the oil well and the jet pump when it is installed at different depths.

Parameters	Depth of Jet Pump Installation, m					
	610	650	700	800	850	890
Input parameters						
Pressure at the inlet of jet pump p_w, (MPa)	3.21	3.51	3.91	4.75	5.20	5.56
Density of working liquid ρ_w, (kg/m^3)	724.7	763.1	806.4	876.7	904.0	922.5
Consumption gas content β_w	0.208	0.169	0.125	0.052	0.024	0.004
Velocity of the working liquid before entering the nozzle w_w, (m/s)	3.3	3.1	2.9	2.7	2.6	2.6
Density of free gas ρ_g, (kg/m^3)	26.4	29.1	32,5	40.1	44.2	47.5
Calculated parameters						
Diameter of the nozzle d_n, (mm)	4.3	4.0	3.7	3.1	2.9	2.8
Diameter of the mixing chamber d_{mc}, (mm)	10.7	9.9	9.1	7.8	7.3	6.9
Outlet diameter of the diffuser d_d, (mm)	28.6	26.6	24.4	21.0	19.6	18.6
Density of the mixed flow at the outlet of the jet pump ρ_m, (kg/m^3)	324.4	339.7	360.2	404.6	428.1	447.4
Pressure of the mixed flow at the outlet of the jet pump p_m, (MPa)	2.12	2.20	2.31	2.58	2.73	2.87
Wellhead pressure p_{wh}, (MPa)	0.47	0.42	0.39	0.42	0.47	0.52
Density of the mixed flow at the wellhead $\rho_m{}^{(wh)}$, (kg/m^3)	33.8	30.6	28.3	30.0	33.9	37.9

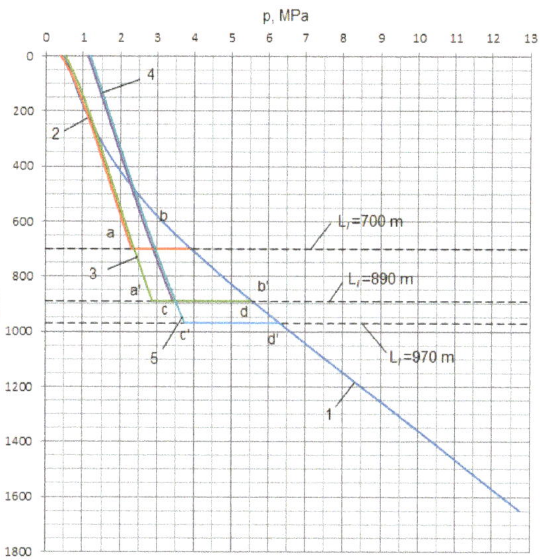

Figure 3. Pressure distribution along the wellbore between jet pumps and wellhead: 1—wellbore pressure distribution; 2, 3—pressure distribution after using jet pump with f_3/f_1 = 6.25; 4, 5—pressure distribution after using jet pump with f_3/f_1 = 4.

Curves 2 and 3 correspond to jet pumps with the cross-sections ratio f_3/f_1 = 6.25 and curves 4 and 5 correspond to jet pumps with the cross-sections ratio f_3/f_1 = 4. In addition, the pressure distribution in oil well between the sucker-rod pump and the wellhead (curve 1) without using jet pump is additionally depicted. The pressure in all four cases decreases almost linearly and much more slowly than in the well without jet pump. The only exception is for the part of the wellbore (approximately 200 m) straight below the mouth, where the pressure drop with and without jet pump is almost the same.

A smaller pressure change gradient is caused by the presence of a significant amount of free gas from the annulus in the tubing. The actual constancy of the pressure change gradient can be explained by the fact that, on the one hand, an increase in the amount of gas in the production liquid should cause a decrease in the pressure change gradient, and, on the other hand, as the amount of gas in the mixture increases, its velocity also increases and this causes the growth in friction losses. These two variable losses almost compensate each other.

The reason for the small wellhead pressure, which low-pressure jet pumps with cross-section areas ratio f_3/f_1 = 6.25 can provide, when the depth of their placement in the well changes in the range from 610 m to 890 m, is a significant pressure drop between inlet and outlet of the jet pump. In Figure 3, these pressure drops are represented by the segments *ab* and *a'b'*. Jet pumps with cross-section areas ratio f_3/f_1 = 4, as it can be seen from Figure 3, allow to obtain significantly greater wellhead pressure than the previous two.

If these jet pumps are installed in the oil well 753-D, then placing the jet pump at a depth of 700 m (f_3/f_1 = 6.25) will entail a slight increase in inlet pressure compared to the pressure in this cross-section of the well without a jet pump. At the same time, the jet pump, located at a depth of 890 m (f_3/f_1 = 6.25), on the contrary, will give a slight decrease in inlet pressure compared to the pressure in this cross-section of the well without a jet pump. Jet pumps with cross-section areas ratio f_3/f_1 = 4 will provide a noticeable decrease in pressure in front of the jet pump, and, consequently, a substantial decrease in outlet pressure of the sucker-rod pump.

Thus, it follows from the above that an increase of the installation depth for such jet pumps will have a little affect on the pressure change in front of it. With increasing of jet pump placement depth in the well, the inlet pressure of it will slightly reduce. It turned out that even greater influence on inlet pressure decreasing has the cross-section areas ratio f_3/f_1.

It would be possible to determine to which level low-pressure jet pumps with the ratio f_3/f_1 = 4 can reduce the outlet pressure of the sucker rod pump. However, since we have proven that the inlet pressure of the jet pump substantially depends on the value f_3/f_1, it is necessary to consider installation of high-pressure jet pumps, for which the value of cross-section areas ratio is $f_3/f_1 < 4$. Such jet pumps provide the highest possible pressure at their outlet.

Table 2 shows the input and some calculated values for such jet pumps, taking into account that input parameters in this case are equal to the corresponding parameters of the well at depths of 700, 800, 850 and 870 m.

Table 2. Input and calculated parameters of the high-pressure oil-gas jet pumps during their installation in oil well at different depths [84].

Parameters	Depth of Jet Pump Installation, m			
	700	800	850	870
Input parameters				
Pressure at the inlet of jet pump p_w, (MPa)	3.91	4.75	5.20	5.38
Density of working liquid ρ_w, (kg/m^3)	806.4	876.7	904.0	913.6
Consumption gas content β_w	0.125	0.052	0.024	0.014
Density of free gas ρ_g, (kg/m^3)	32.5	40.1	44.2	45.8
Velocity of the working liquid before entering the nozzle w_w, (m/s)	2.9	2.7	2.6	2.6
Calculated parameters				
Density of the mixed flow at the outlet of the jet pump ρ_m, (kg/m^3)	418.7	485.9	523.6	541.9
pressure of the mixed flow at the outlet of the jet pump p_m, (MPa)	2.99	3.64	4.09	4.34
Injection ratio before entering the mixing chamber u_2	1.36	1.57	1.69	1.74
Velocity of the working liquid before outflowing from the nozzle w_w, (m/s)	64.9	76.6	82.1	84.2

After that, for these four jet pumps, the pressure distribution between them and the wellhead was determined using previously mentioned computer program for the Baksendel method. In Figures 4 and 5 the dotted line (curves 2) shows the pressure distribution between the jet pump, installed at depths of 700 and 870 m in the well, and its mouth, obtained from the results of this program.

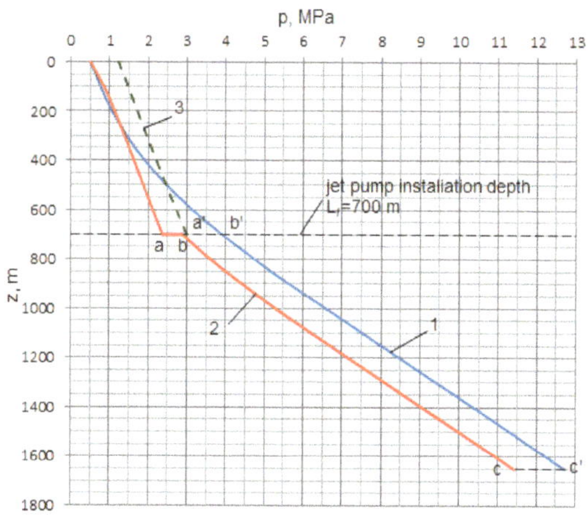

Figure 4. Pressure distribution between the sucker-rod pump and the wellhead: 1—without the jet pump; 2—with the jet pump (f_3/f_1 = 3); 3—with the jet pump after decreasing inlet pressure p_w = 3.908 MPa.

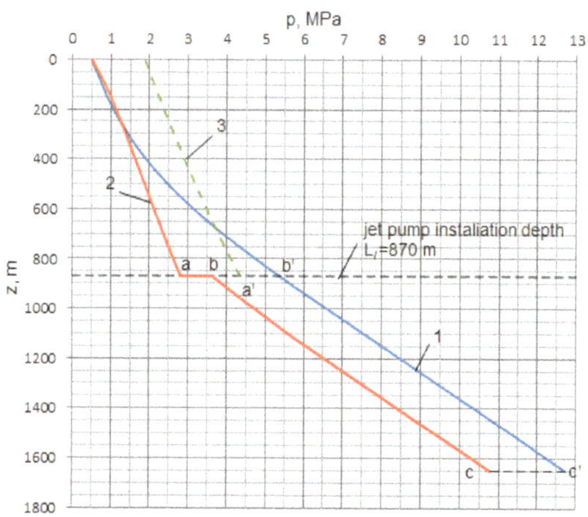

Figure 5. Pressure distribution between the sucker-rod pump and the wellhead: 1—without the jet pump; 2—with the jet pump (f_3/f_1 = 3); 3—with the jet pump after decreasing inlet pressure p_w = 5.381 MPa.

As seen in Figures 4 and 5, if the inlet pressure of the jet pump is equal to the pressure that exists in the well in this cross-section (in the figures it is point b'), then the pressure at the wellhead will be greater than required p_{wh} = 0.5 MPa. Therefore, the next task was to find such inlet pressure of the jet pump, at which the wellhead pressure would be 0.5 MPa. The geometry and all other parameters of the jet pump were determined when its inlet pressure was changing, as well as the pressure distribution along the wellbore from the jet pump to the wellhead.

By successively reducing the inlet pressure of the jet pumps, such values of it were found at which the wellhead pressure are equal to 0.5 MPa. Knowing the inlet pressure of the jet pumps and the temperatures at their installation points, calculations were made to determine the pressure distribution along the wellbore from jet pumps to the sucker-rod pump. According to the results of using created computer programs, pressure distribution curves for the oil well 753-D were built for cases of jet pumps installation at depths of 700 and 870 m respectively (curves 3) (Figures 4 and 5).

Figures 4 and 5 show that the usage of high-pressure jet pumps in the well can significantly reduce the outlet pressure of the sucker-rod pump. This reduction is from 1.32 MPa, for the case of jet pump installing at a depth of 700 m and up to 1.96 MPa, when jet pump is installed at a depth of 870 m. Reducing the outlet pressure of the sucker-rod pump means decreasing the stem load. For the case that is considered, such a reduction in load can reach up to 26% (with a decrease in outlet pressure of sucker-rod pump by 1.96 MPa). In addition, by reducing the load on the beam pumping unit's mechanism, smaller counterbalance can be used [84].

Thus, we may say that it is the most rational to install the high-pressure jet pump with $f_3/f_1 = 3$ at depth of 870 m in the oil well 753-D Dolyna Oil Field "Dolynanaftogaz" PJSC "Ukrnafta".

In the general case, it can be claimed that usage of tandem installation, which consists of the sucker-rod pump and oil-gas jet pump is advisable when the gas–oil ratio of the well is around 100 m^3/t and higher. However, even then the efficiency of such tandem installation will be determined by many other factors.

It is proposed to use an electrochemical chrome coating formed in a flowing electrolyte to increase the wear and corrosion resistance of the polished rod of the rod pump unit.

The cost of applying the chrome coating on parts in a flowing electrolyte does not exceed the cost of applying a chrome coating on parts in a quiet electrolyte—the traditional way. It should be noted that the application of a thick layer of electrochemical chrome coating in the flowing electrolyte reduces the amount of allowance for diamond grinding of parts compared to the quiet electrolyte [88], which ultimately reduces the cost of manufacturing the suitable chrome parts that have high microhardness to our technology (Figure 6).

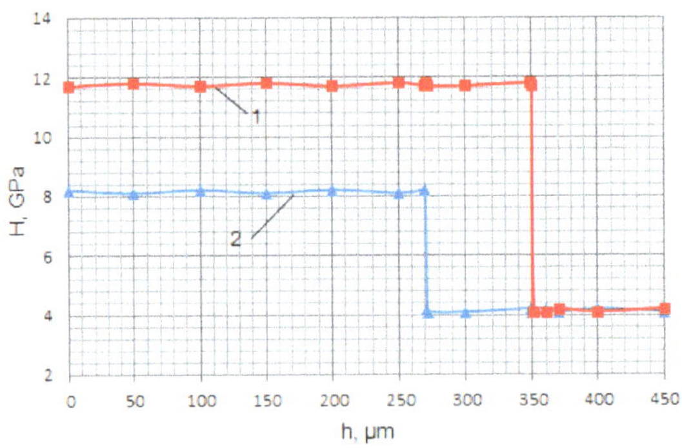

Figure 6. Change in microhardness according to the thickness of the chromium electrochemical coating formed in the electrolyte: 1—flow; 2—calm.

The use of chromium plating in a flowing electrolyte provides thicker coatings compared to chromium plating in a quiet electrolyte. In this case, the coatings applied in the flowing electrolyte evenly cover the steel base, have a lower surface roughness and higher microhardness (11.8 GPa) compared to chromium plating in the quiet electrolyte (8.1 GPa). This is achieved by creating a stable flow of electrolyte in the annular interelectrode space of the electrochemical cell and intensive removal of gases formed during electrolysis on the surface of the part and the electrode. In addition, the saturation of the chrome coating and the steel base with hydrogen is reduced.

Tests of samples with chrome coatings were performed on an upgraded installation that simulates the reciprocating motion (reversible friction) of the polished rod of the rocking machine and allows you to continuously record the value of the coefficient of friction. The results are presented in Figure 7.

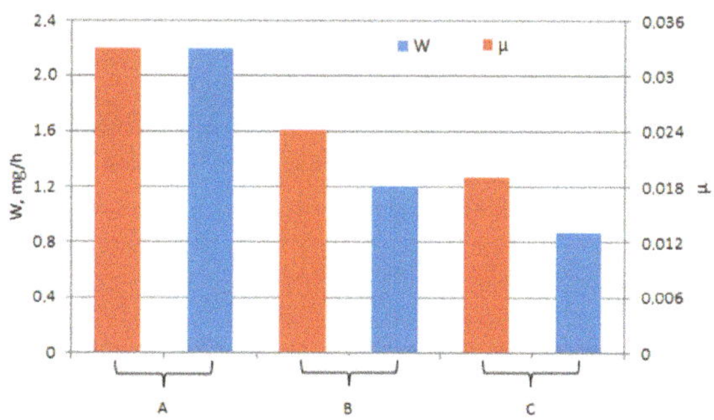

Figure 7. Wear rate of chrome-plated coatings in the friction pair with stuffing box seal and the value of the coefficient of friction: A—steel 40KHN; B—chrome coating (quiet electrolyte); C—chrome coating (flowing electrolyte).

The results of tests of samples with chrome coatings in reciprocating motion showed that the coating reduces the wear rate of the chrome rod, which is operated in pair with the stuffing box seal of the rod installation. The combination of high microhardness of chrome coatings, as well as the specifics of the structure provide high wear resistance of chrome coatings applied from a flowing electrolyte. Chromium plating in a flowing electrolyte also reduces the coefficient of friction in the metal-polymer friction pair as 20 percent compared to chromium plating in a quite electrolyte. It is established that the minimum wear for reversible friction pair of electrochemical chrome coating—stuffing box seal is provided by the surface roughness of the coating Ra from 0.16 to 0.32 μm. All this ultimately increases the service life of pump parts, reduces the load on the mechanical part of the rod installation of the rocking machine, the cost of electricity to overcome the forces of thorns in the installation for rod production of oil from wells. In addition, the combination of ejector and plunger pumps for joint work increases the energy efficiency of oil production and ensures the rational use of potential energy of oil gas, as well as improving the environmental situation in oil fields, as it eliminates the combustion of oil gas.

The usage of the developed method to specify the rational placement of jet pumps in oil wells allows us to find the optimal operation mode of simultaneous operation of the sucker-rod pumps and the oil-gas jet pumps. This method will provide offtaking of the whole amount of free oil gas to enter the annulus of the well, and in turn it will provide the lowest possible stem load.

Results of this research are formulated in the form of a methodology for calculating the working parameters of a downhole ejection system, which is implemented in OGPD "Dolynanaftogaz" PJSC "Ukrnafta" and the methodology for determining the rational placement of oil-gas jet pumps in oil wells, approved by the PJSC "Ukrnafta", and the main design parameters of suggested jet pump for tandem installation in well 753-D as well as the exact place of its location in the well have been submitted for implementation.

This article does not really mention the effect of non-uniformity of the rod pump supply on the operation of the downhole ejector. We have conducted such studies, and they indicate that even at the minimum speed and flow rate of the working flow in the ejector (corresponding to the movement of the rods up) can suck the injected flow from the annulus of the well. In addition, for some wells it is advisable to use periodic ejector operation. This means that the gas will be removed from the annulus only when a certain critical pressure value is reached there.

The performed hydrodynamic calculations prove the efficiency of the proposed scheme of operation of the downhole jet pump and the rod pump in the well. Of course, the authors understand that this method of operation cannot be used for all oil wells and in each case it is necessary to justify the possibility of joint operation of the rod pump and ejector.

The novelty of the proposed study lays in a systematic approach to rod pump and our patented ejector pump operation in the oil well [85] and chrome plating of pump parts. The result of scientific research is a sound method of determining the rational location of the ejector in the oil well and the calculation of its geometry, which will provide a complete selection of petroleum gas released into the annulus of the oil well.

This will make possible obtaining the positive results, namely:

- effectively using the potential energy of free oil gas from annulus;
- making sucker-rod pump operation more stable;
- decreasing dynamic level fluctuations in the well (to avoid dangerous fluctuations in the case of a small immersion of the sucker-rod pump);
- increasing the production liquid to the surface by the mixed flow density reduction (after the jet pump);
- reducing the stem load, which in turn will extend their overhaul life;
- decreasing electricity consumption during well operation and decrease investments in oil production.

Especially this technology can show its efficiency in conditions of high gas–oil ratios and in regions with predominantly low ambient temperatures.

The results of industrial tests of jet and plunger pumps with reinforced chrome-plated parts showed an increase in their service life of about 27%.

In further research, the authors plan to provide calculations considering the curvature of wellbores.

4. Conclusions

The study showed the following: the operation capability and efficiency of oil-gas jet pumps was proved, and the following conclusions were made:

It was established that the usage of low-pressure jet pumps with a cross-section areas ratio f_3/f_1 = 6.25 cannot provide a significant decrease in the outlet pressure of the sucker-rod pump even with an increase in the installation depth due to the large difference inlet/outlet pressure of the jet pumps.

Furthermore, the minimum tubing load is specified by the combination of two factors: the minimum possible cross-section areas ratio f_3/f_1 and the maximal installation depth. This way, the ratio f_3/f_1 for the jet pumps is a factor to decrease significantly tubing loads. Depth of installation makes a much smaller effect if increased.

According to studies, the best operation mode for the 753-D oil well lays in the tandem installation (sucker-rod pump and the jet pump; f_3/f_1 = 3 and placed at depth 870 m). This mode provides decreasing of outlet pressure of the sucker-rod pump up to 1.96 MPa.

This way, the authors recommend the following for the 753-D oil well OGPD "Dolynanaftogaz" PJSC "Ukrnafta":

- to install the high-pressure oil-gas jet pump (f_3/f_1 = 3) at a depth of 870 m to take off the whole amount of free annulus gas. This means to aim at reducing the stem load by 26%. The aforementioned jet pump has the following geometrical options: nozzle diameter d_n = 3.88 mm, mixing chamber diameter d_{mc} = 6.72 mm, and the diffuser outlet diameter d_d = 18.01 mm.

Electrochemical chromium plating according to the developed technological process using an installation equipped with an automated control system of technological parameters of electrochemical chromium plating process in flowing electrolyte with nanoadditives, provided uniform coating with low surface roughness, high microhardness and wear resistance. The proposed innovations ensure the high quality of the working surfaces of parts and increase the performance of ejector and plunger pumps during oil production.

In addition, the usage of jet pumps will also improve the ecological situation in the oil production fields through the recycling of associated oil gas.

Author Contributions: Conceptualization, O.D., L.R. and O.B.; methodology, O.D. and M.S.; software, O.D. and T.P.; validation, O.D., L.R. and M.S.; formal analysis, O.D. and M.S.; investigation, O.D. and V.L.; resources, O.D. and L.R.; data curation, O.D. and M.S.; writing—original draft preparation, O.D., M.S. and V.L.; writing—review and editing, O.D., L.R., M.S. and V.L.; visualization, O.D., M.S., T.P. and V.L.; supervision, O.D. and L.R.; project administration, O.B. and L.R.; funding acquisition, O.B. All authors have read and agreed to the published version of the manuscript.

Funding: This study was carried out as part of the project "Belt and Road Initiative Institute for Chinese-European studies (BRIICES)" and was funded by the Guangdong University of Petrochemical Technology.

Institutional Review Board Statement: Not applicable.

Informed Consent Statement: Not applicable.

Data Availability Statement: Data are contained within the article.

Acknowledgments: The team of authors express their gratitude to the reviewers for valuable recommendations that have been taken into account to improve significantly the quality of this paper. The authors are grateful to the Ministry of Science and Education of Ukraine for the grant to implement the project D 8-21-P (RK 0121U109591).

Conflicts of Interest: The authors declare no conflict of interest.

References

1. Rajović, V.; Kiss, F.; Maravić, N.; Bera, O. Environmental flows and life cycle assessment of associated petroleum gas utilization via combined heat and power plants and heat boilers at oil fields. *Energy Convers. Manag.* **2016**, *118*, 96–104. [CrossRef]
2. IEA. *World Energy Outlook. Flagship Report—2019*; International Energy Agency: Paris, France, 2019; Available online: https://www.iea.org/reports/world-energy-outlook-2019 (accessed on 20 May 2021).
3. Saik, P.; Petlovanyi, M.; Lozynskyi, V.; Sai, K.; Merzlikin, A. Innovative approach to the integrated use of energy resources of underground coal gasification. *Solid State Phenom.* **2018**, *277*, 221–231. [CrossRef]
4. Koltun, P.; Klymenko, V. Cradle-to-gate life cycle assessment of the production of separated mix of rare earth oxides based on Australian production route. *Min. Miner. Depos.* **2020**, *14*, 1–15. [CrossRef]
5. Pivnyak, G.; Bondarenko, V.; Kovalevska, I. *New Developments in Mining Engineering 2015: Theoretical and Practical Solutions of Mineral Resources Mining*; Taylor & Francis: Abingdon, UK, 2015; 607p.
6. Le Billon, P.; Kristoffersen, B. Just cuts for fossil fuels? Supply-side carbon constraints and energy transition. *Environ. Plan. A Econ. Space* **2020**, *52*, 1072–1092. [CrossRef]
7. Krichevskiy, S. Evolution of technologies, "green" development and grounds of the general theory of technologies. *Philos. Cosmol.* **2015**, *14*, 120–139.
8. Khomenko, O.; Barna, T. Zonal-and-wave structure of open systems on micro, mega- and macrolevels of the universe. *Philos. Cosmol.* **2019**, *22*, 24–32. [CrossRef]
9. Rozin, V. From engineering and technological process to post-cultural technology. *Future Hum. Image* **2020**, *15*, 99–109. [CrossRef]
10. Akchiche, M.; Beauquin, J.-L.; Serra, S.; Sochard, S. Exergoeconomic Optimization of Oil and Gas Production Systems. In Proceedings of the SPE Europec Featured at 82nd EAGE Conference and Exhibition, virtual, 1–3 December 2020; pp. 1–15. [CrossRef]
11. Rabaev, R.U.; Belozerov, V.V.; Molchanova, V.A. Associated annular gas utilization methods. *Pet. Eng.* **2019**, *17*, 88–93. [CrossRef]
12. Anosike, N.; El-Suleiman, A.; Pilidis, P. Associated Gas Utilization Using Gas Turbine Engine, Performance Implication—Nigerian Case Study. *Energy Power Eng.* **2016**, *8*, 137–145. [CrossRef]
13. Pavlychenko, A.; Kovalenko, A. The investigation of rock dumps influence to the levels of heavy metals contamination of soil. In *Mining of Mineral Deposits*; CRC Press: Boca Raton, FL, USA, 2013; pp. 237–238. [CrossRef]
14. Pujihatma, P.; Hadi, S.P.; Rohmat, T.A. Combined heat and power–multi-objective optimization with an associated petroleum and wet gas utilization constraint. *J. Nat. Gas Sci. Eng.* **2018**, *54*, 25–36. [CrossRef]
15. Law, B.E.; Ulmishek, G.F.; Clayton, J.L.; Kabyshev, B.P.; Pashova, N.T.; Krivosheya, V.A. Basin-centered gas evaluated in Dnieper-Donets basin, Donbas foldbelt, Ukraine. *Oil Gas J.* **1998**, *96*, 74–78.
16. Fyk, M.; Biletskyi, V.; Abbood, M.; Al-Sultan, M.; Abbood, M.; Abdullatif, H.; Shapchenko, Y. Modeling of the lifting of a heat transfer agent in a geothermal well of a gas condensate deposit. *Rozrobka Rodovyshch* **2020**, *14*, 66–74. [CrossRef]

17. Panevnyk, D.O.; Panevnyk, O.V. Investigation of the joint work of a jet and plunger pump with a balancing crank-rod drive. *Neftyanoe Khozyaystvo Oil Ind.* **2020**, *2*, 58–61. [CrossRef]
18. Panevnyk, O.V.; Dubei, O.Y. Determination of the pressure and temperature distribution along the oil well bore. *Nauk. Visnyk Natsionalnoho Hirnychoho Universytetu* **2015**, *148*, 98–103.
19. Bekbergenov, D.; Jangulova, G.; Kassymkanova, K.K.; Bektur, B. Mine technical system with repeated geotechnology within new frames of sustainable development of underground mining of caved deposits of the Zhezkazgan field. *Geod. Cartogr.* **2020**, *46*, 182–187. [CrossRef]
20. Drozdov, A.; Gorbyleva, Y. Improving the Operation of Pump-ejector Systems at Varying Flow Rates of Associated Petroleum Gas. *J. Min. Inst.* **2019**, *238*, 415–422. [CrossRef]
21. Gabdrkhmanova, K.; Izmailova, G.; Samigullina, L. Solution of the Problem of Annular Space Gas Utilization in Wells Operated by Walking-Beam Pumping Unit. *IOP Conf. Ser. Earth Environ. Sci.* **2020**, *459*, 042080. [CrossRef]
22. Telkov, V.P. Improvement of oil recovery by jet and electrical centrifugal pumping technology of water/gas influence. In Proceedings of the SPE Annual Technical Conference and Exhibition, Anaheim, CA, USA, 11–14 November 2007; pp. 4–9. [CrossRef]
23. Yelemessov, K.; Krupnik, L.; Bortebayev, S.; Beisenov, B.; Baskanbayeva, D.; Igbayeva, A. Polymer concrete and fibre concrete as efficient materials for manufacture of gear cases and pumps. *E3S Web Conf.* **2020**, *168*, 00018. [CrossRef]
24. Prokopenko, D.; Shatskyi, I.; Vorobiov, M.; Ropyak, L. Cyclic deformation of separating tape in electromagnetic rolling pump. *J. Phys. Conf. Ser.* **2021**, *1741*, 012029. [CrossRef]
25. Tarel'nik, V.B.; Konoplyanchenko, E.V.; Kosenko, P.V.; Martsinkovskii, V.S. Problems and Solutions in Renovation of the Rotors of Screw Compressors by Combined Technologies. *Chem. Petrol Eng.* **2017**, *53*, 540–546. [CrossRef]
26. Sun, T.; Zhang, X.; Liu, S.; Cao, Y.; Xie, R. Annular Pressure Buildup Calculation When Annulus Contains Gas. *Chem. Technol. Fuels Oils* **2018**, *54*, 484–492. [CrossRef]
27. Jianxin, S. Application of composite jet-rod pumping system in a system in a deep heavy-oil field in Tarim China. In Proceedings of the SPE Annual Technical Conference and Exhibition, Florence, Italy, 19–22 September 2010; p. 8. [CrossRef]
28. Drozdov, A.N.; Drozdov, N.A. Prospects of development of jet pump's well operation technology in Russia. In Proceedings of the SPE Russian Petroleum Technology Conference, Moscow, Russia, 26–28 October 2015; pp. 1–14. [CrossRef]
29. Fedorov, A.E.; Verbitsky, V.S.; Goridko, K.A. Experimental studies and analysis of gas-jet device's operation characteristics for oil and gas production in abnormal operating conditions. In Proceedings of the SPE Russian Petroleum Technology Conference and Exhibition, Moscow, Russia, 24–26 October 2016. [CrossRef]
30. Skitsa, L.; Yatsyshyn, T.; Liakh, M.; Sydorenko, O. Ways to improve safety of a pumping-circulatory system of a drilling rig. *Min. Miner. Depos.* **2018**, *12*, 71–79. [CrossRef]
31. Velichkovich, A.S. Shock absorber for oil-well sucker-rod pumping unit. *Chem. Pet. Eng.* **2005**, *41*, 544–546. [CrossRef]
32. Moysyshyn, V.M.; Lyskanych, M.V.; Borysevych, L.V.; Zhovniruk, R.A. Multifactorial Empirical Model of Three-Cone Bit Gear Oscillation Energy. *Metallofiz. Noveishie Tekhnol.* **2020**, *42*, 1729–1752. [CrossRef]
33. Velychkovych, A.; Petryk, I.; Ropyak, L. Analytical study of operational properties of a plate shock absorber of a sucker-rod string. *Shock Vib.* **2020**, *2020*, 3292713. [CrossRef]
34. Zhang, T.; Zhang, N. Vibration Modes and the Dynamic Behaviour of a Hydraulic Plunger Pump. *Shock Vib.* **2016**, *2016*, 9679542. [CrossRef]
35. Moisyshyn, V.M.; Lyskanych, M.V.; Borysevych, L.V.; Kolych, N.B.; Zhovniruk, R.A. Integral Indicators of Change of Drilling Column Vibration-Criterion for Assessing of Roller Cone Bit Wear. *Metallofiz. Noveishie Tekhnol.* **2019**, *41*, 1087–1102. [CrossRef]
36. Dutkiewicz, M.; Gołębiowska, I.; Shatskyi, I.; Shopa, V.; Velychkovych, A. Some aspects of design and application of inertial dampers. *MATEC Web Conf.* **2018**, *178*, 06010. [CrossRef]
37. Velichkovich, A.S.; Velichkovich, S.V. Vibration-impact damper for controlling the dynamic drillstring conditions. *Chem. Pet. Eng.* **2001**, *37*, 213–215. [CrossRef]
38. Shatskyi, I.; Velychkovych, A. Increase of compliance of shock absorbers with cut shells. *IOP Conf. Ser. Mater. Sci. Eng.* **2019**, *564*, 012072. [CrossRef]
39. Ropyak, L.Y.; Velychkovych, A.S.; Vytvytskyi, V.S.; Shovkoplias, M.V. Analytical study of "Crosshead—Slide rail" wear effect on pump rod stress state. *J. Phys. Conf. Ser.* **2021**, *1741*, 012039. [CrossRef]
40. Li, W.; Dong, S.; Sun, X. An Improved Sucker Rod Pumping System Model and Swabbing Parameters Optimized Design. *Math. Probl. Eng.* **2018**, *2018*, 4746210. [CrossRef]
41. Konoplianchenko, I.; Tarelnyk, V.; Martsynkovskyy, V.; Gaponova, O.; Lazarenko, A.; Sarzhanov, A.; Mikulina, M.; Zhengchuan, Z.; Pirogov, V. New technology for restoring Babbitt coatings. *J. Phys. Conf. Ser.* **2021**, *1741*, 012040. [CrossRef]
42. Kalwar, S.A.; Awan, A.Q.; Qureshi, F.A. Optimum selection and application of hydraulic jey pump for well-1A: A case study. In Proceedings of the Abu Dhabi International Petroleum Exhibition & Conference, Abu Dhabi, United Arab Emirates, 13–16 November 2017. [CrossRef]
43. Zhautikov, B.A.; Aikeyeva, A.A. Development of the system for air gap adjustment and skip protection of electromagnetic lifting unit. *J. Min. Inst.* **2018**, *229*, 62–69. [CrossRef]

44. Kumar, S.; Kalwar, S.A.; Farouque, K.; Qureshi, S.; Anjum, Z.A. Production enhancement by installation of jet pump in SSD in Pakistan Ghauri Oil Field. In Proceedings of the SPE Middle East Artificial Lift Conference and Exhibition, Manama, Bahrain, 30 November–1 December 2016. [CrossRef]
45. Kurkjian, A.L. Optimazing jet-pump production in the presence of gas. *SPE Prod. Oper.* **2018**, *34*, 373–384. [CrossRef]
46. Panevnik, D.A.; Velichkovich, A.S. Assessment of the stressed state of the casing of the above-bit hydroelevator. *Neftyanoe Khozyaystvo—Oil Ind.* **2017**, *1*, 70–73.
47. Shatskyi, I.; Popadyuk, I.; Velychkovych, A. Hysteretic Properties of Shell Dampers. In *Dynamical Systems in Applications*; Springer Proceedings in Mathematics & Statistics Book Series; Springer: Berlin/Heidelberg, Germany, 2018; Volume 249, pp. 343–350. [CrossRef]
48. Velichkovich, A.; Dalyak, T.; Petryk, I. Slotted shell resilient elements for drilling shock absorbers. *Oil Gas Sci. Technol.—Rev. IFP Energ. Nouv.* **2018**, *73*, 34. [CrossRef]
49. Shats'kyi, I.P.; Makoviichuk, M.V. Contact Interaction of Crack Lips in Shallow Shells in Bending with Tension. *Mater. Sci.* **2005**, *41*, 486–494. [CrossRef]
50. Velichkovich, A.S.; Popadyuk, I.I.; Shopa, V.M. Experimental study of shell flexible component for drilling vibration damping devices. *Chem. Petrol. Eng.* **2011**, *46*, 518–524. [CrossRef]
51. Mikhlin, Y.V.; Zhupiev, A.L. An application of the ince algebraization to the stability of non-linear normal vibration modes. *Int. J. Non-Linear Mech.* **1997**, *32*, 393–409. [CrossRef]
52. Kopei, V.; Onysko, O.; Panchuk, V. The Application of the Uncorrected Tool with a Negative Rake Angle for Tapered Thread Turning. In *Advances in Design, Simulation and Manufacturing II*; Springer: Cham, Switzerland, 2020; pp. 149–158. [CrossRef]
53. Onysko, O.R.; Kopey, V.B.; Panchuk, V.G. Theoretical investigation of the tapered thread joint surface contact pressure in the dependence on the profile and the geometric parameters of the threading turning tool. *IOP Conf. Ser. Mater. Sci. Eng.* **2020**, *749*, 012007. [CrossRef]
54. Pryhorovska, T.; Ropyak, L. Machining Error Influnce on Stress State of Conical Thread Joint Details. In Proceedings of the IEEE 8th International Conference on Advanced Optoelectronics and Lasers (CAOL), Sozopol, Bulgaria, 6–8 September 2019; Volume 9019544, pp. 493–497. [CrossRef]
55. Onysko, O.; Kopei, V.; Medvid, I.; Pituley, L.; Lukan, T. Influence of the Thread Profile Accuracy on Contact Pressure in Oil and Gas Pipes Connectors. In *Advances in Design, Simulation and Manufacturing III*; Lecture Notes in Mechanical Engineering Book Series; Springer: Cham, Switzerland, 2020; pp. 432–441. [CrossRef]
56. Tutko, T.; Dubei, O.; Ropyak, L.; Vytvytskyi, V. Determination of Radial Displacement Coefficient for Designing of Thread Joint of Thin-Walled Shells. In *Advances in Design, Simulation and Manufacturing IV*; Lecture Notes in Mechanical Engineering Book Series; Springer: Cham, Switzerland, 2021; pp. 153–162. [CrossRef]
57. Onysko, O.; Kopei, V.; Panchuk, V.; Medvid, I.; Lukan, T. Analytical study of kinematic rake angles of cutting edge of lathe tool for tapered thread manufacturing. In *Advanced Manufacturing Processes*; Lecture Notes in Mechanical Engineering Book Series; Springer: Cham, Switzerland, 2020; pp. 236–245. [CrossRef]
58. Shatskyi, I.; Ropyak, L.; Velychkovych, A. Model of contact interaction in threaded joint equipped with spring-loaded collet. *Eng. Solid Mech.* **2020**, *8*, 301–312. [CrossRef]
59. Ropyak, L.Y.; Vytvytskyi, V.S.; Velychkovych, A.S.; Pryhorovska, T.O.; Shovkoplias, M.V. Study on grinding mode effect on external conical thread quality. *IOP Conf. Ser. Mater. Sci. Eng.* **2021**, *1018*, 012014. [CrossRef]
60. Piotter, V.; Klein, A.; Plewa, K. Development of a ceramic injection molding process for liquid jet nozzles to be applied for X-ray free-electron lasers. *Microsyst. Technol.* **2018**, *24*, 1247–1252. [CrossRef]
61. Zhang, F.L. Machining Mechanism of Abrasive Water Jet on Ceramics. *Key Eng. Mater.* **2010**, *426*, 212–215. [CrossRef]
62. Ropyak, L.; Ostapovych, V. Optimization of process parameters of chrome plating for providing quality indicators of reciprocating pumps parts. *East.-Eur. J. Enterp. Technol.* **2016**, *80*, 50–62. [CrossRef]
63. Snizhko, L.O.; Yerokhin, A.; Gurevina, N.L.; Ciba, A.V.; Matthews, A. Voltastatic studies of magnesium anodising in alkaline solutions. *Surf. Coat. Technol.* **2010**, *205*, 1527–1531. [CrossRef]
64. Dzyubyk, A.; Sudakov, A.; Dzyubyk, L.; Sudakova, D. Ensuring the specified position of multisupport rotating units when dressing mineral resources. *Rozrobka Rodovyshch* **2019**, *13*, 91–98. [CrossRef]
65. Shatskii, I.P. Tension of a plate containing a rectilinear cut with hinged rims. *J. Appl. Mech. Techn. Phys.* **1989**, *30*, 828–830. [CrossRef]
66. Shatskyi, I.P.; Makoviichuk, M.V.; Shcherbii, A.B. Equilibrium of cracked shell with flexible coating. In *Shell Structures: Theory and Applications*; Taylor & Francis Group: Abingdon, UK, 2018; Volume 4, pp. 165–168. [CrossRef]
67. Ropyak, L.Y.; Shatskyi, I.P.; Makoviichuk, M.V. Influence of the Oxide-Layer Thickness on the Ceramic-Aluminium Coating Resistance to Indentation. *Metallofiz. Noveishie Tekhnologii* **2017**, *39*, 517–524. [CrossRef]
68. Ropyak, L.; Schuliar, I.; Bohachenko, O. Influence of technological parameters of centrifugal reinforcement upon quality indicators of parts. *East.-Eur. J. Enterp. Technol.* **2016**, *79*, 53–62. [CrossRef]
69. Shatskyi, I.P.; Ropyak, L.Y.; Makoviichuk, M.V. Strength optimization of a two-layer coating for the particular local loading conditions. *Strength Mater.* **2016**, *48*, 726–730. [CrossRef]
70. Sudakov, A.; Dreus, A.; Kuzin, Y.; Sudakova, D.; Ratov, B.; Khomenko, O. A thermomechanical technology of borehole wall isolation using a thermoplastic composite material. *E3S Web Conf.* **2019**, *109*, 00098. [CrossRef]

71. Sudakov, A.; Dreus, A.; Ratov, B.; Delikesheva, D. Theoretical bases of isolation technology for swallowing horizons using thermoplastic materials. *News Natl. Acad. Sci. Repub. Kazakhstan Ser. Geol. Tech. Sci.* **2018**, *428*, 72–80.
72. Sudakov, A.K.; Khomenko, O.Y.; Isakova, M.L.; Sudakova, D.A. Concept of numerical experiment of isolation of absorptive horizons by thermoplastic materials. *Nauk. Visnyk Natsionalnoho Hirnychoho Universytetu* **2016**, *5*, 12–16.
73. Sudakov, A.; Dreus, A.; Sudakova, D.; Khamininch, O. The study of melting process of the new plugging material at thermomechanical isolation technology of permeable horizons of mine opening. *E3S Web Conf.* **2018**, *60*, 00027. [CrossRef]
74. Bloch, H.P.; Geitner, F.K. Protecting machinery parts against the loss of surface. In *Machinery Component Maintenance and Repair*; Elsevier: Amsterdam, The Netherlands, 2019; pp. 551–633. [CrossRef]
75. Ropyak, L.Y.; Shatskyi, I.P.; Makoviichuk, M.V. Analysis of interaction of thin coating with an abrasive using one-dimensional model. *Metallofiz. Noveishie Tekhnologii* **2019**, *41*, 647–654. [CrossRef]
76. Shatskyi, I.P.; Perepichka, V.V.; Ropyak, L.Y. On the influence of facing on strength of solids with surface defects. *Metallofiz. Noveishie Tekhnologii* **2020**, *42*, 69–76. [CrossRef]
77. Bazaluk, O.; Lozynskyi, V.; Falshtynskyi, V.; Saik, P.; Dychkovskyi, R.; Cabana, E. Experimental Studies of the Effect of Design and Technological Solutions on the Intensification of an Underground Coal Gasification Process. *Energies* **2021**, *14*, 4369. [CrossRef]
78. Falshtynskyi, V.S.; Dychkovskyi, R.O.; Lozynskyi, V.G.; Saik, P.B. Determination of the Technological Parameters of Borehole Underground Coal Gasification for Thin Coal Seams. *J. Sustain. Min.* **2013**, *12*, 8–16. [CrossRef]
79. Ovetska, O.; Ovetskyi, S.; Vytiaz, O. Conceptual principles of project management for development of hydrate and other unconventional gas fields as a component of energy security of Ukraine. *E3S Web Conf.* **2021**, *230*, 01021. [CrossRef]
80. Bazaluk, O.; Slabyi, O.; Vekeryk, V.; Velychkovych, A.; Ropyak, L.; Lozynskyi, V. A technology of hydrocarbon fluid production intensification by productive stratum drainage zone reaming. *Energies* **2021**, *14*, 3514. [CrossRef]
81. Mei, W. Surface jet ump trial test in Safania Field: Evaluation and case study. In Proceedings of the SPE Kingdom of Saudi Arabia Annual Technical Symposium and Exhibition, Dammam, Saudi Arabia, 24–27 April 2017. [CrossRef]
82. Dubei, O.Y.; Panevnyk, O.V. Basic equation of the low-pressure oil-gas jet. *Mod. Eng. Innov. Technol.* **2017**, *2*, 7–15. [CrossRef]
83. Humeniuk, A.I.; Popadyuk, O.Y.; Ropyak, L.Y.; Kostyk, V.V. Method for Cleaning Air from Mechanical Particles, Microbes and Viruses. Ukrainian Patent Application No. u202002929; International Patent Application No. IPC/B01D47/02. 15 May 2020. Available online: https://sis.ukrpatent.org/en/search/detail/1447618/ (accessed on 15 October 2021).
84. Dubei, O.Y. Analytical study of the efficient operation modes of oil-gas jet pumps used in oil wells. *AGH Drill. Oil Gas* **2019**, *36*, 19–29. [CrossRef]
85. Panevnik, O.V.; Dubey, O.Y.; Yaremko, I.Y.; Lyakh, M.M. Well Ejector. Ukrainian Patent Application No. a201307030; International Patent Application No. IPC/F05B47/02; Ukrainian Patent No. 105135. 4 June 2013. Bul. No. 7/2014, Published on 10 April 2014. Available online: https://sis.ukrpatent.org/uk/search/detail/1018049/ (accessed on 15 October 2021).
86. Dubei, O.Y. Improving the Efficiency of the Sucker Rod Pump Units by Using Oil-Gas Jets. Dissertation for Technical Sciences Candidate's Degree in Specialty 05.05.12—Machines of Oil and Gas Industry, Ivano-Frankivsk National Technical University of Oil and Gas, Ivano-Frankivsk, Ukraine, 2016; 218p. Available online: https://old.nung.edu.ua/files/attachments/disertaciya_dubey.pdf (accessed on 15 October 2021).
87. Dubei, O.Y. Rivnyannya visokonapirnogo naftogazovogo ezhektora. In Proceedings of the Internet Symposium "Scientific Answers to the Challenges of Modernity—2016"; Volume 2, pp. 107–126. Available online: https://www.sworld.com.ua/simpoz6/25.pdf (accessed on 15 October 2021).
88. Ropyak, L.Y.; Shovkoplias, M.V.; Vytvytskyi, V.S. Determination of machining allowance for parts with chimney. *Bull. Cherkasy State Technol. Univ.* **2021**, *2*, 117–127. [CrossRef]

energies

MDPI

Article

Enzymatic and Oxidizing Agents for the Decomposition of Polymers Used in the Composition of Drilling Muds

Małgorzata Uliasz, Grzegorz Zima, Sławomir Błaż and Bartłomiej Jasiński *

Oil and Gas Institute—National Research Institute, 31-503 Krakow, Poland; uliasz@inig.pl (M.U.); zima@inig.pl (G.Z.); blaz@inig.pl (S.B.)
* Correspondence: jasinski@inig.pl; Tel.: +48-1343-689-41

Abstract: The article presents the results of research on the influence of enzymatic and oxidizing agents on polymers used in drilling mud technology. The research used calcium hypochlorite, urea peroxide, sodium persulfate, amylase and cellulase. This allowed us to determine the optimal concentrations of these agents in order to reduce the rheological parameters of drilling muds. The ability of enzymes and oxidants to decompose colloids used in the composition of drilling muds was determined on the basis of measurements of their rheological parameters and the influence of oxidants and enzymes on the properties of filter cakes. The final effect of the presented research was the development of a drilling mud treatment method before the cementing procedure and the development of washer fluids containing enzymatic or oxidizing agents to be used to prepare the borehole for the cementing procedure. The drilling muds selected for the research contained starch and cellulose polymer colloids, which allowed for the selection of appropriate agents depending on the type of polymer set used in the drilling mud. Positive results were obtained regarding the possibility of thinning the drilling mud immediately before cementing, as well as the possibility of using oxidants and enzymes in the composition of washer fluids.

Keywords: drilling mud; oxidants; enzymes; filter cake

Citation: Uliasz, M.; Zima, G.; Błaż, S.; Jasiński, B. Enzymatic and Oxidizing Agents for the Decomposition of Polymers Used in the Composition of Drilling Muds. *Energies* **2021**, *14*, 5032. https://doi.org/10.3390/en14165032

Academic Editor: Vijay Kumar Thakur

Received: 2 July 2021
Accepted: 12 August 2021
Published: 16 August 2021

1. Introduction

In order to improve the quality of cementing of casing, the drilling mud is treated to lower its rheological parameters. Obtaining low rheological parameters of the drilling mud facilitates displacement of the drilling mud solidified at the wall of the borehole and partial leaching of the clay cake that hinders the contact of the cement grout with the rock. The treatment of drilling muds with thinning agents changes the rheological parameters without significantly changing the properties of the drilling mud [1,2].

The filter cake on the borehole wall is formed by filtering the water phase from the drilling mud. It is a mixture of a polymer solution, weighting agent and finely drilled solid phase. The filter cake produced during drilling increases the stability of the borehole walls and reduces the permeability of the porous layers, therefore there are no major losses of the water phase from the drilling mud, which can significantly reduce the strength of the rocks. When drilling cracked rocks, the filter cake is also deposited in the crevices, binding individual rock fragments, and increasing their strength. In addition, the filter cake contributes to an even distribution of hydrostatic pressure on the borehole wall, which prevents its damage. In weakly compact rocks, the filter cake protects the borehole wall against the erosive impact of the drilling mud stream and ensures a good technological process until the casing is inserted into the borehole [3–6].

The use of agents which decompose colloids, which are components of the drilling mud, thins the drilling mud and removes clay deposits stuck by these colloids. Polymers can be decomposed in several ways, by biological, mechanical, chemical or radiation decomposition [7,8].

Water-soluble polysaccharides used as components of drilling muds have unique, stable rheological properties in drilling operations, but their structure and properties also have a decisive influence on the formation of a polymer film on the surface of the borehole walls, which in combination with the solid phase becomes difficult to remove from the walls of the borehole. Polysaccharides contain many glycosidic bonds (acetals). Glycoside bonds are sensitive to hydrolysis which leads to degradation of the polymer. The chemical decomposition process of the polymer must be properly designed for specific borehole conditions: temperature, pH, salinity, and exposure time. On the basis of laboratory tests, the most optimal and balanced conditions for polymer decomposition should be determined. Chemical degradation of polymers can be carried out by the action of oxidants and enzymes [3,7,9,10].

Oxidizing agents decompose the polymers contained in the filter cake. Oxidation processes take place as a result of the exchange of electrons between the oxidizing substance and the reducing substance, as a result of which the atoms of the elements taking part in the reaction change their oxidation state. According to the electron interpretation of oxidation-reduction processes, oxidants are atoms, ions, or molecules capable of accepting electrons, i.e., receiving them from other atoms or groups of atoms, thus causing their oxidation. Thus, oxidants are themselves reduced by oxidizing other substances. The most popular oxidants are the most electronegative elements, i.e., fluorine F_2, chlorine Cl_2, bromine Br_2 and oxygen O_2. Among chemical compounds, oxidants are those with certain elements in the highest oxidation states, e.g., potassium permanganate $KMnO_4$, potassium dichromate $K_2Cr_2O_7$, hydrogen peroxide H_2O_2, potassium nitrate KNO_3, nitric acid HNO_3 and others [11–14].

The principle action of oxidants is based on the gradual interaction of compounds with high electronegativity on the main polymer chains, breaking them into smaller fragments and, consequently, a decrease in the molecular weight of the polymer. The advantage of oxidants is their wide range of action at a pH of 3 to 14. Oxidizers react quickly with the polymer at high temperatures. Oxidizers cannot be used in an acidic environment as it is possible to oxidize H_2S to elemental sulfur and clog the pores of reservoir rock [12].

Enzymes are another type of agent used to decompose the filter cake. Enzymes are biological catalysts, classified as proteins, which accelerate specific chemical reactions as a result of lowering their activation energy. Due to their protein nature, the activity of enzymes depends on the physical and chemical parameters of the environment, i.e., temperature, pH. The rate of enzymatic reactions rises with increasing temperature. After exceeding the optimal temperature, the thermal denaturation of enzymes progresses, and their activity drops sharply. Most enzymes also have an optimal operating pH. The pH optimum, next to the temperature optimum, is the second most important environmental parameter characterizing the activity of enzymes. Enzymes are most active in their optimal pH range and the activity declines with increasing or decreasing pH beyond this range. In most cases, the effects of enzymes heavily depend on their substrates [15–20].

In the case of using an invert drilling mud [21], a different approach should be applied to the problem of filter cake removal prior to cementing. The main task of the washer fluid is to remove the hydrophobic sediment and make the borehole wall hydrophilic, which requires the use of specially selected surfactants, which can possibly be combined with enzymes.

Enzymes such as cellulase can also be used to modify the properties of cellulose without decomposing it, for example to increase the amount of the crystalline phase in it [22], which may find application in the design of nanocomposites containing cellulose fillers.

The article presents the results of research carried out at the Oil and Gas Institute—National Research Institute, which aim to improve the quality of casing cementing. Due to insufficient quality of cementation manifested by the occurrence of gas exhalation, research is being carried out on the development of a methodology for the use of oxidants and enzymatic agents that break down polymers used in the composition of drilling muds. These results can be used in the preparation of the borehole prior to cementing to remove filter cake which contribute to reducing the bond between the cement stone and the borehole wall. Most of the research carried out in various research centers concerns the use of enzymes or oxidants in treatment fluids, the purpose of which is to remove drilling mud residues from rock pores in order to reduce damage to rock permeability. The authors [8–11,15,20] use enzymatic agents in their research to remove mud sediments from the pore space of rocks. For the same purpose, ultrasound was used in [13]. Part of the work carried out [15,16,19] concerns the evaluation of enzyme activity under various conditions of temperature, pH, and salt content (chlorides, bromides, formates). The activity of the enzymes was assessed on the basis of the degree of their degradation. The paper [18] presents the results of research on the efficiency of filter cake removal with fluids containing enzymatic, oxidizing, and chelating agents depending on the used weighting agent.

2. Materials and Methods

2.1. Raw Materials

In the laboratory tests, natural and semi-synthetic polymers used in the drilling mud technology as well as enzymes and chemical compounds with an oxidizing effect were used. The chemicals and drilling mud materials used in the tests were:

- Biopolymer (xanthan), obtained in the process of fermentation of polysaccharides with microorganisms (Xanthomonas campestris bacteria). It is used to increase the viscosity of drilling muds of various degrees of salinity and to hold the weighting agents in suspension.
- Gelatinized starch, a natural starch that has been modified by thermal gelatinization to render it water-soluble and to form a colloidal solution.
- Carboxymethylcellulose, a semi-synthetic anionic polymer adsorbing on clay minerals. It is one of the most active agents in reducing filtration. It is formed by treating cellulose activated with sodium hydroxide with monochloroacetic acid.
- The partially hydrolyzed polyacrylamide (PHPA) forms a thin layer on the borehole wall and cuttings, which acts as a barrier to the migration of filtrate from the drilling mud to the rock.
- Carbonate blocker ($CaCO_3$), a weighting agent obtained as a result of the mechanical processing of marble. Due to its solubility in acids, it is used primarily for weighting drilling muds and workover fluids used in opening the deposit and in workover works in boreholes [23,24].
- Chemical compounds with an oxidizing effect: calcium hypochlorite, urea peroxide, sodium persulfate.
- Enzymes: amylase, cellulase.

2.2. Research Methodology

The research was carried out on muds with different density, salinity, and a set of colloids and polymers used for their preparation. To determine the efficiency of oxidants and enzymes in the drilling mud environment, three fluid compositions were selected for drilling holes in the Carpathian Foredeep and the Polish Lowlands. The efficiency of oxidizing agents and enzymes was assessed on the basis of changes in rheological parameters, filtrate volume, and changes in the pH value.

The drilling muds prepared according to the compositions used by the drilling mud services were used for the tests. The drilling muds marked as A and B correspond to the compositions of the drilling muds used in the Carpathian Foredeep. These drilling muds differ in the type of polymers used for their preparation. Drilling mud A contains a starch agent and a biopolymer, drilling mud B contains a cellulose agent, a biopolymer and a PHPA encapsulating polymer. The drilling muds marked with symbols A and B also differed in densities, which were, respectively, 1130 kg/m^3 and 1330 kg/m^3. The drilling mud marked as C corresponded to the compositions of drilling muds in the area of the Polish Lowlands, therefore it contained NaCl in its composition in the amount corresponding to the concentration of the saturated solution. A biopolymer, a cellulose agent and a PHPA polymer were used to prepare drilling mud C. Drilling mud C has a density of 1535 kg/m^3. All drilling muds used in the test (A, B, C) additionally comprised of milled Miocene shale imitating cuttings, selected muds contained barite as a material for adjusting the density, and chemical contamination in the form of calcium and magnesium chlorides was additionally introduced into drilling mud C.

The efficiency of polymer decomposition by particular agents (oxidants and enzymes) in the drilling mud environment was assessed on the basis of changes in the rheological parameters of the drilling muds after the introduction of a given oxidant or enzyme. The tests were carried out in various temperature conditions enabling the assessment of the efficiency of the tested agents in borehole conditions. The efficiency of the enzymes was tested at the temperatures of 20 °C, 40 °C, 60 °C and 80 °C due to the decrease in their activity at temperatures above 65 °C. The optimal temperature for the application of the tested enzymes is determined in the range of 50–65 °C.

The efficiency of the filter cake removal was determined by the percentage loss of the filter cake due to the treatment fluid used, which differed in the type of the measures used. Filter cakes were formed on ceramic filter discs (OFITE No. 170–530) with a defined pore size of 50 µm and a fixed permeability of 15 D using a dynamic High Temperature High Pressure (HTHP) filter press. The weight method was used to assess the efficiency of cake removal.

2.2.1. Weight Method of the Efficiency of Removing Filter Cake from Ceramic Discs

Cake Formation

For 15 min the ceramic disc was soaked in a salt solution corresponding to the salinity of the tested drilling mud. The soaked ceramic disc was weighed, and its mass was determined as W_1. Then, the drilling mud filtration was measured on a dynamic HPHT filter press on a previously soaked ceramic disc, at a pressure of 0.7 MPa of N_2 gas, a set temperature, and the speed of the stirrer in the filtration chamber at 200 rpm. After reaching the set temperature, the drilling mud filtration and formation of the cake were measured for a period of 1 h.

Removal of Cake with Fluids of Various Compositions

After cooling, the measuring chamber was disassembled and the ceramic disk with the cake was carefully removed and washed with water. Then the disc with the cake was weighed and the mass was recorded as W_2. The previously prepared treatment fluid was slowly poured into the filtration chamber and the ceramic disc with the cake was placed back in the filtration chamber. At the pressure of 0.7 MPa, the set temperature was achieved and the stirrer's rotation in the filtration chamber was set to 500 rpm. Then, the filtration of the treatment fluid was measured for a period of 30 min. After cooling, the ceramic disc was removed and, after washing with water, weighed and the mass was determined as W_3. The efficiency of filter cake removal was calculated from the following formula:

$$E_U = \frac{W_2 - W_3}{W_2 - W_1} \cdot 100 \ (\%) \tag{1}$$

where:

E_U—efficiency of filter cake removal

W_1—the weight of a ceramic disc soaked with a salt solution corresponding to the salinity of the tested drilling mud (g)

W_2—the weight of the ceramic disc with the generated filter cake (g)

W_3—the weight of the ceramic disk after the effect of the treatment fluid (g).

3. Results and Discussion

3.1. *Influence of Oxidants and Enzymes on the Properties of Potassium-Polymer Drilling Mud A*

A biopolymer and gelatinized starch were used to prepare potassium-polymer drilling mud A. The drilling mud was salted with 5% KCl and a 7% carbonate blocker was added to it. Then the drilling mud was weighted with barite to a density of 1100 kg/m^3 and contaminated with ground Miocene clay simulating cuttings. The drilling mud prepared in this way had a density of 1130 g/cm^3 and a plastic viscosity of 18 mPa·s, an apparent viscosity of 28.5 mPa·s and a yield point of 10 Pa.

The research on the influence of oxidants and the enzyme on the properties of drilling mud A are presented in Tables 1 and 2 and in Figures 1–4. Various amounts of oxidants and enzymes were added to the drilling mud in order to select the optimal dose at which the greatest thinning effect of the drilling mud can be achieved.

Table 1. The properties of potassium-polymer drilling mud A with the addition of oxidizing agents and amylase.

<table>
<tr><th rowspan="2">No.</th><th rowspan="2" colspan="2">Drilling Mud Composition (%)</th><th colspan="6">Reading on the Viscometer (Pa)</th><th colspan="2">Viscosity η_{pl} η_s</th><th>Yield Point τ_y</th><th>Filtration</th><th rowspan="2">pH</th></tr>
<tr><th>600</th><th>300</th><th>200</th><th>100</th><th>6</th><th>3</th><th colspan="2">(mPa·s)</th><th>(Pa)</th><th>(cm^3)</th></tr>
<tr><td>1</td><td>XCD
Gelatinized starch
KCl
Carbonate blocker
Barite up to 1.1 g/cm3
Cuttings (miocene)</td><td>0.2
3.0
3.0
7.0

5.0</td><td>29.1</td><td>19.9</td><td>15.8</td><td>10.7</td><td>2.6</td><td>1.5</td><td>18</td><td>28.5</td><td>10.0</td><td>2.4</td><td>9.0</td></tr>
<tr><td rowspan="3">2</td><td rowspan="3">Drilling mud No. 1 + calcium hypochlorite</td><td>0.3</td><td>27.1</td><td>19.4</td><td>16.3</td><td>11.7</td><td>3.6</td><td>3.1</td><td>15</td><td>26.5</td><td>11.0</td><td>2.2</td><td>8.5</td></tr>
<tr><td>0.5</td><td>27.1</td><td>19.4</td><td>16.3</td><td>11.7</td><td>3.6</td><td>3.1</td><td>15</td><td>26.5</td><td>11.0</td><td>2.4</td><td>8.3</td></tr>
<tr><td>1.0</td><td>19.4</td><td>13.3</td><td>11.7</td><td>8.7</td><td>2.0</td><td>1.5</td><td>12</td><td>19</td><td>6.7</td><td>3.2</td><td>8.0</td></tr>
<tr><td rowspan="3">3</td><td rowspan="3">Drilling mud No. 1 + sodium persulfate</td><td>0.3</td><td>23.5</td><td>16.9</td><td>13.8</td><td>10.2</td><td>3.1</td><td>2.6</td><td>13</td><td>23</td><td>9.6</td><td>2.4</td><td>8.8</td></tr>
<tr><td>0.5</td><td>21.4</td><td>15.3</td><td>12.3</td><td>8.7</td><td>3.1</td><td>2.0</td><td>12</td><td>21</td><td>8.6</td><td>2.8</td><td>8.5</td></tr>
<tr><td>1.0</td><td>23.5</td><td>16.9</td><td>13.8</td><td>9.7</td><td>3.1</td><td>2.6</td><td>13</td><td>23</td><td>9.6</td><td>3.0</td><td>8.2</td></tr>
<tr><td rowspan="3">4</td><td rowspan="3">Drilling mud No. 1 + urea peroxide</td><td>0.3</td><td>28.1</td><td>20.4</td><td>17.4</td><td>13.3</td><td>4.6</td><td>4.1</td><td>15</td><td>27.5</td><td>12.0</td><td>2.2</td><td>8.5</td></tr>
<tr><td>0.5</td><td>21.4</td><td>14.3</td><td>11.7</td><td>8.2</td><td>2.0</td><td>1.5</td><td>14</td><td>21</td><td>6.7</td><td>2.4</td><td>8.3</td></tr>
<tr><td>1.0</td><td>19.9</td><td>13.8</td><td>11.2</td><td>8.2</td><td>2.0</td><td>1.5</td><td>12</td><td>19.5</td><td>7.2</td><td>2.8</td><td>8.1</td></tr>
<tr><td>5</td><td>Drilling mud No. 1 + amylase</td><td>0.3</td><td>11.2</td><td>7.1</td><td>6.1</td><td>4.6</td><td>1.5</td><td>1.0</td><td>8</td><td>11</td><td>2.9</td><td>34</td><td>8.7</td></tr>
</table>

Table 2. The rheological parameters of potassium-polymer drilling mud A with thinning additives at various temperature values.

Type of Added Agent	Plastic Viscosity (mPa·s)				Apparent Viscosity (mPa·s)				Yield Point (Pa)			
	20 °C	40 °C	60 °C	80 °C	20 °C	40 °C	60 °C	80 °C	20 °C	40 °C	60 °C	80 °C
+ 1.0% calcium hypochlorite	12	8	7	5	19	14	11.5	8.5	6.7	5.7	4.3	3.4
+1.0% urea peroxide	12	9	9	5	19.5	15	14.5	10.5	7.2	5.7	5.3	5.3
+ 1.0% sodium persulfate	13	13	12	13	23	19.5	18.5	17	9.6	6.2	6.2	3.8
+0.3% amylase	8	6	5	4	11	9	7.5	6.5	2.9	2.9	2.4	2.4

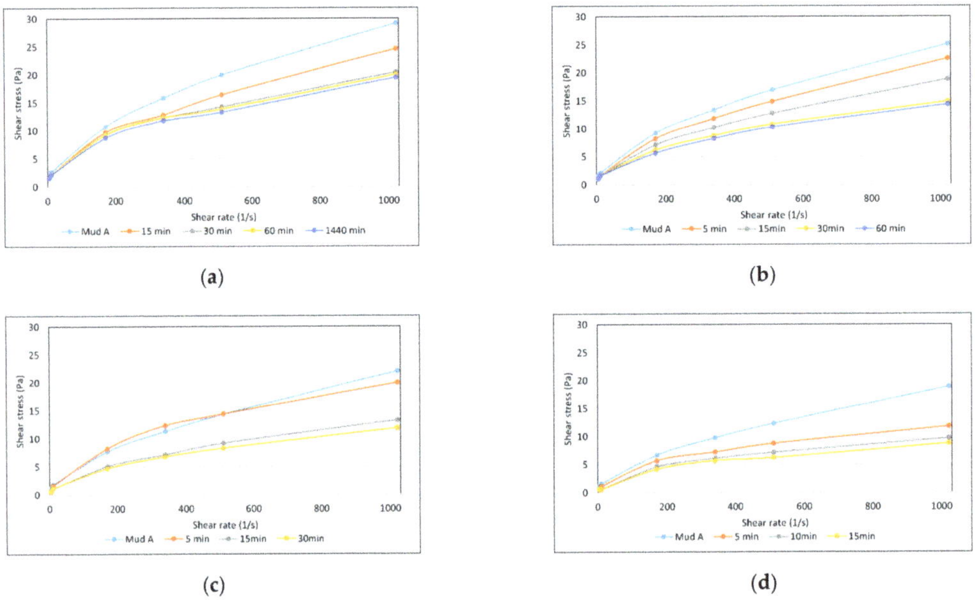

Figure 1. Flow curves of drilling mud A with 1.0% calcium hypochlorite: (**a**) at a temperature of 20 °C, (**b**) at a temperature of 40 °C, (**c**) at a temperature of 60 °C, (**d**) at a temperature of 80 °C.

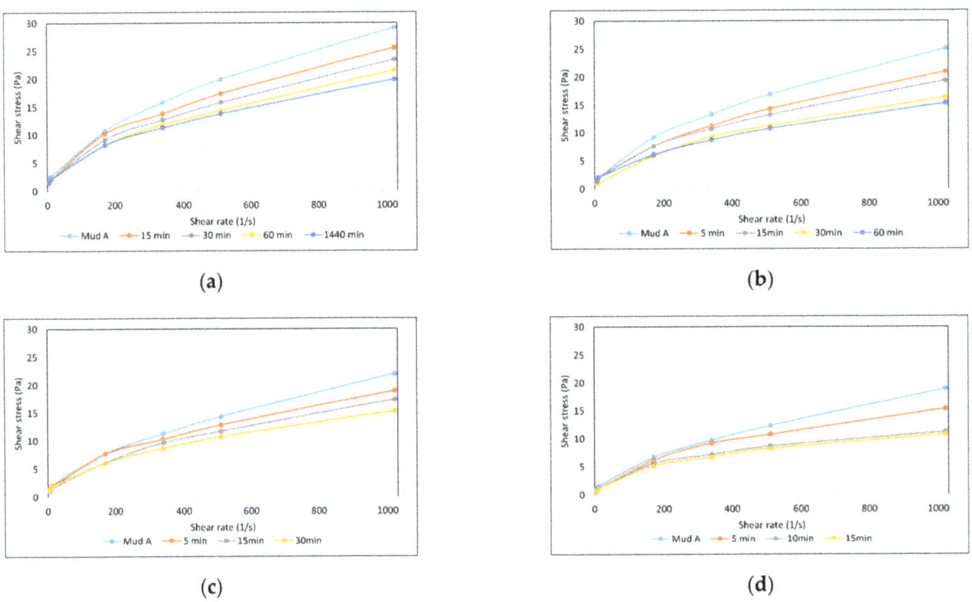

Figure 2. Flow curves of drilling mud A with the addition of 1.0% urea peroxide: (**a**) at 20 °C, (**b**) at 40 °C, (**c**) at 60 °C, (**d**) at 80 °C.

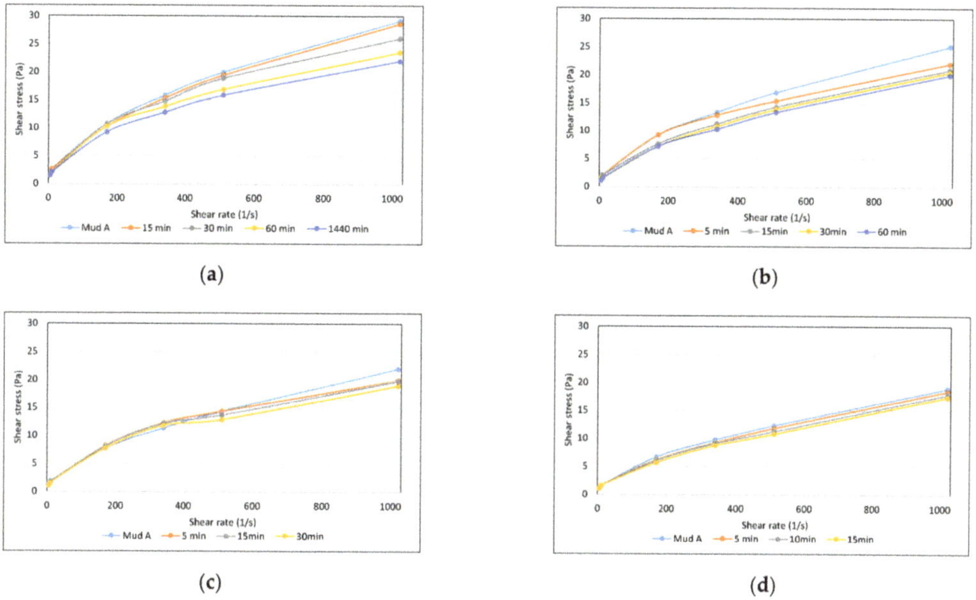

Figure 3. Flow curves of drilling mud A with 1.0% sodium persulfate: (**a**) at 20 °C, (**b**) at 40 °C, (**c**) at 60 °C, (**d**) at 80 °C.

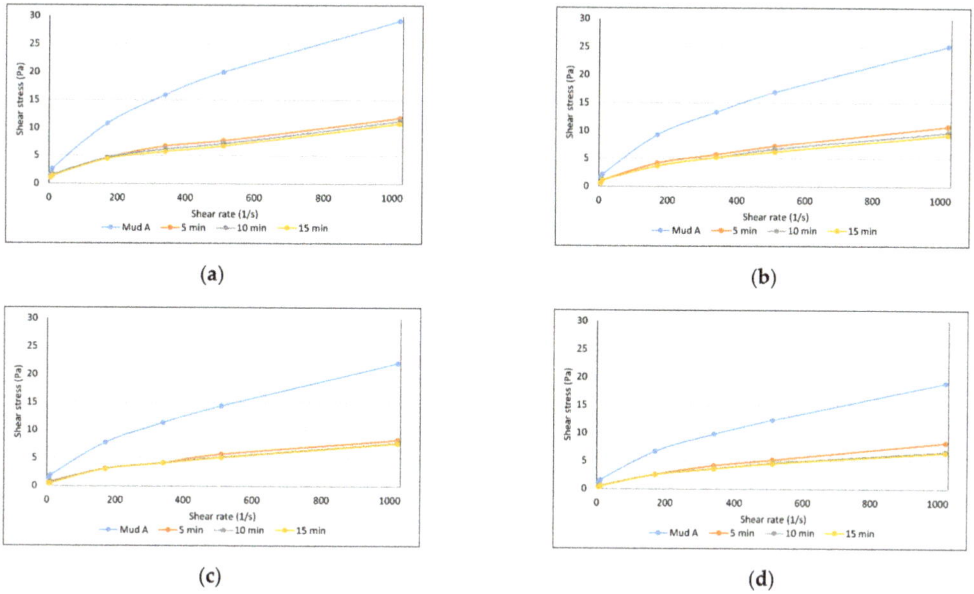

Figure 4. Flow curves of drilling mud A with the addition of 0.3% amylase: (**a**) at 20 °C, (**b**) at 40 °C, (**c**) at 60 °C, (**d**) at 80 °C.

Potassium-polymer drilling mud A at 20 °C was resistant to the action of oxidants, the addition of 10 g/dm^3 of calcium hypochlorite to the drilling mud caused its plastic viscosity to decrease from 18 to 12 mPa·s, the yield point from 10 to 6.7 Pa, and increased its filtration from 2.4 to 3.2 cm^3 (Table 1). The rate of the oxidation reaction at 20 °C did not change significantly over time (Figure 1), the greatest decrease in the viscosity of the drilling mud took place after about 1 h, and the further influence of calcium hypochlorite did not significantly affect a further decrease in rheological parameters. Increasing the temperature to 40 °C accelerated the process of decomposition of polymers contained in the drilling mud, the oxidation process was much faster, and the maximum thinning of the drilling mud took place after about 30 min of the effects of calcium hypochlorite. At the temperature of 80 °C, the process of decomposition of the polymers contained in the drilling mud was turbulent, after about 5 min the drilling mud thinned as a result of the oxidant and the further extension of the time of the oxidant did not significantly change the rheological parameters.

The effect of a 1% addition of urea peroxide on the properties of drilling mud A was marked by a reduction in the plastic viscosity down to 12 mPa·s and the yield point down to 7.2 Pa. The time required for the process of thinning the drilling mud at room temperature is about 24 h, and at the temperature of 40 °C the decomposition of the polymers contained in the drilling mud took place after about 30 min. However, at the temperature of 80 °C, the same thinning effect of the drilling mud was achieved after about 5 min.

Sodium persulfate showed a much weaker effect than urea peroxide and calcium hypochlorite. For the effective oxidation of the drilling mud with sodium persulfate, it was necessary to increase the temperature to about 60 °C. The drilling mud decomposition process at 60 °C was much faster and more effective. After the effect of 15 min of sodium persulfate, the plastic viscosity of the drilling mud decreased to 12 mPa·s, and the yield point to 6.7 Pa.

The action of amylase addition on the properties of the potassium-polymer drilling mud A is shown in Table 1. The decomposition of the polymers contained in the drilling mud A by means of amylase at a temperature of 20 °C was very fast, after about 5 min there was almost complete decomposition of the starch component of the drilling mud, which resulted in the reduction of the rheological parameters of the drilling mud. Extending the time of the enzyme's action and increasing the temperature did not significantly increase the decomposition effect of the drilling mud.

3.2. *Influence of Oxidants and Enzymes on the Properties of Potassium-Polymer Drilling Mud B*

Drilling mud B XCD on the biopolymer matrix and the cellulose polymer colloid weighted with barite to a density of 1330 kg/m^3 has a plastic viscosity of 41 mPa·s, a yield point of 18.2 Pa and a filtration of 4.4 cm^3. The influence of oxidants on the rheological parameters of potassium-polymer drilling mud B at 20 °C is shown in Tables 3 and 4.

The highest degree of thinning of the drilling mud with calcium hypochlorite was obtained by using its addition in the amount of 10 g/dm^3. With such a concentration of calcium hypochlorite in the drilling mud, the plastic viscosity decreased from 41 to 20 mPa·s, and the yield point from 18.2 to 8.1 Pa. The drilling mud filtration increased from 4.4 to 6.2 cm^3, and the pH value decreased to a value of 7.5. Under the conditions of elevated temperature, i.e., 40 and 60 °C, the oxidation process took place to a greater extent. The best results of the drilling mud thinning were obtained by carrying out the oxidation process at 80 °C for 15 min (Figure 5). The plastic viscosity of the drilling mud decreased to 6 mPa·s and the yield point up to 3.3 Pa (Table 4).

The most effective oxidizing action in relation to drilling mud B at the temperature of 20 °C was shown by urea peroxide (Figure 6), which, with the addition of 10 g/dm^3 after an exposure time of 24 h caused the reduction of its plastic viscosity to 13 mPa·s and the yield point to 5.7 Pa. Increasing the temperature improved the efficiency of the agent in relation to oxidation carried out at room temperature.

Table 3. The properties of potassium-polymer drilling mud B with the addition of oxidizing agents and cellulase.

No.	Drilling Mud Composition (%)		Reading on the Viscometer (Pa) 600	300	200	100	6	3	Viscosity η_{pl} (mPa·s)	η_s (mPa·s)	Yield Point τ_y (Pa)	Filtration (cm^3)	pH
1	XCD CMC KCl PHPA Carbonate blocker Barite up to 1.3 g/cm^3 Cuttings (miocene)	0.2 2.0 5.0 0.2 7.0 5.0	61.3	40.3	34.2	20.4	4.1	3.1	41	60	18.2	4.4	9.6
2	Drilling mud No. 1 + calcium hypochlorite	0.3	46.5	30.1	23.5	15.3	3.6	2.6	32	45.5	12.9	4.4	8.0
		0.5	38.3	25.0	19.4	12.3	2.6	2.0	26	37.5	11.0	4.8	7.7
		1.0	29.1	18.9	14.8	9.7	2.6	2.0	20	28.5	8.1	6.2	7.5
3	Drilling mud No. 1 + sodium persulfate	0.3	43.4	28.6	22.5	14.8	3.6	2.6	29	42.5	12.9	4.6	8.6
		1.0	39.3	25.5	19.9	13.3	3.6	2.6	27	38.5	11.0	4.8	8.5
		1.5	44.4	28.6	22.5	14.8	3.6	2.6	31	43.5	12.0	5.0	7.8
4	Drilling mud No. 1 + urea peroxide	0.8	22.0	14.3	10.7	7.7	2.0	1.5	15	21.5	6.2	4.8	7.5
		1.0	19.4	12.8	9.7	6.6	1.5	1.0	13	19	5.7	5.2	8.2
		1.5	14.3	9.2	7.7	5.1	1.5	1.0	10	14	3.8	5.8	8.1
5	Drilling mud No. 1 + cellulase + citric acid	 0.3 0.1	19.4	12.8	10.2	6.6	1.7	1.4	13	19	5.7	8.4	8.7

Table 4. The rheological parameters of potassium-polymer drilling mud B with thinning additives at various temperature values.

Type of Added Agent	Plastic Viscosity (mPa·s) 20 °C	40 °C	60 °C	80 °C	Apparent Viscosity (mPa·s) 20 °C	40 °C	60 °C	80 °C	Yield Point (Pa) 20 °C	40 °C	60 °C	80 °C
+ 1.0% calcium hypochlorite	20	14	11	6	28.5	21	13.5	9.5	8.1	6.7	2.4	3.3
+1.0% urea peroxide	13	9	8	6	19	13.5	12.5	10.5	5.7	4.3	4.3	4.3
+ 1.0% sodium persulfate	27	13	11	8	38.5	18	14.5	11.5	11.0	4.8	3.3	3.3
+0.3% cellulase	13	8	7	8	19	13.5	12	12	5.7	5.3	4.8	3.8

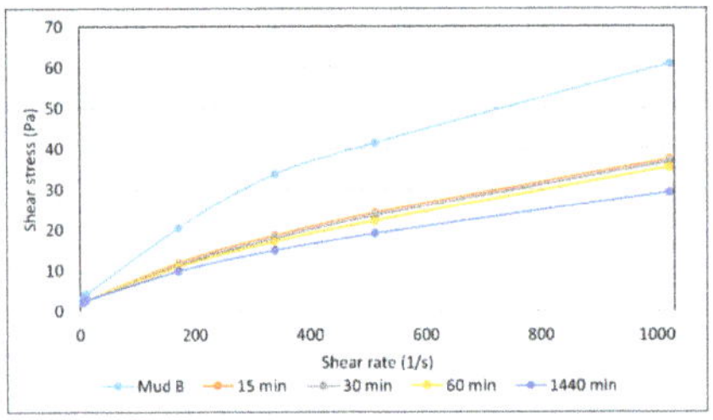

(a)

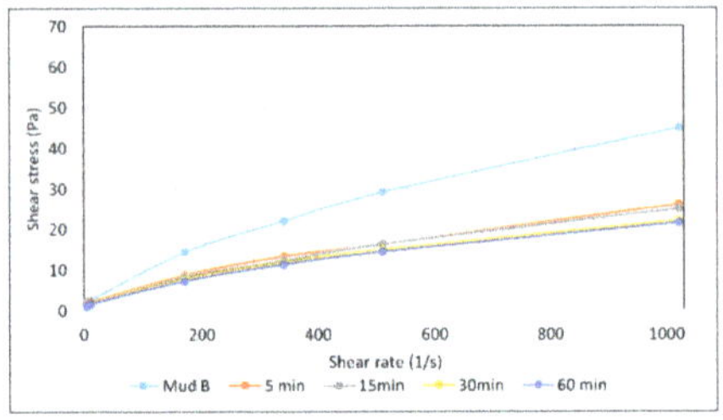

(b)

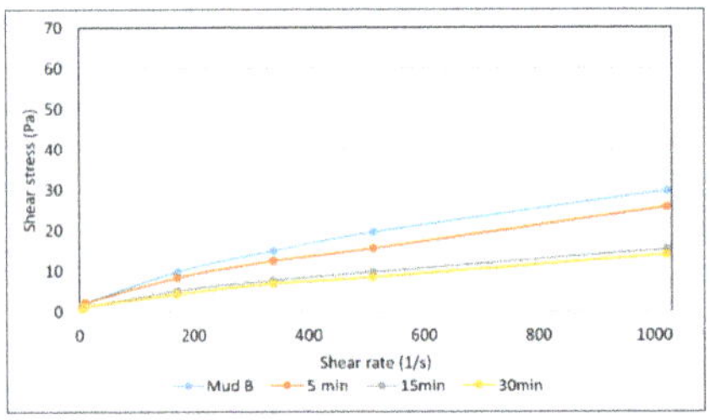

(c)

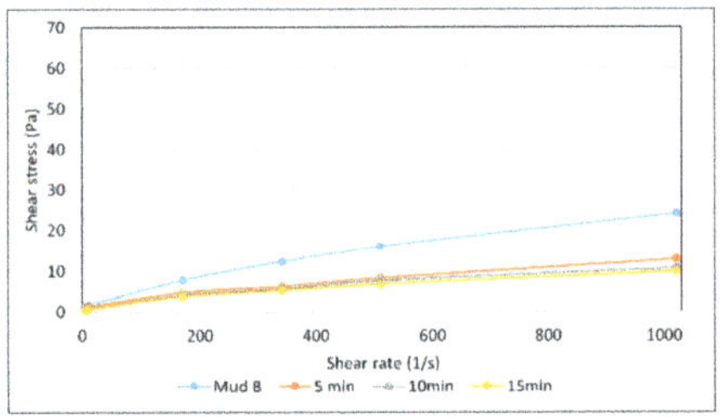

(d)

Figure 5. Flow curves of drilling mud B with 1.0% calcium hypochlorite: (**a**) at a temperature of 20 °C, (**b**) at a temperature of 40 °C, (**c**) at a temperature of 60 °C, (**d**) at a temperature of 80 °C.

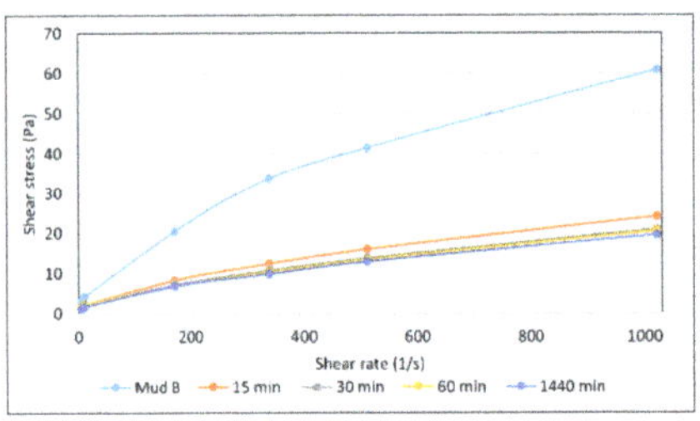

(a)

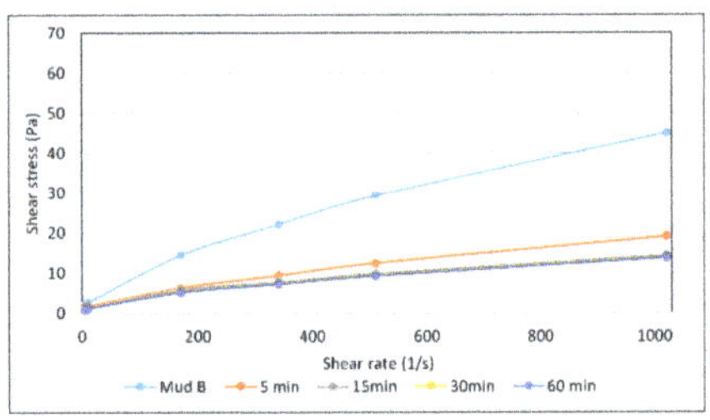

(b)

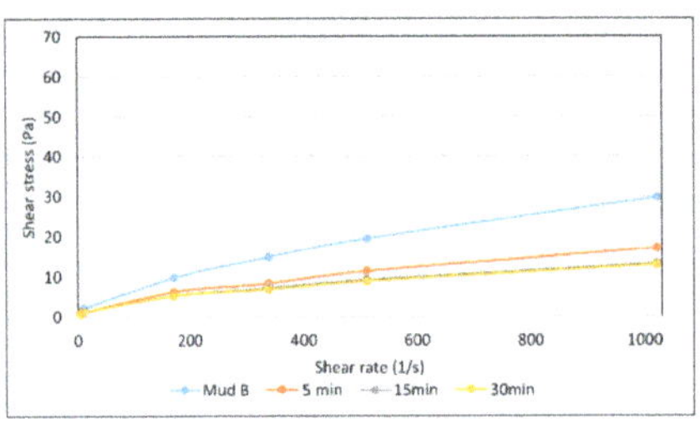

(c)

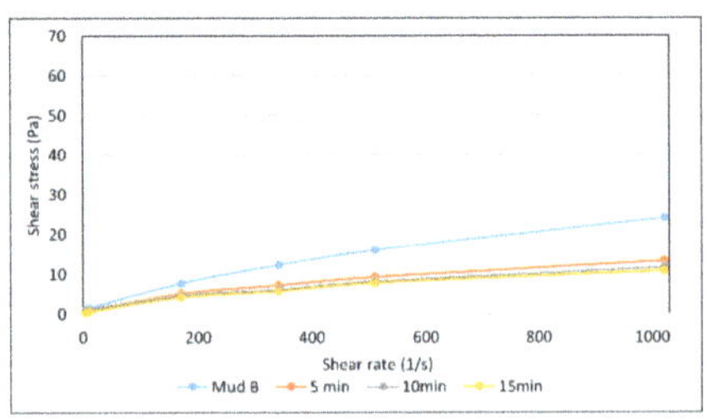

(d)

Figure 6. Flow curves of drilling mud B with the addition of 1.0% urea peroxide: (**a**) at 20 °C, (**b**) at 40 °C, (**c**) at 60 °C, (**d**) at 80 °C.

The process of oxidating the drilling mud at 20 °C with sodium persulfate does not significantly change its rheological parameters (Figure 7). The introduction of 10 g/dm^3

of sodium persulfate at 20 °C reduces the plastic viscosity of the drilling mud from 41 to 27 mPa·s and the yield point from 18.2 to 11 Pa (Table 3).

The action of cellulase on the drilling mud only partially thins it by decomposing cellulose compounds (Figure 8). The addition of 0.3 vol.% of the enzyme reduces the plastic viscosity from 41 to 13 mPa·s and the yield point from 18.2 to 5.7 Pa (Table 3).

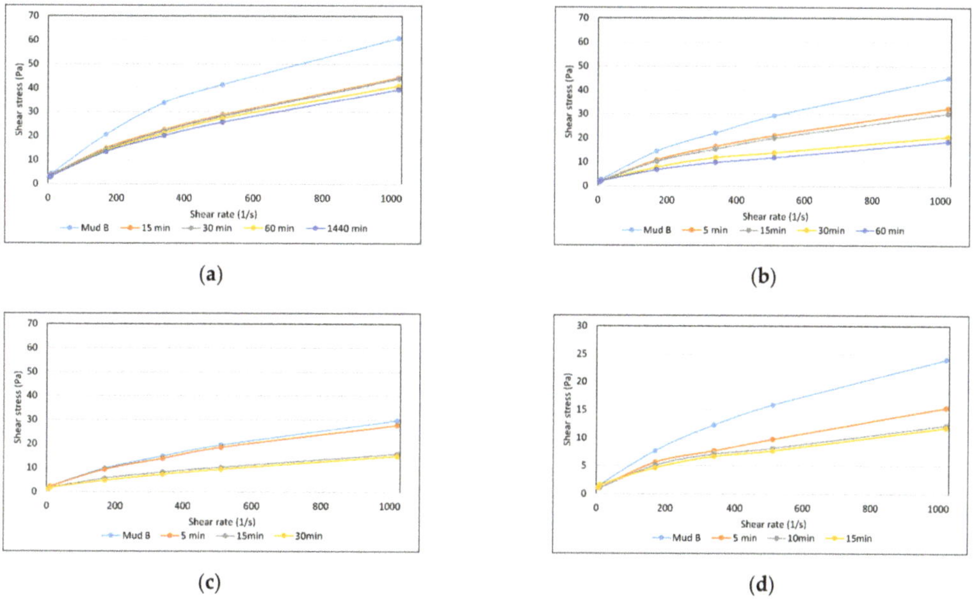

Figure 7. Flow curves of drilling mud B with 1.0% sodium persulfate: (**a**) at 20 °C, (**b**) at 40 °C, (**c**) at 60 °C, (**d**) at 80 °C.

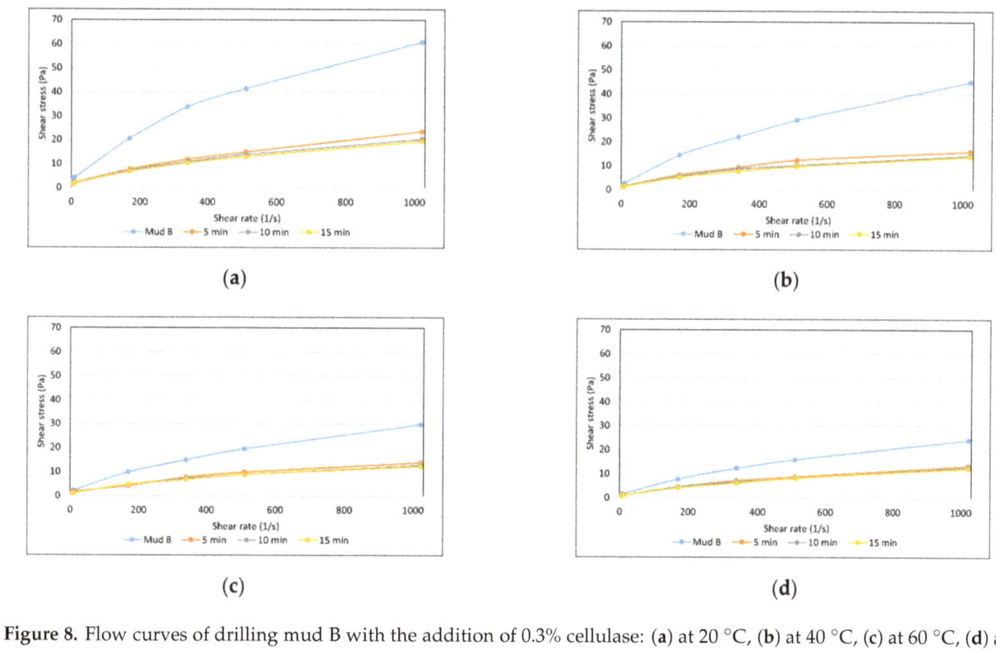

Figure 8. Flow curves of drilling mud B with the addition of 0.3% cellulase: (**a**) at 20 °C, (**b**) at 40 °C, (**c**) at 60 °C, (**d**) at 80 °C.

3.3. The Influence of Oxidants and Enzymes on the Properties of Salt-Barite Drilling Mud C

Salt-barite drilling mud C based on the XCD biopolymer and cellulose colloid was salted with 33% of NaCl, weighted with barite to a weight of 1500 kg/m^3 and contaminated with calcium and magnesium ions. The prepared drilling mud has a plastic viscosity of 57 mPa·s and a yield point of 23 Pa. The influence of polymer degrading agents on the rheological parameters of the drilling mud at 20 °C is shown in Tables 5 and 6.

Table 5. The properties of salt-barite drilling mud C, density 1530 kg/m^3, with the addition of oxidizing agents and cellulase enzyme.

No.	Drilling Mud Composition (%)		Reading on the Viscometer						Viscosity		Yield Point	Filtration	pH
									η_{pl}	η_s	τ_y		
			600	300	200	100	6	3	(mPa·s)		(Pa)	(cm^3)	
1	NaCl XCD CMC PHPA Carbonate blocker Barite up to 1.5 g/cm^3 Cuttings (miocene) CaCl2 MgCl2	33.0 0.2 2.0 0.2 7.0 5.0 1.0 0.5	82.7	53.6	41.4	26.6	5.1	3.6	57	81	23.0	1.6	9.2
2	Drilling mud No. 1 + calcium hypochlorite	0.3	44.4	27.6	20.9	12.8	2.6	2.0	33	43.5	10.1	1.6	9.0
		0.5	39.8	25.0	18.9	11.7	2.6	1.5	29	39	9.6	1.6	7.1
		1.0	34.2	20.9	15.8	9.7	2.0	1.5	26	33.5	7.2	2.4	6.9
3	Drilling mud No. 1 + sodium persulfate	0.3	46.5	29.6	22.5	14.3	3.1	2.0	33	45.5	12.0	1.8	7.6
		0.5	45.5	28.6	22.0	13.8	3.1	2.0	33	43.5	11.0	2.0	6.5
		1.0	44.9	27.6	21.4	12.8	3.1	2.0	34	44	9.6	2.2	6.3
4	Drilling mud No. 1 + urea peroxide	0.5	23.0	14.8	11.2	8.7	1.5	1.0	16	22.5	6.2	32	7.2
		0.8	18.4	11.7	9.2	5.1	1.0	1.0	13	17	4.8	58	6.9
		1.0	14.8	9.7	6.6	3.1	1.0	0.5	10	14.5	4.3	66	6.3
5	Drilling mud No. 1 + cellulase	0.3	37.3	23.5	17.9	11.2	2.3	1.8	27	36.5	9.1	4.2	9.1
6	Drilling mud No. 1 + cellulase + citric acid	0.3 0.1	32.2	20.9	15.8	9.7	2.0	1.5	22	31.5	9.1	6.4	8.3

Table 6. The rheological parameters of salt-barite drilling mud C with thinning additives at various temperature values.

Type of Added Agent	Plastic Viscosity (mPa·s)				Apparent Viscosity (mPa·s)				Yield Point (Pa)			
	20 °C	40 °C	60 °C	80 °C	20 °C	40 °C	60 °C	80 °C	20 °C	40 °C	60 °C	80 °C
+ 1.0% calcium hypochlorite	26	15	13	14	33.5	19.5	17	18	7.2	4.3	3.8	3.8
+1.0% urea peroxide	10	9	8	5	14.5	11.5	9.5	6	4.3	2.4	1.4	1.0
+ 1.0% sodium persulfate	34	21	15	13	44	27.5	21	18	9.6	6.2	5.7	4.8
+1.0% cellulase	22	22	14	13	31.5	31	20	17.5	9.1	8.6	5.7	4.3

After adding calcium hypochlorite in an amount of 1.0 vol.% there is a significant decrease in the rheological parameters of the drilling mud. The plastic viscosity is reduced to 26 mPa·s and the yield point is reduced to 7.2 Pa. The filtration of the drilling mud increased to 2.4 cm^3 and the pH value decreased to 6.9. The oxidation reaction at an elevated temperature (Figure 9) is accompanied by a greater reduction in the drilling mud viscosity and the decomposition of polymers contained in the drilling mud occurs much faster than at 20 °C. Increasing the temperature of the reaction environment to 40 °C accelerates the process of decomposition of the drilling mud. At 60 °C, the oxidation process is similar, the maximum thinning of the drilling mud takes place after 15 min of the effect of calcium hypochlorite (Figure 9).

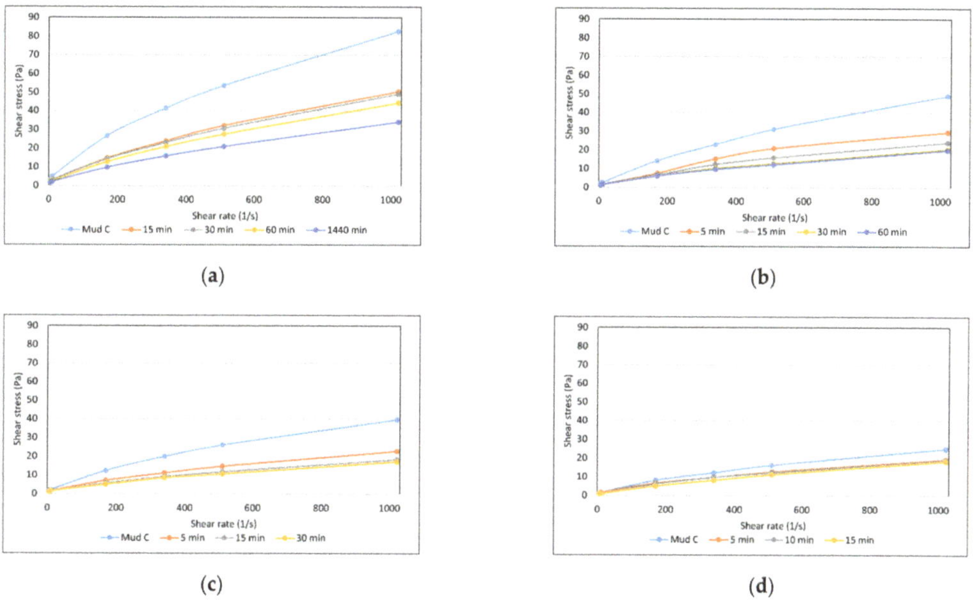

Figure 9. Flow curves of drilling mud C with 1.0% calcium hypochlorite: (**a**) at a temperature of 20 °C, (**b**) at a temperature of 40 °C, (**c**) at a temperature of 60 °C, (**d**) at a temperature of 80 °C.

The effectiveness of sodium persulfate in a salt-barite drilling mud at 20 °C is much lower than that of calcium hypochlorite. The addition of sodium persulfate to the drilling mud C in the amount of 10 g/dm^3 reduces the plastic viscosity of the drilling mud from 57 to 34 mPa·s and the yield point from 23 to 9.6 Pa (Table 5). The course of the oxidation reaction does not change significantly over time (Figure 10), the greatest decrease in the viscosity of the drilling mud takes place after approx. 1 h, and the further effect of sodium persulfate does not significantly affect a further decrease in rheological parameters. The use of sodium persulfate in the degradation of the polymers contained in drilling mud salt-barite C necessitates a treatment at 60 or 80 °C, the oxidation of the drilling mud at a temperature of 20 °C is inefficient and very slow (Figure 10).

An effective means for the distribution of polymers in salt-barite drilling mud C was urea peroxide, which with the addition of 10 g/dm^3 to the drilling mud after approx. 30 min (Figure 11) causes almost complete decomposition of polymers, which lowers its rheological parameters: plastic viscosity from 57 to 10 mPa·s, and yield point from 20 to 4.3 Pa. Carrying out the process of decomposition of the drilling mud at higher temperatures slightly increases the effectiveness of its operation (Figure 11). However, the use of urea in the environment of full salinity does not require raising the temperature and extending the time of its effect. The process can be carried out at a temperature of about

20 °C. The disadvantage of thinning the drilling mud with urea peroxide is that it causes a significant increase in filtration. After the action of the agent on the drilling mud, it is completely degraded, thus the filtration value increases from 1.6 to 66 cm^3 (Table 5).

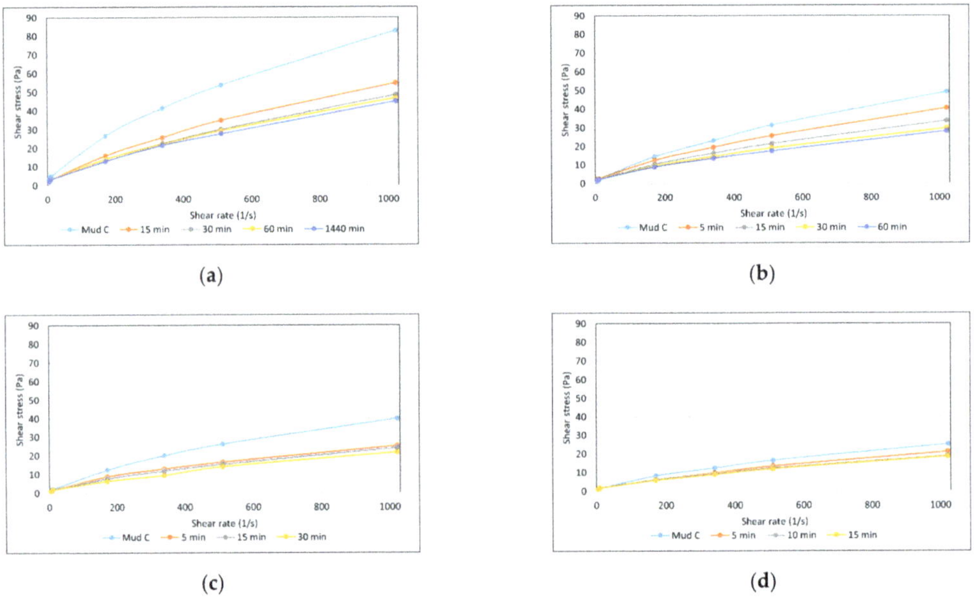

Figure 10. Flow curves of drilling mud C with the addition of 1.0% sodium persulfate: (**a**) at 20 °C, (**b**) at 40 °C, (**c**) at 60 °C, (**d**) at 80 °C.

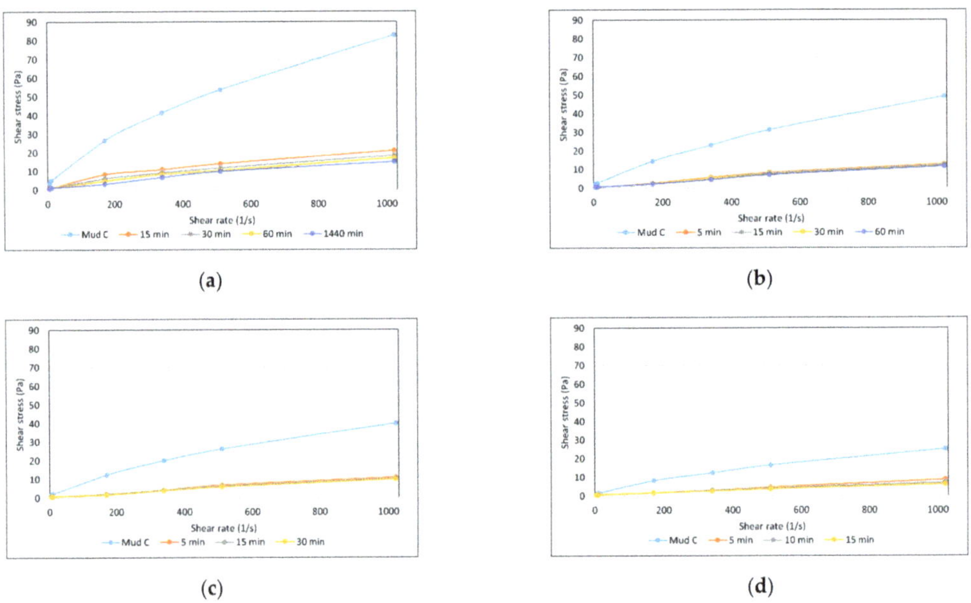

Figure 11. Flow curves of drilling mud C with 1.0% urea peroxide: (**a**) at 20 °C, (**b**) at 40 °C, (**c**) at 60 °C, (**d**) at 80 °C.

The use of the enzyme to decompose salt-barite drilling mud C allows for partial thinning of the drilling mud (Figure 12), the addition of 0.3% cellulase enzyme at 20 °C reduces the plastic viscosity of the drilling mud to 27 mPa·s and the yield point to 9.1 Pa (Table 5). When the pH of the drilling mud drops below 9, the efficiency of the enzyme activity increases. After its action, the plastic viscosity drops to 22 mPa·s, and the apparent viscosity to 31.5 mPa·s.

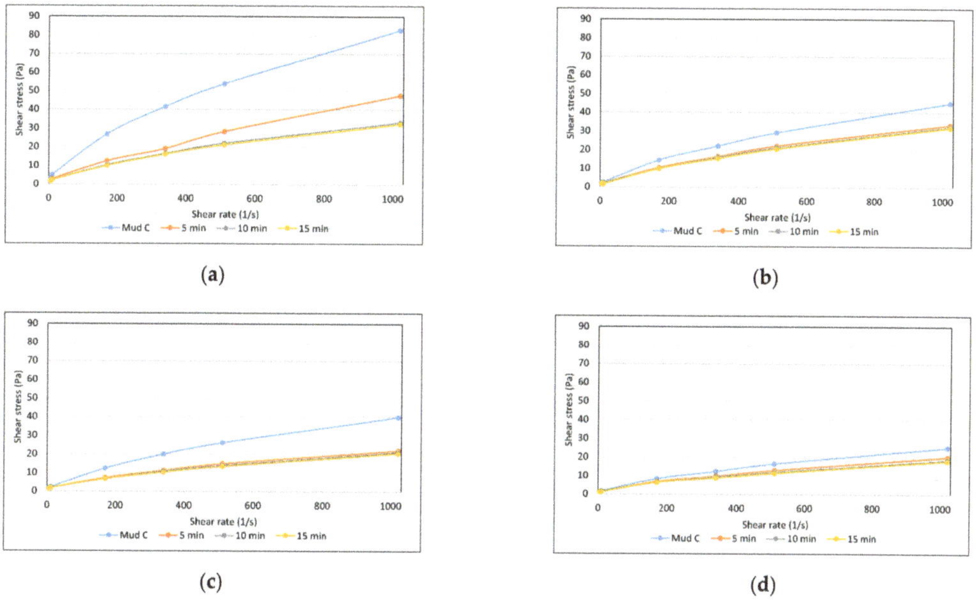

Figure 12. Flow curves of drilling mud C with the addition of 0.3% cellulase: (**a**) at 20 °C, (**b**) at 40 °C, (**c**) at 60 °C, (**d**) at 80 °C.

The cake produced from the potassium-polymer drilling mud taken from the K-1 borehole was characterized by a density of 2.21 g/cm^3 and a hydration level of 52.88% (Figure 13a). Attempts were made to remove the cakes produced from the K-1 drilling mud with the fluids shown in Table 7. On the basis of the obtained results (Figure 14), it was found that the most effective agents in removing the cakes generated from the K-1 drilling mud were oxidants (urea, calcium hypochlorite). Eddy eroding at 50 °C and a differential pressure of 0.7 MPa with the help of fluid No. 2 with the addition of urea peroxide, it allowed to remove 86% of the cake (Figure 13c), while the efficiency of cake removal under the same conditions with fluid No. 1 with the addition of calcium hypochlorite allowed for the removal of 90% of the filter cake (Figure 13b).

Table 7. Selected fluid compositions for eroding filter cakes produced from drilling mud taken from the K-1 borehole.

Fluid 1	Fluid 2	Fluid 3
Water KCl—5% calcium hypochlorite—1%	Water KCl—5% urea peroxide—1%	Water KCl—5% cellulase—1%

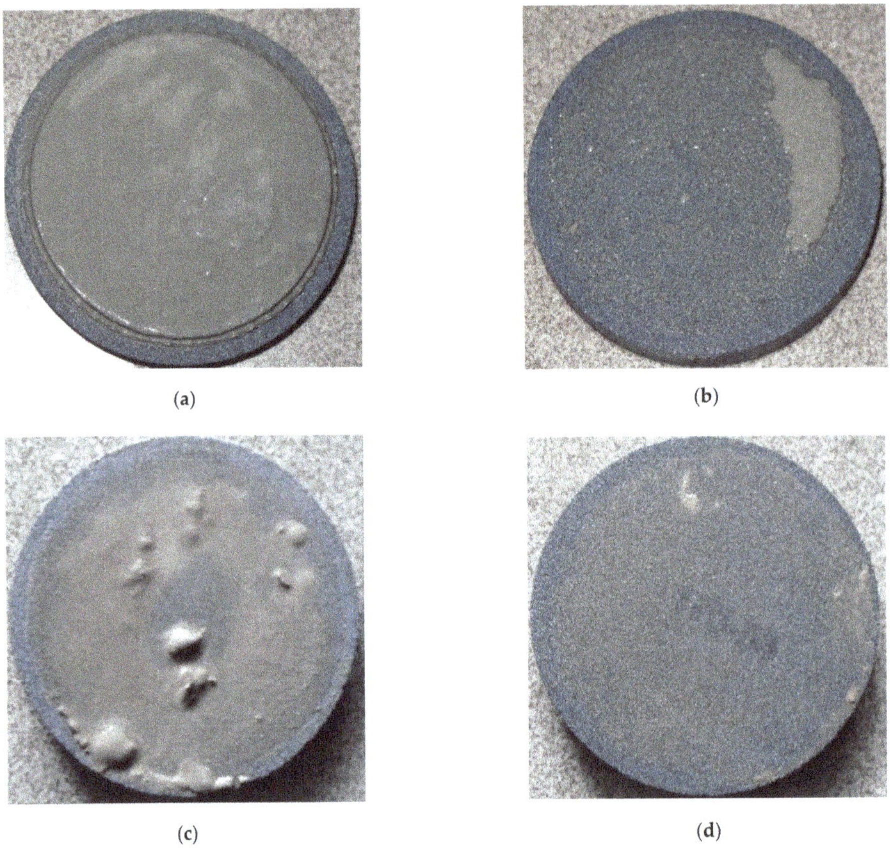

Figure 13. Filter cake produced from the drilling mud taken from the K-1 borehole: (**a**) before eroding, (**b**) eroded with fluid No. 1, (**c**) eroded with fluid No. 2, (**d**) eroded with fluid No. 3.

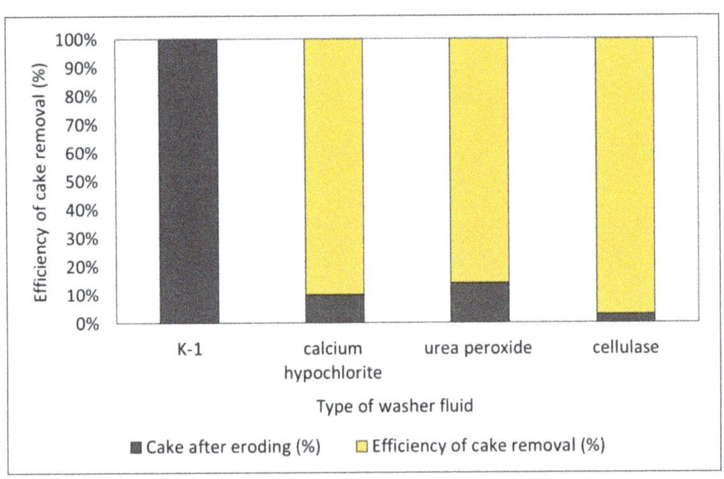

Figure 14. The efficiency of the removal of filter cakes produced from the drilling mud taken from the K-1 borehole, depending on the fluids used for eroding.

High efficiency of cake removal was achieved with fluid No. 3 with the addition of the enzyme, with this type of fluid 97% of the cake was removed (Figure 13d).

4. Conclusions

The research concerned the recognition of the possibility of using enzymatic agents and oxidants for the decomposition of polymers used in drilling mud compositions. The obtained results of laboratory tests indicate the possibility of selecting, depending on the borehole conditions, the appropriate type of agents (enzymes or oxidants) that cause the decomposition of agents that impart viscosity to drilling muds. The effectiveness of selected enzymes and oxidants has been confirmed in the conducted tests, both in polymer solutions and drilling muds prepared according to the compositions used by the drilling services. Studies have shown that at low temperatures it is more advantageous to use enzyme agents, while oxidizing agents should be used at elevated temperatures.

Cellulase and amylase enzyme agents used in the research can be used to decompose colloids used in the composition of drilling muds. In the case of using enzymes in drilling muds at a pH level above 9.0 and with significant salinity, it may be necessary to increase the concentration of the enzyme or adjust the pH with citric acid. The presence of NaCl salt in the drilling muds in a concentration corresponding to a saturated solution, or calcium and magnesium chloride in the concentrations usually found in drilling muds, does not cause a significant reduction in enzyme activity at pH levels below 9. Among the oxidants used in the research, the most effective polymer decomposition was recorded in the case of calcium hypochlorite, urea peroxide and sodium persulfate. The effectiveness of the oxidizing agents in drilling muds does not depend on the pH value, and they show unchanged activity over a wide pH range. The increase in temperature and extension of the exposure time increases the effectiveness of the oxidants. For the decomposition of polymers in the composition of drilling muds, oxidizing agents and enzymes should be selected, on the basis of laboratory tests, according to the type of colloids and other polymers used in the drilling muds, and the most optimal conditions for their decomposition should be determined. The most effective thinning agent for salt-barite drilling muds is urea peroxide, which, regardless of the sodium chloride concentration in the drilling mud, showed the highest oxidizing efficiency. The oxidation reaction carried out in an environment of full salinity does not require increasing the temperature and extending the exposure time of urea peroxide. Calcium hypochlorite shows less efficiency as its effect depends on the sodium chloride content in the drilling mud, while sodium persulfate shows the weakest oxidizing effect. The use of cellulase and amylase enzyme solutions, as well as oxidant solutions (urea peroxide, sodium persulfate and calcium hypochlorite), as shown by the results of leaching of clay cakes in laboratory conditions, significantly facilitates the removal of filter cakes from drilling muds. The performed tests allowed for the development of methods of preparing the borehole for the cementing procedure with the use of oxidants or enzymes:

- Thinning of the drilling muds by the addition of enzymatic or oxidizing agents before cementing. Due to the application of such treatment, apart from thinning the drilling mud, the polymers contained in the solidified drilling mud at the borehole wall and clay cake are partially decomposed, which will contribute to better displacement of the drilling mud by the cement paste.
- Application of washer fluids in the form of 1% solutions of enzymes (cellulase) or oxidants (urea peroxide and calcium hypochlorite) prior to the cementing procedure in order to disassemble and then remove the remains of drilling mud and cake from the borehole walls.

In the holes where the expected temperature does not exceed 80 °C, it is recommended to use enzymatic agents (amylase, cellulase) to thin the drilling muds and remove cakes, while in the holes where the temperature exceeds 60 °C, oxidizing agents (urea peroxide, calcium hypochlorite) should be used.

Author Contributions: Conceptualization, M.U., G.Z., S.B. and B.J.; methodology, G.Z. and B.J.; validation, G.Z., S.B. and B.J.; formal analysis, M.U, G.Z. and S.B.; investigation, M.U., G.Z., B.J.; resources, G.Z. and S.B.; data curation, G.Z. and B.J.; writing—original draft preparation, M.U. and G.Z.; writing—review and editing, B.J. and S.B.; visualization, G.Z. and B.J.; supervision, M.U.; project administration, M.U. and G.Z. All authors have read and agreed to the published version of the manuscript.

Funding: This research was funded by the Ministry of Science and Higher Education—Archival no.: 0003/KW, and Oil and Gas Institute—National Research Institute—Archival no.: 1286/KW.

Institutional Review Board Statement: Not applicable.

Informed Consent Statement: Not applicable.

Data Availability Statement: Not applicable.

Conflicts of Interest: The authors declare no conflict of interest.

References

1. Uliasz, M.; Zima, G.; Błaż, S.; Jasiński, B. Ocena właściwości cieczy wiertniczych w aspekcie zapobiegania migracji gazu w otworach na przedgórzu Karpat. *Nafta-Gaz* **2015**, *1*, 11–17.
2. Gawlik, P.; Szymczak, M. Migracje gazowe w przestrzeniach międzyrurowych otworów realizowanych na przedgórzu Karpat. *Nafta-Gaz* **2006**, *62*, 349–358.
3. Zima, G. Rozkład koloidów ochronnych płuczek wiertniczych środkami enzymatycznymi. *Nafta-Gaz* **2011**, *4*, 249–253.
4. Błaż, S. Dobór środków chemicznych do degradacji polimerów i koloidów ochronnych w płuczkach wiertniczych. *Nafta-Gaz* **2009**, *5*, 371–383.
5. Block, P.A.; Brown, R.A.; Robinson, D. Novel Activation Technologies for Sodium Persulfate In Situ Chemical Oxidation. In Proceedings of the Fourth International Conference on the Remediation of Chlorinated and Recalcitrant Compounds, Monterey, CA, USA, 24–27 May 2004.
6. Kameda, E.; Langone, M.A.; de Queiroz Neto, J.C.; Zarur Coelho, M.A. Removal of Polymeric Filter Cake in Petroleum Wells by Enzymatic Treatment. In Proceedings of the 4th Mercosur Congress on Process Systems Engineering and 2nd Mercosur Congress on Chemical Engineering, Rio de Janeiro, Brazil, 14–18 August 2005.
7. Jasiński, B. Wpływ oksydantów na wielkość sedymentacji fazy stałej w zasolonych płuczkach wiertniczych. *Nafta-Gaz* **2012**, *9*, 602–610.
8. Leal, A.B.; Barroso, A.L.; Miranda, X.; Flores, L.; Medeiros, G.; Marcelino, C.; Moranezi, L.; Pereira, R.A.; Oliveira, P.S.; Cândido, H.B. Reservoir Drilling and Completion Best Practices: Well Productivity Assessment Applying Drill in Fluid, Chelant/Enzyme Breaker System and Stimulation Design on Onshore Well BHT Scenario in Brazil. In Proceedings of the Offshore Technology Conference, Houston, TX, USA, 6–9 May 2019. [CrossRef]
9. Nasr-El-Din, H.A.; Al-Otaibi, M.B.; Al-Qahtani, A.A.; Al-Fuwaires, O.A. Filter-cake cleanup in MRC wells using enzyme/surfactant solutions. In Proceedings of the SPE International Symposium and Exhibition on Formation Damage Control, Lafayette, LA, USA, 15 February 2006. [CrossRef]
10. Zhang, Y.; Wang, H.; Liu, G.; Zhu, Z.; Duan, Y.; Pang, Q.; Zhou, X. Applying Biological Enzyme to Remove Plugging in Screen Pipe Completed Horizontal Wells in Volcanic Gas Reservoir. In Proceedings of the International Petroleum Technology Conference, Beijing, China, 28 March 2013. [CrossRef]
11. Al-khaldi, M.H.; Ghosh, B.; Ghosh, D. A Novel Enzyme Breaker for Mudcake Removal in High Temperature Horizontal and Multi-lateral Wells. In Proceedings of the SPE Asia Pacific Oil and Gas Conference and Exhibition, Jakarta, Indonesia, 20–22 September 2011. [CrossRef]
12. Yong, M.; Rong, C.M.; Yang, G.; Qing, S.; Li, L. How to Evaluate the Effect of Mud Cake on Cement Bond Quality of Second Interface? In Proceedings of the SPE/IADC Middle East Drilling and Technology Conference, Cairo, Egypt, 22–24 October 2007. [CrossRef]
13. Vakilinia, B. Experimental Investigation of Formation Damage Reduction: Mud Cake Removal and Mud Filtration Treatment using Ultrasonic Wave Radiation. In Proceedings of the SPE Annual Technical Conference and Exhibition, San Antonio, TX, USA, 8–10 October 2012. [CrossRef]
14. Saldungaray, P.M.; Troncoso, J.C.; Santoso, B.T. Simultaneous Gravel Packing and Filter Cake Removal in Horizontal Wells Applying Shunt Tubes and Novel Carrier and Breaker Fluid. In Proceedings of the SPE Middle East Oil Show, Manama, Bahrain, 17–20 March 2001. [CrossRef]
15. Battistel, E.; Cobianco, S.; Bianchi, D.; Fornaroli, M.; Guglielmetti, G. Enzyme Breakers for Chemically Modified Starches. In Proceedings of the SPE European Formation Damage Conference, Sheveningen, The Netherlands, 25–27 May 2005. [CrossRef]
16. Kameda, E.; de Queiroz Neto, J.C.; Langone, M.A.; Coelho, M.A.Z. Removal of polymeric filter cake in petroleum wells: A study of commercial amylase stability. *J. Pet. Sci. Eng.* **2007**, *59*, 263–270. [CrossRef]

17. Chen, E.; Wang, C.; Meng, R. A new type of cementation flushing fluid for efficiently removing wellbore filter cake. *Nat. Gas. Ind. B* **2015**, *2*, 455–460. [CrossRef]
18. Siddig, O.; Mahmoud, A.A.; Elkatatny, S. A review of different approaches for water-based drilling fluid filter cake removal. *J. Pet. Sci. Eng.* **2020**, *192*, 107346. [CrossRef]
19. Battistel, E.; Bianchi, D.; Fornaroli, M.; Cobianco, S. Enzymes breakers for viscosity enhancing polymers. *J. Pet. Sci. Eng.* **2011**, *77*, 10–17. [CrossRef]
20. Nasr-El-Din, H.A.; Al-Otaibi, M.B.; Al-Qahtani, A.A.; Samuel, M.M. An Effective Fluid Formulation to Remove Drilling Fluid Mud Cake in Horizontal and Multi-Lateral Wells. *SPE Drill. Complet.* **2007**, *22*, 26–32. [CrossRef]
21. Błaż, S.; Zima, G.; Jasiński, B.; Kremieniewski, M. Invert Drilling Fluids with High Internal Phase Content. *Energies* **2021**, *14*, 4532. [CrossRef]
22. Zielińska, D.; Rydzkowski, T.; Thakur, V.K.; Borysiak, S. Enzymatic engineering of nanometric cellulose for sustainable polypropylene nanocomposites. *Ind. Crop. Prod.* **2020**, *161*, 113188. [CrossRef]
23. Uliasz, M.; Zima, G.; Rzepka, M.; Dębińska, E. Role of chemicals in development of technological properties of the drilling fluids Rola środków chemicznych w kształtowaniu właściwości technologicznych cieczy wiertniczych. *Przemysł Chem.* **2015**, *5*, 715–722. [CrossRef]
24. Bielewicz, D. *Płyny Wiertnicze*; Wydawnictwo Akademii Górniczo-Hutniczej: Kraków, Poland, 2009.

MDPI
St. Alban-Anlage 66
4052 Basel
Switzerland
Tel. +41 61 683 77 34
Fax +41 61 302 89 18
www.mdpi.com

Energies Editorial Office
E-mail: energies@mdpi.com
www.mdpi.com/journal/energies

www.ingramcontent.com/pod-product-compliance
Ingram Content Group UK Ltd.
Pitfield, Milton Keynes, MK11 3LW, UK
UKHW060124010826
14089UKWH00006B/130

9 783036 543758